# Evolutionary Tinkering in Gene Expression

# NATO ASI Series

## Advanced Science Institutes Series

*A series presenting the results of activities sponsored by the NATO Science Committee,
which aims at the dissemination of advanced scientific and technological knowledge,
with a view to strengthening links between scientific communities.*

The series is published by an international board of publishers in conjunction with the
NATO Scientific Affairs Division

| | | |
|---|---|---|
| A | **Life Sciences** | Plenum Publishing Corporation |
| B | **Physics** | New York and London |
| | | |
| C | **Mathematical** | Kluwer Academic Publishers |
| | **and Physical Sciences** | Dordrecht, Boston, and London |
| D | **Behavioral and Social Sciences** | |
| E | **Applied Sciences** | |
| | | |
| F | **Computer and Systems Sciences** | Springer-Verlag |
| G | **Ecological Sciences** | Berlin, Heidelberg, New York, London, |
| H | **Cell Biology** | Paris, and Tokyo |

*Recent Volumes in this Series*

*Volume 164*—Cochlear Mechanisms: Structure, Function, and Models
edited by J. P. Wilson and D. T. Kemp

*Volume 165*—The Guanine–Nucleotide Binding Proteins: Common Structural
and Functional Properties
edited by L. Bosch, B. Kraal, and A. Parmeggiani

*Volume 166*—Vascular Dynamics: Physiological Perspectives
edited by N. Westerhof and D. R. Gross

*Volume 167*—Human Apolipoprotein Mutants 2: From Gene Structure to
Phenotypic Expression
edited by C. R. Sirtori, G. Franceschini,
H. B. Brewer, Jr., and G. Assmann

*Volume 168*—Techniques and New Developments in Photosynthesis Research
edited by J. Barber and R. Malkin

*Volume 169*—Evolutionary Tinkering in Gene Expression
edited by Marianne Grunberg-Manago, Brian F. C. Clark,
and Hans G. Zachau

*Volume 170*—*ras* Oncogenes
edited by Demetrios Spandidos

*Series A: Life Sciences*

# Evolutionary Tinkering in Gene Expression

Edited by

## Marianne Grunberg-Manago

Institut de Biologie Physico-Chimique
Paris, France

## Brian F. C. Clark

Aarhus University
Aarhus, Denmark

and

## Hans G. Zachau

University of Munich
Munich, Federal Republic of Germany

Plenum Press
New York and London
Published in cooperation with NATO Scientific Affairs Division

Proceedings of a NATO/FEBS Advanced Research Workshop on
Evolutionary Tinkering in Gene Expression,
held August 29–31, 1988,
in Spetsai, Greece

ISBN 978-1-4684-5666-0          ISBN 978-1-4684-5664-6 (eBook)
DOI 10.1007/978-1-4684-5664-6

Library of Congress Cataloging in Publication Data

NATO/FEBS Advanced Research Workshop on Evolutionary Tinkering in Gene
Expression (1988: Nisos Spetsai, Greece)
    Evolutionary tinkering in gene expression / edited by Marianne Grunberg-
Manago, Brian F. C. Clark, and Hans G. Zachau.
        p.      cm. — (NATO ASI series. Series A, Life sciences: vol. 169.)
    "Proceedings of a NATO/FEBS Advanced Research Workshop on Evolutionary
Tinkering in Gene Expression, held August 29–31, 1988, in Spetsai, Greece"—T.p.
verso.
    "Published in cooperation with NATO Scientific Affairs Division."
    Includes bibliographies and index.
    ISBN 978-1-4684-5666-0
    1. Gene expression—Congresses. 2. Evolution—Congresses. I. Grunberg-
Manago, Marianne, 1921-      II. Clark, Brian Frederic Carl. III. Zachau, Hans G.
IV. Title. V. Series: NATO ASI series. Series A, Life sciences; v. 169
QH450.N36   1988                                                        89-8493
574.87'322—dc20                                                              CIP

PREFACE


The Workshop on "Evolutionary Tinkering in Gene Expression" which was held at the end of August 1988, was planned to celebrate 20 successful Advanced Study Institutes (A. S. I. ) in Molecular and Cell Biology. The first Institute was held in 1966 on the Island of Spetsai, after a N. A. T. O. suggestion and was entirely financed by N. A. T. O. The success was immediately so great that the Institute grew very rapidly and in the following years, N. A. T. O. , E. M. B. O. (since 1972) and F. E. B. S. (since 1981) co-sponsored it. Since the start of the ASI, the U. S. National Science Foundation has granted travel money for a limited number of American participants each year. In addition, the course was supported by minor industrial subsidies of varying amounts which enabled the organizers to improve some of the local facilities particularly with respect to the lecture hall. In particular, Boehringer Mannheim has contributed since 1966. Furthermore, the Greek Ministry of Science and Culture has provided support at least for a social event during each ASI.

There were a number of different events that contributed to the origin of this Institute. In the middle of the 1960's, there was a shortage of Advanced Study Institutes in Molecular Biology and N. A. T. O. representatives approached me to organize one. They were keen on holding it in a country where molecular biology was not taught at the University. I thought the time was appropriate for an Institute on Molecular Biology and that it would be nice to organize it on a Greek island. I believed that a pleasant location would contribute to a successful School and I felt that the environment would help to attract the best lecturers and encourage them to stay the whole time at the School (around two weeks). They would be relaxed and in a mood to interact socially and scientifically with the students. I never believed, as some did at that time, that there was a danger of limited attendance at the lectures: good lecturers would compete advantageously with the nice environment. Leslie Orgel (La Jolla, USA) who happened to visit the Institute of Biology in Paris encouraged me to organize it the way I thought and promised to help me. The same year, I was invited to lecture at Harvard (Cambridge, USA) and I had a discussion with A. Rich concerning my ideas about an ideal Summer School and my wish to have it on a Greek island. He became very enthusiastic and told me that he knew the right man who could help me: a Greek working at M. I. T. , who was returning to Athens and who wanted to do something to help science in his country : that was Thanos Evangelopoulos. We therefore decided to coorganize the first Summer School. Finally, a Greek named E. Bricas, who was working in France at that time, suggested to me the island of Spetsai. The location looked ideal. There was a college (with facilities for sporting activities) large enough to acommodate students and a hotel at a short distance from it, both of them adjacent to a good beach. The island was small enough to facilitate contacts between students and professors and it was nevertheless, large enough to provide peace and quiet.

The first meeting was very successful: Dr. Zarvas and Dr. Pullman (who previously organized two N.A.T.O. Schools: one in Squaw Valley and one in Ravenna) were coorganizers. The inhabitants of the island were very proud to have an international course. F. Crick came on his boat and J. Monod also ; although he didn't take part in the course, he introduced me to the inhabitants of the island that he already knew. That time, we had a chance to meet J. Dassin and Melina Mercouri. P. Doty and J. Watson were thrilled to know her and she had a preview of the "double helix". Already during this first course, informal discussions were organized. The course immediately achieved a good reputation and was organized from that time on every year with two interruptions : 1967 and 1968 because of the "colonels". Two Institutes (1971 and 1974) were organized in Erice but after that, the course came back to Spetsai. After 1972, the directorship of the schools was rotated between Dr. H. G. Zachau (München, F.R.G.), Dr. B.F.C. Clark (Aarhus, Denmark) and myself (Paris, France). Since 1983, however the directorship has included Dr. C.T. Caskey (Houston, U.S.A.) to make a four-yearly rotation. All of the Advanced Study Institutes held in Spetsai have involved Dr. A.E. Evangelopoulos of the National Hellenic Foundation in Athens as a local representative of the organizing committee. In addition, each year, the organizing committee and the set of lecturers have changed according to the topic chosen. I would like to underline the role played by the secretaries* in the organization of these ASI, particularly, Morfo Houlis (Greek secretary) who has helped all of us for 20 years now. Little by little, the lectures have been supplemented with research seminars and discussions, advanced tutorials in small groups, and poster sessions followed by discussions, thus generating a summer school atmosphere, all the more because they took place on the beach or more frequently on the terrace of the hotel. Extensive interaction of teachers and participants is encouraged by eating together. At the beginning, the topic of the School was general molecular biology but now the teaching varies every year to cover more specialized topical fast moving subjects at a level not available elsewhere in Europe or USA.

Spetsai participants form a community. Some of them decided to work in one of the lecturers' laboratories after their participation in the course and one can see from the list in the appendix that some of the students became very well known scientists (for instance in 1966: M. van Montagu, J. Abelson, R. Cape who is a founder and chairman of the Board of the Cetus Corporation) and came back to Spetsai as lecturers in the more recent courses. The course has been a great success over the years and only one third of the applicants can be accepted. The participants are usually late pre-doctoral students or postdoctoral workers in the same field, or those in other fields who come to gain an insight into the topic of the course for possible career re-directioning. Some of them come from developing countries but we find that, now, they are better trained than a few years ago. The summer school represents one of the very important events of their life.

The participants of the 1988 Workshop were chosen among the lecturers of the previous Spetsai Summer Schools who particularly interacted with the students. The goal was to put into the perspective of evolution the variety of mechanisms of gene expression and regulation. A large part of the Workshop was devoted to analysing the recently discovered mechanisms of gene regulation that differ from the classical operon model of Jacob and Monod. The originality of the Workshop was that it forced molecular biologists to think about their particular regulatory mechanism from an evolutionary point of view and to reflect on the overall field of gene regulation, contrasting the field as it existed when the Spetsai Schools started 20 years or so ago and as it exists in its multiple forms today.

---

* In recent years, the secretariat has been in the capable hands of L. Heilesen and A. Soerensen (Denmark), Mrs. Oppermann (F.R.G.), N. Franzos and L. Paineau (France), and L. Tanagho (U.S.A.)

ACKNOWLEDGEMENTS

The Organizing Committee of the NATO/FEBS Advanced Research Workshop on "Evolutionary Tinkering in Gene Expression" gratefully acknowledges additional sponsorship from the following sources:

Boehringer Mannheim GmbH
Cetus Corporation
Institut Choay
Integrated Genetics
Senetek
Smith Kline Beckman Corporation

# CONTENTS

Conformational Polymorphism in DNA
A. Rich   1

Translational Control of Gene Expression in
E. coli: The Case of Threonyl-tRNA Synthetase
M. Grunberg-Manago, H. Moine, M. Springer, P. Romby
J.-P. Ebel, C. Ehresmann and B. Ehresmann   17

Structural and Functional tRNA Mimicry of the 3'-end
of Turnip Yellow Mosaic Virus RNA
J.-P. Ebel, R. Giégé and C. Florentz   29

Protein Engineering of Elongation Factor Tu
M. Jensen, K.K. Mortensen, H.U. Petersen, T.F.M. la Cour
J. Nyborg, M. Kjeldgaard, S. Thirup and B.F.C. Clark   41

Translational Control by Phosphorylation of
Mammalian Initiation Factors
J.W.B. Hershey, R.F. Duncan, S.C. Milburn,
V.K. Pathak, S.Y. Choi and R.J. Kaufman   49

$Ca^{2+}$-Calmodulin-Dependent Protein Kinases and
Protein Kinase C: Functional Similarities
T.G. Sotiroudis, S.M. Kyriakidis, L.G. Baltas,
V.G. Zevgolis and A.E. Evangelopoulos   59

Translocation of Proteins into Mitochondria
G. Schatz   71

tRNA Genes – Tinkering in Organization and Expression ?
H. Feldmann   79

Transcriptional Control by Retroviral LTR Regions
N.O. Kjeldgaard, A.J. Baekgaard, H. Yan Dai,
M. Etzerodt, P. Jørgensen, S. Lovmand,
H. S. Olsen and F.S. Pedersen   87

Negative Regulation of Cell Growth
C. Schneider, W. Ansorge and L. Philipson   101

The Human Immunoglobulin K Locus and its Evolution
H.G. Zachau   111

Murine Ornithine Transcarbamylase : Structure
and Expression
S.E. Scherer, G. Veres, W. J. Craigen,
S.N. Jones and C.T. Caskey                                    121

Compositional Patterns in Vertebrate Genomes :
Conservation and Change in Evolution
G. Bernardi, D. Mouchiroud, C. Gautier and G. Bernardi        133

Organization and Evolution of Genomes as Seen
from a Megabase Perspective
C.R. Cantor, L. Pevny and C.L. Smith                          143

Line-1 Sequences : Human Transposable Elements
M. F. Singer                                                  155

The Evolutionary Origin of Glycosomes : How Glycolysis Moved
from Cytosol to Organelle in Evolution
P. Borst and B.W. Swinkels                                    163

Can Plant RNA Viruses Exchange Genetic Material ?
J. -C. Boyer, M. -D. Morch and A. -L. Haenni                  175

Yeast Pre-mRNA Splicing Mutants
U. Vijayraghavan and J. Abelson                               193

Alternative Splicing to Tissue Specific Splicing -
an Evolutionary Pathway ?
E. Brody, J. Marie, M.S. Goux-Pelletan
and B. Clouet d'Orval                                         203

Was RNA the First Genetic Polymer ?
L.E. Orgel                                                    215

Sequence Space, Quasispecies and Statistical Geometry
New Concepts in Molecular Biology
M. Eigen and R. Winkler-Oswatitsch                            225

Chronicle of the Summer Schools on Molecular Biology
on the Occasion of the 20th Anniversary                       235

Contributors                                                  271

Index                                                         274

# CONFORMATIONAL POLYMORPHISM IN DNA

Alexander Rich

Department of Biology
Massachusetts Institute of Technology
Cambridge, Massachusetts

Biological systems are characterized by the abundant participation of macromolecules in carrying out most of their activities. These activities are varied and include information transfer in which nucleic acids play a specialized role. The proteins form organized structures including the catalytically active molecules that are responsible for channelling the metabolic activity of the cell.

There are some generalizations that can be made regarding the macromolecules. Almost all of them, proteins and nucleic acids alike, are polymorphic, which means that they can form in more than one conformation. These conformations may not be equally stable, but each of them represents a local energy minimum that is often utilized in biological systems. The phenomenon of allostery in proteins is well known. Proteins can often adopt different conformations either as single subunits or in the manner in which subunits of proteins assemble together to make larger entities. Much of the biology and detailed chemistry of proteins is associated with changes in conformation. The proteins are essentially machines, and the machines work by changes in three-dimensional structure. X-ray crystallographic studies have clearly revealed these changes in increasing numbers, and this perspective leads us to understand the way that these different structural changes play a role in developing biological activity. Quite often enzymes are catalytically active because the conformation of the protein in the initial state differs significantly from the conformation of the protein in the final state.

Knowledge about the conformational differences in proteins has been widely appreciated for many years. It is only relatively recently that we have been able to gain information about conformational polymorphism or structural changes in the nucleic acids. The physical basis for conformational changes in the proteins is associated with the fact that the total energy of the protein is made up of a large number of interactions between its component pieces. Proteins tend to have many different ways of assembling. Although there are certain structural motifs that are frequently found, such as the alpha helix or pleated sheet, the manner in which these assemble together can vary and can be influenced by such things as the binding of small substrate molecules or other changes nearby. The level of complexity of the system is great enough that it is easy to understand how the interactions of its various components can be modified by having a localized change in one part of the molecule influence its three-dimensional organization at a distance removed from it.

1

Our knowledge of the conformational changes in the nucleic acids is much more limited than in proteins. It initially seemed as if their potential for conformational changes was more limited, due in part to the fact that the basic motif of the nucleic acids, the double helix, is fairly simple in its structure. Thus, it was not readily obvious that many different conformations could be adopted. However, we are beginning to accumulate much knowledge of conformational changes in DNA.

## CONFORMATIONAL INFORMATION IN GENETIC MATERIAL

DNA is the central information storehouse in biological systems. The information contained in it is encoded in a sequence of nucleotides that has the ability to direct a variety of biological reactions. The most prominent of these are the reactions carried out by the coding nucleotides in RNA transcription. In thinking about DNA, we usually stress the informational content in terms of coding sequences. However, to understand the complete role of DNA, it is important to remember that transcription must be carried out in a regulated fashion. The information that may be essential in regulating RNA transcription is not confined solely to coding sequences. Some of it may be found in sequences of nucleotides that induce the formation of specialized conformations in DNA itself. These conformations may then play a direct role in modifying or regulating the expression of genetic information in coding sequences. Conformational changes may also play a role in a wide variety of other biolotical activities, such as replication, recombination, mutation, transposition, and so forth.

## CODING INFORMATION AND CONFORMATIONAL INFORMATION

It is generally assumed that the major conformation of DNA in biological systems is B-DNA. Here we consider that other conformations also exist and that they have a significant biological role.

Proteins that bind to sequence-specific DNA may do so by two distinct mechanisms. One is a mechanism in which the protein binds to the B-DNA conformation and sequence specificity is determined almost entirely through the interaction of the protein with the bases along the double helix. The binding of different proteins, as in a number of repressors, for example, would be differentiated by the particular sequences of bases that are embedded in a B-DNA conformation.

A second type of protein binding is one in which the sequence of nucleotides may itself facilitate a conformation different from B-DNA. When the protein binds to this alternative conformation, it may recognize the sequence by interacting with the alternative conformation as well as with the bases in the sequence. In a more extreme form, the protein may simply recognize the modified conformation of the backbone alone that is stabilized by the particular nucleotide sequence.

The distinction between these two types of protein recognition of nucleic acid sequences has been made more evident to us by the discovery of left-handed Z-DNA, in which the nucleotide sequence has a significant role in stabilizing an alternative conformation of the double helix (Wang et al. 1979). However, this is not likely to be the only example found in biological systems, and we may come to recognize a large number of cases. Additional examples include conformational changes that can occur in homopurine-homopyrimidine sequences, in poly dA-poly dT segments or in chromosome telomere sequences. Explicitly, we are discussing the possibility of having nucleotide sequences that stabilize altered conformations of the sugar-phosphate backbone so that it is no longer B-DNA. If the alterations are great

2

enough, for example, producing changes of more than 1 Å in distances between phosphates, a protein could recognize this altered form.

We can describe those segments of DNA that stabilize an altered conformation as containing "conformational information" in contrast to the "coding information" found in sequences that code for RNA strands.

Altered DNA conformations may be widely distributed, but there is only limited information about them at present. The most extreme form of a sequence-dependent conformational change is seen in left-handed Z-DNA. Here, a somewhat specialized nucleotide sequence stabilizes a conformation in which the double helix actually twists in the opposite sense! More is known about the detailed molecular geometry of this system than of the others cited above. It may be that understanding Z-DNA will provide some insight into what may be anticipated for many other systems with altered DNA conformation.

## LEFT-HANDED Z-DNA

Z-DNA was discovered when the crystal structure of a self-complementary hexanucleotide, d(CpGpCpGpCpG), was solved at atomic resolution using X-ray diffraction techniques (Wang et al, 1979). In the crystal, the molecules adopted the form of a left-handed double helix (Fig. 1) in which the successive hexamer fragments stacked upon each other in a manner that made it appear as a continuous molecule running through the lattice. There are Watson-Crick base pairs between the strands, but the asymmetric unit of the helix is now a dinucleotide (pCpG) instead of a mononucleotide as in right-handed B-DNA. Although right-handed B-DNA has a major and minor groove, Z-DNA has only one deep groove, which is analogous to the minor groove of B-DNA. The concave major groove of B-DNA forms the convex outer surface of the Z-DNA molecule. Z-DNA has a diameter of 18 Å with 12 bp per turn; B-DNA has a diameter of 20 Å with close to 10 pb per turn. The deoxyguanosine nucleotides of Z-DNA have undergone two conformational changes relative to B-DNA. The pucker of the sugar ring has changed from C3' endo to C2' endo, and, in addition, the guanosine residues are in the syn conformation in Z-DNA, whereas they are in the anti conformation in B-DNA (Fig. 2).

These changes produce significant alterations in the relation between successive base pairs along the helix. Because the guanine residues are in the *syn* conformation in Z-DNA, its C8 is found on the outside of the helix. This is in contrast to guanine C8 in B-DNA, where its hydrogen is in van der Waals contact with the sugar-phosphate chain on the outside of the helix. In Z-DNA the base pairs are on the outside of the molecule. In Z-DNA the twist angle between successive base pairs along the helix depends on the sequence, but this is not generally so in B-DNA. The CpG sequence in Z-DNA has very little twist between the two base pairs, whereas the GpC sequence has a large twist angle. In B-DNA there is an approximately steady twist angle near 36 degrees between each successive base pair along the helix.

Every other residue in the initial Z-DNA crystal was a purine in the *syn* conformation. For some time it has been known that purine residues can adopt either *syn* or *anti* conformations, based on theoretical (Haschemeyer and Rich 1967) as well as experimental studies (Son et al 1972). These studies also suggest that it is more difficult for pyrimidines to form the *syn* conformation. However, it is possible to detect *syn* conformation of pyrimidines in nucleotide solutions. This difference in the relative ease with which purines can adopt the *syn* conformation compared with pyrimidines means that the Z-DNA conformation is likely to be stabilized where there are stretches of DNA that

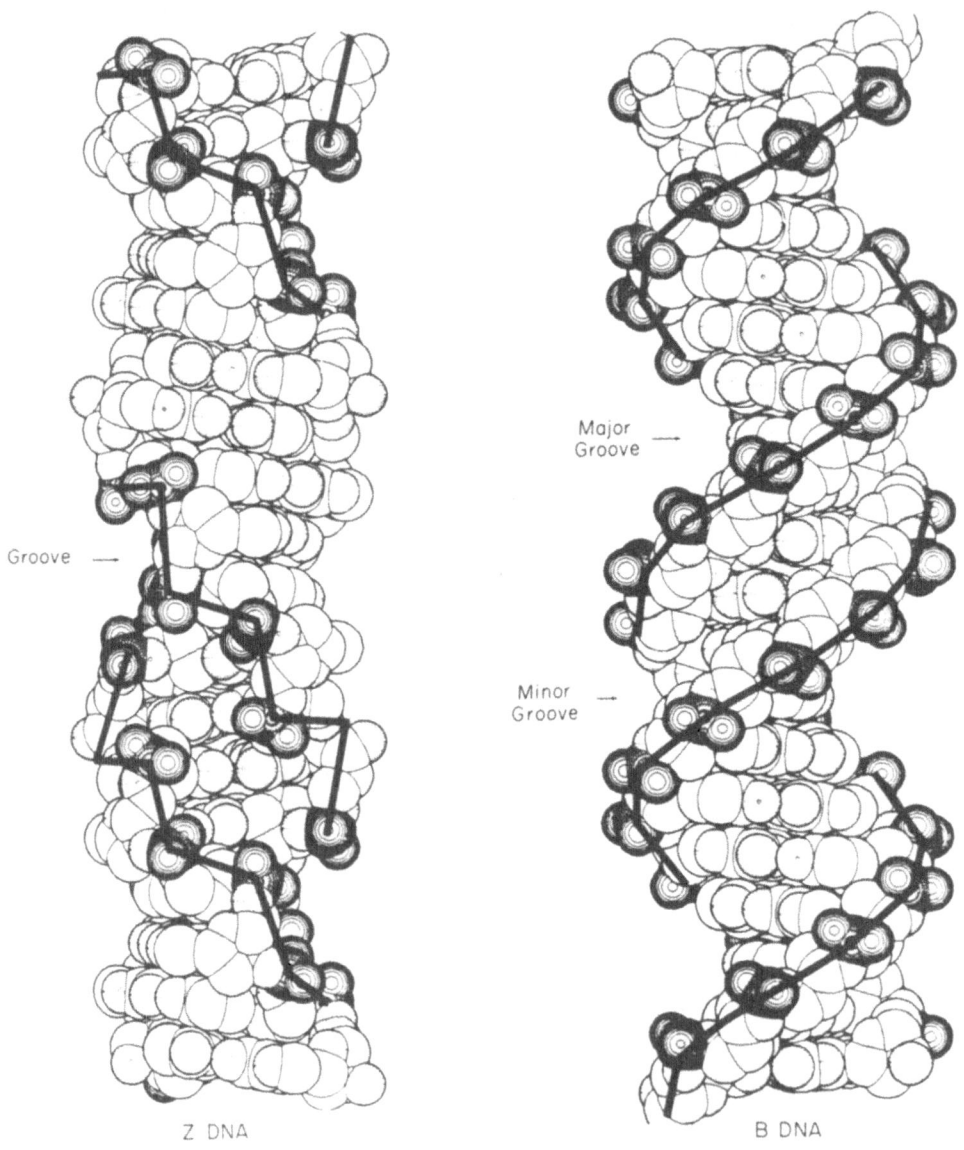

Figure 1. Van der Waals diagrams of Z-DNA and B-DNA. The irregularity of the Z-DNA backbone is illustrated by the heavy lines which go from phosphate to phosphate residue along the chain. Z-DNA with its single groove is shown as it appears in the hexamer crystal. In contrast, B-DNA has a smooth line connecting the phosphate groups and it has two grooves, neither of which extends to the axis of the helical molecule.

have sequences with alternating purines and pyrimidines. The Z-DNA conformation is most favored by alternations in guanine and cytosine residues, but single-crystal studies have revealed the presence of AT base pairs in the Z-DNA conformation. In addition, Z-DNA crystals form in sequences that do not have alternations of purine and pyrimidines (Wang et al, 1985).

Another significant difference between B-DNA and Z-DNA is the orientation of the base pairs relative to the sugar-phosphate backbone. Figure 3 shows a diagram in which the bases are drawn as short, board-like structures, and the sugar-phosphate backbone is drawn as a simple ladder. A segment of 4 bp is converted from B-DNA to Z-DNA. This conversion is brought about by changing the orientation of the base pairs relative to the backbone by "flipping" the bases over. This is indicated by the change in shading. In the case of the guanine residues, the flipping occurs by rotation of guanine about the glycosyl bond from the *anti* to the *syn* conformation. In the case of cytosine residues, the entire unit, sugar plus base, is turned upside down to form Z-DNA. It is this rotation of the cytosine sugar residues that produces the zigzag conformation of the backbone.

The conversion from right-handed B-DNA to left-handed Z-DNA is not brought about by a simple twisting of the double helix in the opposite direction. Instead, it involves a series of complex, internal rearrangements in which there is a basic change in the positioning of the bases relative to the backbone.


CONFORMATIONAL EQUILIBRIUM IN DNA

It is believed that DNA exists largely in the B-DNA conformation in biological systems. However, the data on which this assumption is made are not very extensive. Diffraction patterns of B-DNA have been found in a variety of biological structures ranging from sperm heads to bacteriophages, but the nature of X-ray diffraction from noncrystalline biological sources is such that it will pick up only those elements of structure that are highly repetitive. Elements that are not repetitive will not register in the pattern. Although B-DNA is present in these systems, the experiments would fail to see other conformations that are not there in a regular and periodic manner. Experiments on DNA more than 30 years ago showed that there are two major conformations that can be visualized in X-ray diffraction studies of DNA fibers (Franklin and Gosling 1953). Air-dried fibers produce an A-type diffraction pattern, whereas hydrated fibers yield a B-type pattern. In A-DNA the base pairs are tilted relative to the helix axis, whereas in B-DNA the bases are almost perpendicular to the axis. A-DNA is shorter and fatter than B-DNA (Fig. 4a). The axis of B-DNA runs through the center of the base pairs, whereas in A-DNA the base pair is displaced away from the axis The position of the helix axis relative to the base pair for three major types of DNA double helices is shown in Figure 4b. Because of the displacement of the base pair in Z-DNA away from the axis, the major groove of A-DNA is quite deep, whereas the minor groove is quite shallow. In B-DNA, where the axis passes through the two base pairs, the difference in depth between major and minor grooves is not very pronounced. The axes of both A-DNA and B-DNA are found on the perpendicular bisector of the line going from C1' of one base to C1' of the other base in the pair. This position preserves the pseudodyad or two-fold symmetry found in the plane of the base pair in both A- and B-DNAs. In contrast, the axis for the Z-DNA helix is not constrained to fall on that line because the asymmetric unit contains two bases. In Z-DNA, the helix axis passes close to the O2 position of cytosine as shown in Figure 4. The position of the helix axis in Z-DNA is such that it makes the groove corresponding to the minor groove much deeper in Z-DNA, and the major groove is eliminated entirely since it now becomes the outer part of the molecule.

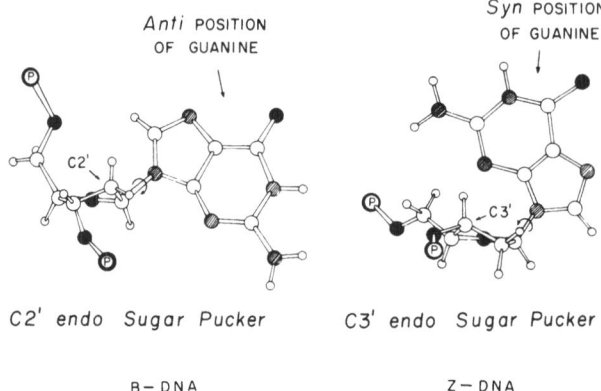

Figure 2. The conformation of deoxyguanosine is shown in B-DNA and in Z-DNA. In the sugar, the plane defined by C1'-O1'-C4' is horizontal. Atoms lying above this plane are in the <u>endo</u> conformation. The C3' is <u>endo</u> in Z-DNA while in B-DNA the C2' is <u>endo</u>. Z-DNA has guanine in the *syn* position, in contrast to the *anti* position in B-DNA. A curved arrow around the glycosyl carbon-nitrogen linkage indicates the site of rotation.

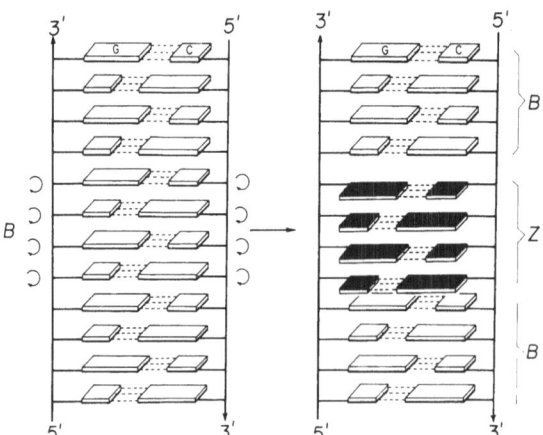

Figure 3. The change in topological relationship is shown when a four base pair segment of B-DNA converts to Z-DNA. The conversion is accomplished by rotation or flipping over the base pairs as indicated by the curved arrows. Rotation of alternate residues about the glycosylic bond puts them into the *syn* conformation.

6

An important point to be made here is that in solution all conformations of DNA are in equilibrium with each other along any particular segment of helix. The distribution of conformations found along any segment of the helix will depend to a significant extent on the nucleotide sequence in that region as well as on the local environment.

The equilibrium between B-DNA and Z-DNA can be observed most simply in solution by measuring the circular dichroism (CD) of poly(dG-dC), which forms B-DNA in a 0.1 M NaCl solution but largely Z-DNA in a 4.0 M NaCl solution (Pohl and Jovin, 1972; Thamann et al, 1981). On lowering the salt concentration from 4.0 molar to 0.1 M, a series of intermediate CD spectra are seen, illustrating a gradual shift in the equilibrium.

Other experiments also illustrate the equilibrium between B-DNA and Z-DNA. For example, in the hexamer crystallization experiments (Wang et al. 1979), the crystals were formed from a low-salt solution. CD measurements of this solution indicate no Z-DNA present, but the crystals that formed were entirely Z-DNA! In this case there were a small number of Z-DNA molecules in equilibrium with B-DNA and they nucleated to form a crystal. As the crystals grew, the equilibrium shifted the molecules from B-DNA to Z-DNA and then into the crystal lattice.

The equilibrium between B-DNA and Z-DNA can also be altered by covalent modification. This is shown most clearly in studies of the methylation of poly(dG-dC) on cytosine C5. When this polymer is methylated, the equilibrium shifts rather strongly toward Z-DNA, so that under physiological salt conditions it is now found entirely as Z-DNA rather than B-DNA (Behe and Felsenfeld 1981). In addition, micromolar amounts of polyamines are sufficient to stabilize this polymer in the Z-DNA conformation. An X-ray crystallographic study of a methylated hexamer with alternating CG sequences (Fujii, et al. 1982) has provided a structural rationalization for Z-DNA stabilization.

In thinking about the equilibrium between different conformational states, it is important to bear in mind that the results of X-ray crystallographic studies generally illustrate one conformation in which there is a local energy minimum. However, this is not necessarily the only energy minimum that the molecule can adopt, but rather the one that happens to form in this particular crystal. The structure of the molecule in a crystal may be likened to a snapshot of the molecule in one particular conformation. The true conformation of the molecule may be thought of as containing a large number of such snapshots, as in a movie reel, in which the molecule moves among a number of different conformations. The actual amount of time spent in a particular conformation will depend on its stability relative to the other coformations. The speed with which the molecule converts from one conformation to the other depends on the energy barriers for the conversion and whether or not there happen to be enzymes that may facilitate the conversion from one form to the other.

It should be pointed out that each of the three major categories of DNA conformation, A, B, and Z (Fig. 4a), represents not a single structure but rather a family of related structures. There are a number of modifications in local twist angle between adjacent base pairs and slight modifications in ring pucker that make each of these helical types exist in slightly different forms. These slight modifications have been seen in a number of single crystal forms of A-, B-, and Z-DNAs. In general, it requires single-crystal X-ray analyses to bring this kind of information forward, since the type of conformational averaging used in interpreting fiber diffraction patterns of necessity loses subtle conformational changes.

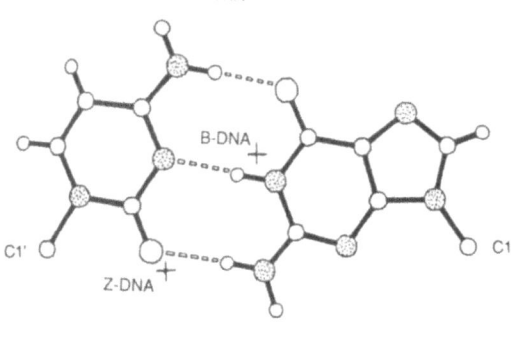

A-DNA          B-DNA          Z-DNA

Figure 4.  Geometry of A-, B- and Z-DNA.
a)  van der Waals drawings of 20 base pairs of DNA in the three different
conformations.
b)  The position of the three helical axes are shown superimposed on a
guanine-cytosine base pair.  The sugar C1' atoms are also shown.  The axes of
A-DNA and B-DNA are found on a line perpendicular to the C1'-C1' line and
halfway between them.  This is necessary because of the pseudodyad found in
both of these structures in the plane of the base pair.  The axis of Z-DNA is not
on that line because the asymmetric unit consists of two bases and there is no
longer a pseudodyad in the plane of the base pairs.  The base pairs are far
removed from the A-DNA axis, so that the major groove is deep and the minor
groove on the external surface in the molecule is shallow.  The Z-DNA axis is
also somewhat removed from the center, so that its groove is deep and the
external surface of Z-DNA is what corresponds to the major-groove side of the
base pair.  In B-DNA the axis goes through the middle of the base pairs, giving
rise to two slightly unequally sized grooves.

We can ask whether alternative left-handed helical forms of DNA exist. An approach to this problem has been made by working on form-V plasmids. Form-V plasmids are made by isolating single-stranded, circular DNA from plasmids and then annealing the two complementary circles together. Segments of these plasmids will form right-handed DNA, and topological constraints will force the remaining segments to adopt left-handed DNA conformations because the two complementary DNA circles are not linked with each other. Spectroscopic studies involving both CD and Raman spectra yield data that suggest that the only left-handed conformation seen in form-V plasmids is that adopted by Z-DNA (Brahms et al, 1982). If these results are borne out by further investigations, it may turn out that the Z-DNA family of structures is the only left-handed conformation that polynucleotide strands can adopt.

The important conclusion to bear in mind is that DNA is a dynamic system that exists in a variety of conformational states. Furthermore, the distribution of conformational states can be altered by modifications of DNA or by changes in the local environment, which includes both small molecules, especially charged ones, as well as the macromolecular proteins.

## Z-DNA IS IMMUNOGENIC

The Z-DNA conformation is strongly immunogenic, and it stimulates the production of antibodies specific for Z-DNA (Lafer et al. 1981). Ths is in marked contrast to B-DNA, which is not immunogenic and for which antibodies cannot normally be produced. Poly(dG-dC) in a high salt solution was stabilized in the Z-DNA form by bromination so that a third of the guanine residues were brominated on the C8 position. The guanine C8 position in Z-DNA is on the outside of the helix. In B-DNA, however, the hydrogen atom attached to guanine C8 is in van der Waals contact with the sugar-phosphate backbone. Substituting the bulkier bromine atom for hydrogen on the C8 position stabilizes the *syn* conformation of guanine, which in turn stabilizes Z-DNA. Injection of this material into rabbits or mice produced a high titer of antibodies that were found to react solely with the Z-DNA conformation (Lafer et al, 1981).

## NEGATIVE SUPERCOILING STABILIZES Z-DNA

In biological systems DNA is generally under torsional constraint, which means that the long molecule is twisted about itself or supercoiled. Supercoiling exists whenever the number of turns of the double helix is not equal to the number of turns the molecule would adopt if it were in a linear or relaxed form. In vivo, DNA is generally not found in a linear form but is either in a circular form as in plasmids or it is constrained in topological domains in the genome. Topoisomerase enzymes can keep DNA in a negatively supercoiled or slightly underwound state.

Negatively supercoiled DNA has a higher free energy than relaxed DNA. The free energy of supercoiling is proportional to the square of the number of superhelical turns or the superhelical density. The free energy of supercoiling can be used to change DNA, since processes that reduce the number of superhelical turns are energetically favored. In negatively supercoiled DNA, these processes include unpairing of bases with strand separation, unwinding of the double helix, cruciform formation, as well as binding to proteins as in nucleosome formation. Another process that is facilitated in a negatively supercoiled plasmid is stabilization of Z-DNA (Singleton et al, 1982, Peck et al, 1982). Converting one turn of right-handed B-DNA into a left-handed Z-DNA segment loses two negative superhelical turns. The negative supercoiling energy of the plasmid effectively stabilizes Z-DNA formation.

We may ask to what extent a macromolecular protein that binds to Z-DNA can influence the equilibrium between B-DNA and Z-DNA. This problem can be addressed most simply by using the antibody raised against poly (dG-dC) in the Z- conformation to study its influence on the B-Z equilibrium. It is easy to show that the antibody has an effect on the equilibrium in the polymer. Poly (dG-dC) converts to Z-DNA in the presence of high concentrations of NaCl (Pohl and Jovin, 1972). There is a direct correlation between the binding constant of the antibody with the polymer and its effectiveness in lowering the required salt concentration (Lafer et al, 1986).

It can be shown that the binding energy of antibodies to Z-DNA is great enough to hold a segment of a plasmid in a Z-DNA conformation in the absence of torsional strain from the plasmid (Lafer et al, 1985). If there were a true equilibrium between B-DNA and Z-DNA, it should be possible to detect the equilibrium starting with linear DNA and Z-DNA binding proteins. Thus, it should be possible to trap the DNA in the Z-conformation.

An experiment of that type is carried out in the following way. Two types of radioactively labelled plasmids are linearized with a restriction endonuclease, either with no insert or with an insert of d(CG/GC)$_{12}$. The linearized plasmids are then incubated with anti Z-DNA antibodies which were raised against poly (dG-dC) in the Z conformation. The results are assayed by nitrocellulose filtration. If the antibody is binding to the linearized plasmid it will retain the plasmid on the filter paper, otherwise, it will pass through. The results of this experiment are shown in Figure 5. It can be seen that incubation of linearized plasmids containing the Z-forming insert are retained on the nitrocellulose filter while no retention is found with the linearized plasmid in the absence of the insert. It should be noted that this reaction

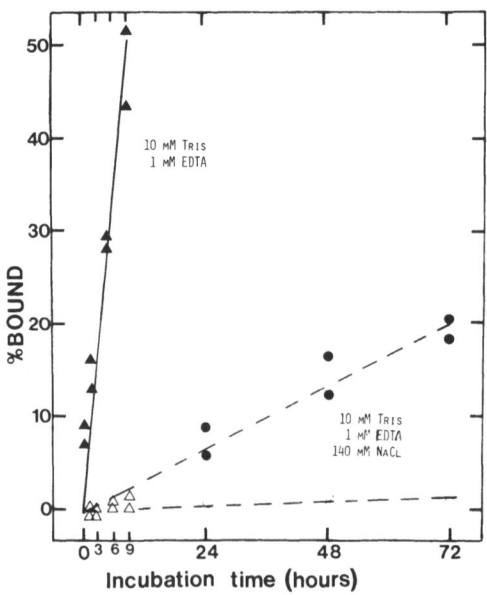

Figure 5. The binding of anti Z-DNA antibodies to linearized plasmids with (△,0) or without (Δ,0) a d(CG/GC)$_{12}$ insert. The DNA is radioactive and the time-dependent binding to nitrocellulose filters is shown at two different salt concentrations.

proceeds very slowly. It takes many hours to establish binding of the linearized fragment by the antibody. This suggests that there is a relatively slow conversion rate from B-DNA to Z-DNA in the absence of torsional strain. Other experiments have demonstrated that the conversion of supercoiled B-DNA to Z-DNA is much more rapid (Peck et al, 1986).

The overall conclusion of experiments described above is that proteins binding to Z-DNA in a sequence specific manner may be associated with large amounts of energy. This energy may, as in the case of the antibody, be great enough not only to hold Z-DNA in that conformation in the absence of torsional strain, but it may also bind to those Z-DNA segments which are found in the normal B-Z equilibrium. In this manner the equilibrium is perturbed and the amount of Z-DNA increases.

## Z-DNA BINDING PROTEINS ARE FOUND IN CHROMATIN

Z-DNA binding proteins have been isolated from the nuclei of Drosophila cells using the method of affinity chromatography (Nordheim et al, 1982). Brominated poly (dG-dC), stable as Z-DNA at physiological salt concentrations, was attached covalently to Sephadex G-25 as an affinity matrix.

Another tissue that has been studied extensively is wheat germ (Lafer et al, 1985). Using techniques similar to those that were employed in the isolation of Drosophila Z-DNA binding proteins, several Z-DNA binding proteins have been isolated with molecular weights that vary from 40,000 to nearly 150,000. Like the Drosophila proteins, these have the ability to bind to brominated poly (dG-dC) (Z-DNA) but not to poly (dG-dC) (B-DNA). It has also been shown that the antibody against Z-DNA competes with the binding of wheat germ Z-DNA binding proteins, suggesting that they are binding at the same Z-DNA sites. Filter binding experiments were also carried out with negatively supercoiled plasmids containing inserts such as $d(CG/GC)_n$ or $d(CA/GT)_n$ where n varies from 10 to 60. The specificity of Z-binding proteins can be tested in this system. When two plasmids are supercoiled to the same negative superhelical density, there is a greater retention of the plasmid with the Z-DNA forming insert than the plasmid without the insert. Such tests clearly demonstrate that the protein binding is due to Z-DNA and is not, for example, due to proteins that bind to supercoiled DNA per se.

In experiments with both Drosophila and wheat germ chromatin, the protein preparations were found to stabilize Z-DNA formation in inserts of $d(CA/GT)_n$ to a greater extent than inserts with $d(CG/GC)_n$. This may reflect the fact that the eukaryotic genome contains many repetitive DNA segments of $d(CA/GT)_n$, where n>50 (Hamada et al, 1982). It is possible that these proteins stabilize those segments.

## SV-40 MINICHROMOSOMES HAVE Z-DNA BINDING PROTEINS

In an attempt to find proteins stabilizing specific DNA sequences, experiments were carried out on the minichromosome formed by the tumor virus SV-40. The SV-40 tumor virus infects monkey kidney cells and transforms the cells. The virus goes through active transcription and replication phases. It can be isolated as a small fragment of chromatin containing 24 nucleosomes associated with each covalently closed double helical SV-40 genome containing slightly over 5 kb of DNA  Experiments using Z-DNA affinity chromatography revealed Z-DNA binding proteins (Azorin and Rich, 1985).

Experiments were carried out to see whether Z-DNA forms in negatively supercoiled SV40 DNA. The negatively supercoiled SV40 was crosslinked to anti-Z-DNA antibodies, the DNA was cleaved by restriction enzymes and some fragments were retained on the nitrocellulose filter. This included segments

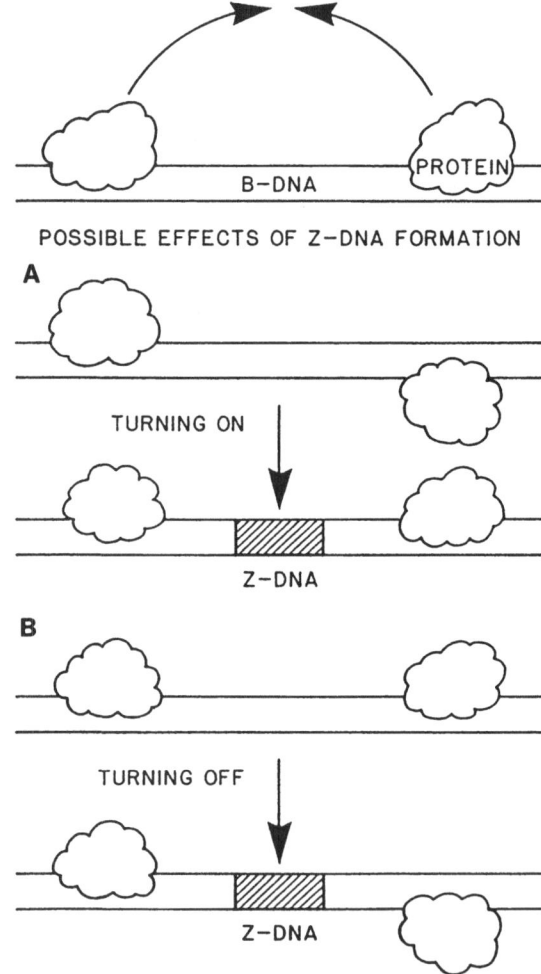

ACTIVATION OF DNA BY BENDING
TO FACILITATE PROTEIN—PROTEIN INTERACTIONS

PROTEIN

B—DNA

POSSIBLE EFFECTS OF Z—DNA FORMATION

A

TURNING ON

Z—DNA

B

TURNING OFF

Z—DNA

Figure 6.   A hypothetical role for Z-DNA in regulating gene expression.   It is assumed that DNA bending or looping results in protein-protein interactions for proteins bound to DNA.   This results in gene activation.   Z-DNA formation between the two proteins would influence their contacts.

of the control region spanning the origin of replication through to the transcriptional enhancer (Nordheim and Rich, 1983). Although the plasmid contains over 5 kb of DNA, Z-DNA formation was found to occur largely in the regulatory region of the virus.

Experiments with Z-DNA binding proteins from SV40 minichromosomes revealed that they bound to SV40 only when it was supercoiled, and they bound so tightly that it is likely that they bind in a sequence specific manner. The major conclusion from positioning experiments is that a Z-DNA binding protein is found that appears to be located near the transcriptional enhancer of SV40 in the control region. Finding a protein in that position suggests that it may participate in some way in regulating the various functions controlled by this segment of DNA in the virus.

## A POSSIBLE ROLE FOR Z-DNA ASSOCIATED WITH GENE ACTIVATION

There is some evidence that proteins bind to DNA and it becomes activated for transcription through bending or looping of the DNA that facilitates protein-protein interaction (Ptashne, 1986). Experiments that support this interpretation of gene activation are coupled to the effect of inserting added nucleotides in the DNA molecule. Thus, for example, we can imagine that two proteins are bound to the same side of the DNA molecule some distance apart (see Fig. 6). Bending or looping of the DNA molecule would facilitate protein-protein interaction and activation of this segment of DNA for transcription. If, however, one inserts added nucleotides between the two proteins, the DNA helix is extended. When five nucleotides have been added between the two proteins bound to right-handed B-DNA, the proteins are then found on the opposite side of the helix and this is associated with inactivation. However, addition of ten base pairs again puts the protein back on the same side, and this is associated with reactivation.

We might ask what the effect would be of forming Z-DNA in a region between the two protein binding sites. If a segment of DNA forms of the appropriate length, it would have the effect of changing the orientation of the two segments of DNA on either side of it. As shown in Fig. 6, formation of the segment of Z-DNA between the two binding sites can bring about a rotation of the DNA segment bound to one protein molecule relative to the segment bound to the other molecule. In effect, formation of Z-DNA can alter the disposition of proteins bound to DNA on either side of the Z-DNA forming segment. In the examples shown in Fig. 6, Z-DNA formation could be associated either with gene activation (bringing the proteins on the same side of the helix) or gene inactivation (putting the proteins on opposite sides of the helix). In this hypothetical example, Z-DNA binding proteins and the formation of Z-DNA might act on transcriptional systems by changing the orientation of DNA segments on either side of the Z-forming region. Only further experimentation will determine whether or not a system of this type has any relevance in our* understanding of transcriptional control.

## Z-DNA AND GENETIC RECOMBINATION

Genetic recombination is a fundamental biological process whereby DNA molecules meet and "cross over," that is, the polynucleotide chains are severed and the DNA molecules are joined in a recombinant mode. The experimental assay for recombination enzymes usually a strand transfer reaction in which a linear double strand sequence is added to a circular single strand. A recombination enzyme usually using ATP has the ability to make a recombinant, that is, to join the circular single strand with its complementary mate obtained from the DNA duplex. This simple assay has been used to demonstrate recombination.

13

Work by Holloman and his co-workers has concentrated on a eukaryotic recombination enzyme Rec I from the corn rust Ustilago (Kmiec et al, 1985; Kmiec and Holloman, 1986). Isolation and purification of that enzyme using the strand transfer assay made it possible to study that enzyme in some detail. The enzyme was found to be a Z-DNA binding protein. The enzyme could carry out a strand transfer reaction and was also able to generate a "paranemic" joint. A paranemic joint is one in which there is no net linking number, i.e., the strands are not linked around each other such as they are, for example, in right-handed B-DNA or in left-handed Z-DNA. In a paranemic joint there are equal numbers of right- and left-handed turns so that there is no net linking.

A human recombination enzyme has been isolated using a Z-DNA affinity column (Fishel et al, 1988). It is possible that many recombination enzymes can be purified by this method. Present evidence thus suggests that Z-DNA may play a role in homologous genetic recombination and sequences may be targeted for recombination if they have the Z-DNA forming potential. Further investigations of these Z-DNA binding proteins is likely to lead to a fuller understanding of the role of these sequences in this fundamental biological process.

## ACKNOWLEDGMENTS

This work was supported by grants from the National Institutes of Health, the American Cancer Society, the National Aeronautics and Space Administration and the Office of Naval Research.

## REFERENCES

Azorin, F. and Rich, A., 1985, Isolation of Z-DNA binding proteins from SV40 minichromosomes, Cell, 41:365.

Behe, M. and Felsenfeld, G., 1981, Effects of methylation on a synthetic polynucleotide: The B-Z transition in poly(dG-m$^5$dC)·poly(dG-m$^5$dC), Proc. Natl. Acad. Sci., 78:1619.

Brahms, S., Vergne, J., Brahms, J.G., DiCapua, E., Bucher, P. and Koller, T., 1982, Natural DNA sequences can form left-handed helices in low salt solution under conditions of topological constraint, J. Mol. Biol., 162:473.

Fishel, R., Detmer, K. and Rich, A., 1988, Identification of homologous pairing and strand-exchange activity from a human tumor cell line based on Z-DNA affinity chromatography, Proc. Natl. Acad. Sci., 85:36.

Franklin, R.E. and Gosling, R.G., 1953, The structure of sodium thymonucleate fibres, Acta Crystallogr., 6:673; 8:151.

Fujii, S., Wang, A.H.-J., van der Marel, G.A., van Boom, J.H, and Rich, A., 1982, Molecular structure of (m$^5$dC-dG)$_3$: The role of the methyl group on 5-methylcytosine in stabilizing Z-DNA, Nucleic Acids Res., 10:7879.

Hamada, H., Petrino, M.G. and Kakunaga, T., 1982, A novel repeated element with Z-DNA-forming potential is widely found in evolutionarily diverse eukaryotic genomes, Proc. Natl. Acad. Sci., 79:6465.

Haschemeyer, A.E.V and Rich, A., 1967, Nucleoside conformations: An analysis of steric barriers to rotation about the glycosidic bond, J. Mol. Biol., 27:369.

Kmiec, E.B., Augelides, D.J. and Holloman, W.K., 1985, Left-handed DNA and the synaptic pairing reaction promoted by Ustilago rec 1 protein, Cell, 40:139.

Kmiec, E.B. and Holloman, W.K., 1986, Homologous pairing of DNA molecules by Ustilago Rec 1 protein is promoted by sequences of Z-DNA, Cell, 44:545.

Lafer, E.M., Moller, A., Nordheim, A., Stollar, B.D. and Rich, A., 1981, Antibodies specific for left-handed Z-DNA, Proc. Natl. Acad. Sci., 78:3546.

Lafer, E.M, Sousa, R. and Rich A., 1985, Anti Z-DNA antibodies can stabilize Z-DNA in relaxed and linear plasmids under physiological conditions, EMBO J., 4:3655.

Lafer, E.M, Sousa, R., Rosen, B., Hsu, A. and Rich, A., 1985, Isolation and characterization of Z-DNA binding proteins from wheat germ, Biochem., 24:5070.

Lafer, E.M., Sousa, R., Rashid, A., Rich, A. and Stollar, B.D., 1986, The effect of anti-Z-DNA antibodies on the B-DNA-Z-DNA equilibrium, J. Biol. Chem., 261:6439.

Nordheim, A., Tesser, P., Azorin, F., Kwon, Y.H., Moller, A. and Rich, A., 1982, Isolation of Drosophila proteins that bind selectively to left-handed Z-DNA, Proc. Natl. Acad. Sci., 79:7729.

Nordheim, A. and Rich, A., 1983, Negatively supercoiled simian virus 40 DNA contains Z-DNA segments within transcriptional enhancer sequences, Nature, 303:674.

Peck, L.J., Nordheim, A., Rich, A. and Wang, J.C., 1982, Flipping of cloned $d(CG)_n \cdot d(CG)_n$ DNA sequences from a right- to a left-handed helical structure by salt, Co(III), or negative supercoiling, Proc. Natl. Acad. Sci. 79:4560.

Peck, L.J., Wang, J.C., Nordheim, A. and Rich, A., 1986, Rate of B to Z structural transition of supercoiled DNA, Mol. Biol., 190:125.

Pohl, F.M. and Jovin, T.M., 1972, Salt-induced co-operative conformational change of a synthetic DNA: Equilibrium and kinetic studies with poly(dG-dC), J. Mol. Biol., 67:375.

Ptashne, M., 1986, Gene regulation by proteins acting nearby and at a distance, Nature, 322:697.

Singleton, C.K., Klysik, J., Stirdivant, S.M. and Wells, R.D., 1982, Left-handed Z-DNA is induced by supercoiling in physiological ionic conditions, Nature, 299:312.

Son, T.-D., Guschlbauer, W., and Gueron, M., 1972, Flexibility and conformations of guanosine monophosphates by the Overhauser effect, J. Am. Chem. Soc. 94:7903.

Thamann, T.J., Lord, R.C., Wang, A.H.-J., and Rich, A., 1981, The high salt form of poly(dG-dC)·poly(dG-dC) is left-handed Z-DNA: Raman spectra of crystals and solutions, Nucleic Acids. Res., 9:5443.

Wang, A.H.-J., Quigley, G.J., Kolpak, F.J. , Crawford, J.L., van Boom, J.H., van der Marel, G., and Rich, A., 1979, Molecular structure of a left-handed DNA fragment at atomic resolution, Nature 282:680.

Wang, A.H.-J., Gessner, R.V., van der Marel, G., van Boom, J.H. and Rich, A., 1985, Crystal structure of Z-DNA without an alternating purine-pyrimidine sequence, Proc. Natl. Acad. Sci. 82:3611.

# TRANSLATIONAL CONTROL OF GENE EXPRESSION IN E. COLI:

## THE CASE OF THREONYL-tRNA SYNTHETASE

M. Grunberg-Manago*, H. Moine+, M. Springer*,
P. Romby+, J.-P. Ebel+, C. Ehresmann+ and B. Ehresmann+

*Institut de Biologie Physico-Chimique
13 rue Pierre et Marie Curie
75005 Paris, France

+Institut de Biologie Moléculaire et Cellulaire
du C.N.R.S.
15 rue René Descartes
67084 Strasbourg Cedex, France

A wide range of functions are devoted to RNA molecules in the cell. Besides their canonical role in the protein biosynthetic machinery, RNAs are involved in a large number of other functions, such as priming of reverse transcription (Kikuchi et al., 1986), self splicing of the rRNA intron of the ciliate Tetrahymena and many other catalytic activities (Cech and Bass, 1986). This makes it likely that life began with RNA serving both as a genome and and as a catalyst for its own replication. Obviously, the RNA three- dimensional structures determine many of their biological activities. Messenger RNAs also must be considered as dynamic molecules whose structure might have a direct effect on their function in the decoding process.

We would like to report that messenger RNA structure is essential in the regulation of the expression of an aminoacyl-tRNA synthetase of E. coli namely threonyl-tRNA synthetase (ThrRS). The control of expression of threonyl-tRNA synthetase, like that of a number of phage and bacterial genes is regulated at the level of translation of its mRNA by a negative feedback mechanism. While we have an increasing number of examples of such regulation, particularly for genes whose products are involved in nucleic acid metabolism, little is known about the mechanism by which translation is controlled or about the structure of the mRNA region involved in the regulatory protein recognition. The study of the molecular mechanism of translational regulation is facilitated in the thrS gene because of the relative facility with which cis and trans acting elements involved in the regulation can be defined. Furthermore, the enzymatic activity of ThrRS is easily tested towards different substrates allowing a biochemical approach.

RESULTS

Three types of evidence demonstrate that the expression of thrS is negatively regulated at the translational level.

1) In vitro studies in a cell free transcription-translation system programmed by thrS DNA have shown an inhibition of ThrRS de novo synthesis by addition of ThrRS to the assay mixture while thrS mRNA synthesis was not affected (Lestienne et al., 1984).

2) In vivo studies with strains carrying various thrS mutant alleles were shown to overproduce a modified form of ThrRS while in these mutants the synthesis rate of thrS mRNA is unchanged compared to the wild type strain (Butler et al., 1986).

3) The third set of experiments proving the translational control of thrS expression was performed with different types of thrS-lacZ fusions. Structural thrS mutants on the chromosome derepress β-galactosidase synthesis from a thrS-lacZ protein fusion cloned in bacteriophage λ and integrated at the lambda attachment site on the chromosome (Springer et al., 1985 and 1986). (In protein fusions the β-galactosidase activity comes from a hybrid THrRS-β galactosidase protein that is expressed from the thrS promoter and the translational initiation site of thrS). The β-galactosidase from the thrS-lacZ fusions is repressed in strains carrying ThrRS overproducing plasmids (Table I). This indicates that thrS expression is inversely related to the active ThrRS concentration. This is not the case for the operon fusions where β-galactosidase is expressed from a thrS promotor but from its own initiation site and therefore do not respond to translation regulation (Springer et al., 1985 and 1986).

Table 1. Repression of β-galactosidase synthesis from a thrS-lacZ fusion (from Springer et al., 1987)

| | Repression caused by overproduction | |
| | W. T. ThrRSase | R$^{++}$4-11-8 ThrRSase |
| --- | --- | --- |
| W. T. type | 13.8 | 3 300 |
| M1 - 2 (T$_{-31}$ → A) | 1.12 | 1.4 |
| L6 - 8 (G$_{-32}$ → A) | 1.07 | 1 |
| M6 - 1 (G$_{-40}$ → A) | 1.1 | 3.1 |
| M4 - 11 (C$_{-16}$ → T) | 0.96 | 30 |
| M2 - 2 (Duplication) | 1.06 | 1.9 |

For each fusion, the repression factor is the ratio of β-galactosidase synthesis in a wild type strain carrying the vector plasmid to that in the same strain carrying a plasmid overproducing either wild type or R$^{++}$ 4-11-8 ThrRSase. W. T. stands for wild type.

GENETIC EVIDENCE FOR THE OPERATOR

The fact that the strains containing the thrS-lacZ protein fusion are sensitive to repression by an excess of ThrRS permits an easy selection of "translational operator" constitutive mutants (Springer et al., 1986). The striking feature of the selection is that all the mutants are neither derepressible (expression enhanced when the host cell contains a mutant form of ThrRS) nor repressible (expression decreased by high levels of wild type ThrRS) (Springer et al., 1986) translational mutants (Table I). The mutants are indeed translational and not transcriptional as the rate of mRNA synthesis is the same in the mutants and in the wild type. The control regions of all the mutants were sequenced from the -35 region of the thrS promoter (the mRNA start is located 163 nucleotides in front of the ATG) to the beginning of the structural gene (Springer et al., 1986). The translational operator mutations are all located upstream of the thrS translational initiation site between -50 and -10 (+1 is the A of the initiation triplet AUG) and downstream of the 5' end of the mRNA. The selection gave mainly point mutations and one small duplication. In addition, one mutation, L18, was constructed in vitro by inserting an oligonucleotide in the stem of the operator. The majority of point mutations occurs either at $G_{-32}$ or $T_{-31}$. These two nucleotides are common to the anticodons of all threonine specific tRNAs.

STRUCTURE OF THE mRNA

A region encompassing nucleotides -172 to +58 was cloned in a vector downstream of an SP6 promoter and transcribed in vitro. Its conformation was investigated using several chemical and enzymatic probes (Moine et al., 1988). The four bases were monitored at one of their Watson-Crick base-pairing positions using dimethylsulfate (DMS) which modifies positions N1-A and N3-C and a carbodiimide derivative (CMCT) which modifies positions N1-G and N3-U. Position N7 of purines involved in tertiary interactions or in stacking was probed with DMS and diethylpyrocarbonate (DEPC) for G and A respectively. Single stranded regions were tested with RNases T1 and U2 and with nuclease S1, and structured regions with RNase V1. Experiments were conducted under native conditions (in the presence of magnesium, as used for the study of the interaction between tRNA and aminoacyl-tRNA synthetase). In addition, in order to test the stability of helices, the reactivity of bases toward the chemical probes was also tested in the absence of magnesium (semi-denaturing conditions). Tertiary interactions which are less stable than canonical base pairings are expected to melt under such experimental conditions. Detection of modified nucleotides and cleavages was done by the primer extension technique. Advantages and limitations of this structure probing technique have been discussed elsewhere (Ehresmann et al., 1987).

Our results indicate that the mRNA region folds in five different structural domains (Fig. 1).

Domain I (between +4 and -12) contains the initiation codon AUG and the Shine and Dalgarno sequence. This domain exhibits a strong reactivity toward the chemical probes indicating that it is essentially unpaired.

Domain II (between -12 and -49) corresponding to the thrS operator shares structure similarities with the tRNA$^{Thr}$ isoacceptors (see Fig. 1) : i) it contains a seven base anticodon-like loop, with four nucleotides idential to those of tRNA$^{Thr}$ isoacceptors. Among the four, two correspond to nucleotides shared by the anticodon sequence of all known isoacceptors ; ii) it contains an anticodon-like helix (helix III) (between -22 and -42) with several base pairs identical to those of the different tRNA$^{Thr}$ isoacceptors. Helix III as well as the neighboring helix II requires the presence of magnesium to be stabilized.

Domain IIIC (between -50 and -73) exhibits reactivity of all nucleotides at Watson-Crick positions under native conditions and also is readily split by single strand specific nucleases, indicating that no Watson-Crick base pairs exist in that region. Some of the purine N7 positions are strongly reactive whereas others are only marginally reactive suggesting their involvment in stacking or in non canonical base pairs.

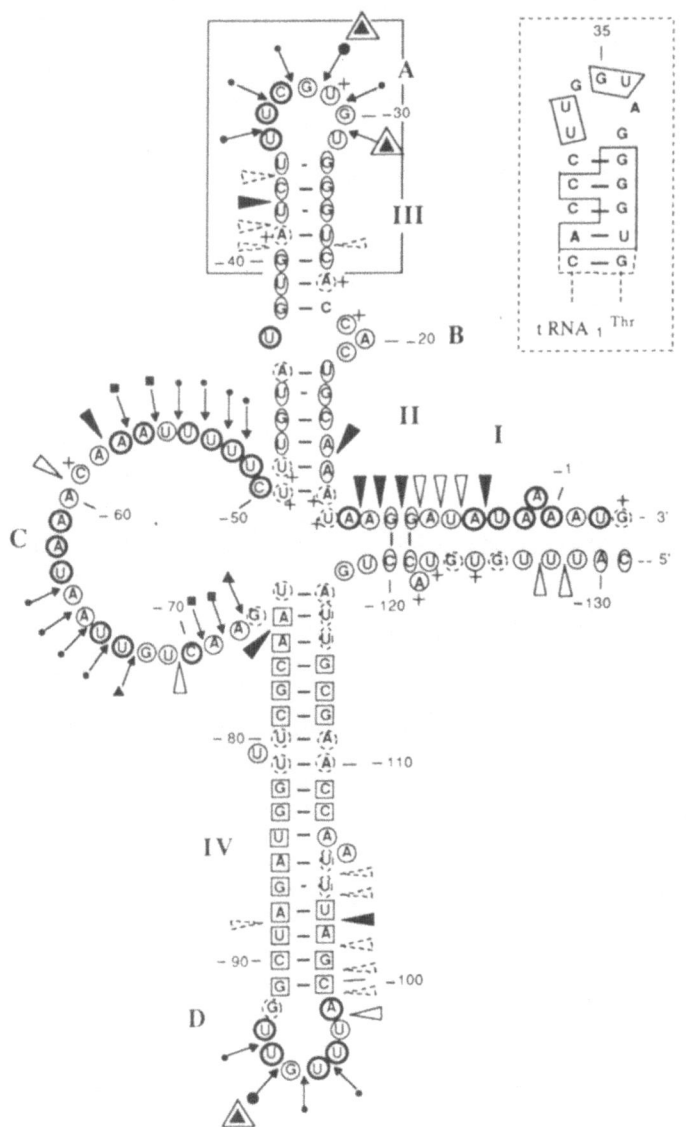

Fig. 1. Reactivity of nucleotides +3 to -131 of the wild-type mRNA towards enzymatic and chemical probes in the absence and the presence of threonine-tRNA ligase. *Accessibility of the naked RNA to the enzymatic probes : RNase T1, ( ▲ ) major cut, ( ▴ ) moderate cut ; RNase U2, ( ■ ) major cut, ( ▪ ) moderate cut ; nuclease S1, ( ● ) major cut, ( • ) moderate cut ; RNase V1, ( ◄ ) major cut, ( ◁ ) moderate cut, ( ◅ ) minor cut. Protection specifically induced by threonine-tRNA ligase towards RNase T1 is denoted by ( ▲ ) for strong protection. Reactivity of Watson-Crick positions towards DMS and CMCT : ( O ) reactive under native conditions (strong reactivity is denoted by bold circle, moderate reactivity by thin circles and marginal reactivity by dotted symbols) ; ( [] ) unreactive under native conditions and reactive under semi-denaturing conditions (low reactivity is denoted by ⫯ ) ; ( □ ) unreactive under both native and semi-denaturing conditions. ( ⫯ ) indicates increase of reactivity after removal of magnesium. The loops are denoted by A to D and helices by I to IV. The anticodon-like arm in the mRNA is enboxed. The inset describes the anticodon region of tRNA$_1^{Thr}$ in which the sequence similarities with the mRNA are framed (from Moine et al., 1988).*

Most of the nucleotides in the helical region of domain IV (between −74 and −117) are unreactive under native or semi-denaturing conditions indicating a high stability of the helix resulting from the high content of G-C pairs. The loop in domain IV is a well defined eight base loop closed by two G-C pairs. Since all nucleotides in the loop are highly reactive to the chemical probes, and exhibit the major sites of cleavage with nuclease S1 and RNase T1, this confirms that all bases are exposed.

Domain V (between −117 and −131) exhibits a strong reactivity to the chemical probes, as domain I indicating that it is essentially unpaired.

A set of deletions and point mutations isolated after site specific mutagenesis indicates that domains I to IV are implicated in the regulation and in the level of thrS expression. The main regulatory domain remains the tRNA-like structure of domain II (Springer et al., unpublished results).

The conformation of three constitutive mutants containing single base changes in the operator region leading to the loss of the regulatory control was also investigated. In mutant L6-8 (Fig. 2), the replacement in the anticodon-like triplet of $G_{-32}$ by an adenine does not lead to any conformational change compared to the wild type mRNA. This suggests that the regulatory control is directly linked to the existence of a guanine at that position in the loop. In the case of mutant M4-11 (Fig. 2), the replacement of $C_{-16}$ involved in a G-C base pair by an uracil results in a decrease of the calculated free energy of helix II from −4.6 to −2.4 Kcal/mol. All nucleotides of helix II of mutant M4-11 become reactive at their Watson-Crick positions as the result of a destabilization of this helix. However, the structure of the anticodon-like region as well as that of domain IV remains unaltered, suggesting that the loss of control in that case is linked to a lack of the correct geometry of the anticodon-like region. In mutant M6-1 (Fig. 3), $G_{-40}$ involved in a G-C base pair within helix III is replaced by an adenine, leading to a drastic decrease of the calculated free energy of the helix (from −6.6 to −0.2 Kcal/mol). Unexpectedly, the destabilization of helix III results in an extensive structural rearrangement of a large region encompassing nucleotides −73 to −13, while the region containing helix IV and loop D remains unaltered. In particular, most of the nucleotides of the anticodon-like loop become unreactive at Watson-Crick positions and are cleaved by RNase V1. From our probing experiments, we propose a secondary structure model of the operator region of this mutant which differs considerably from the wild-type structure, in particular for the anticodon-like stem and loop which are now completely involved in another structure. Note that $G_{-32}$ found essential in feedback control is base paired with U−74 in this mutant.

## mRNA − THREONYL-tRNA SYNTHETASE BINDING

A feedback regulation mechanism can be imagined in which the synthetase would bind to the mRNA at a region located just upstream of the Shine-Dalgarno sequence, sharing sequence and structural similarities with $tRNA^{Thr}$, thus preventing the ribosomes from translating the thrS mRNA. In line with this, modification of a single base in the anticodon loop that is essential for protein binding or a conformational change of the anticodon-like region would prevent the binding of the ThrRS leading to loss of the control. The finding that both the anticodon loop and stem of $tRNA^{Thr}$ interact with ThrRS (Theobald et al., 1988) gives strong support to the reality of an interaction between the mRNA and ThrRS.

In order to provide evidence for such a mechanism, a first set of footprinting experiments using RNase T1 as a probe was conducted on the wild-type mRNA fragment and on the three different mutants in the presence of ThrRS, under conditions used for the formation of the tRNA/synthetase complex. Results

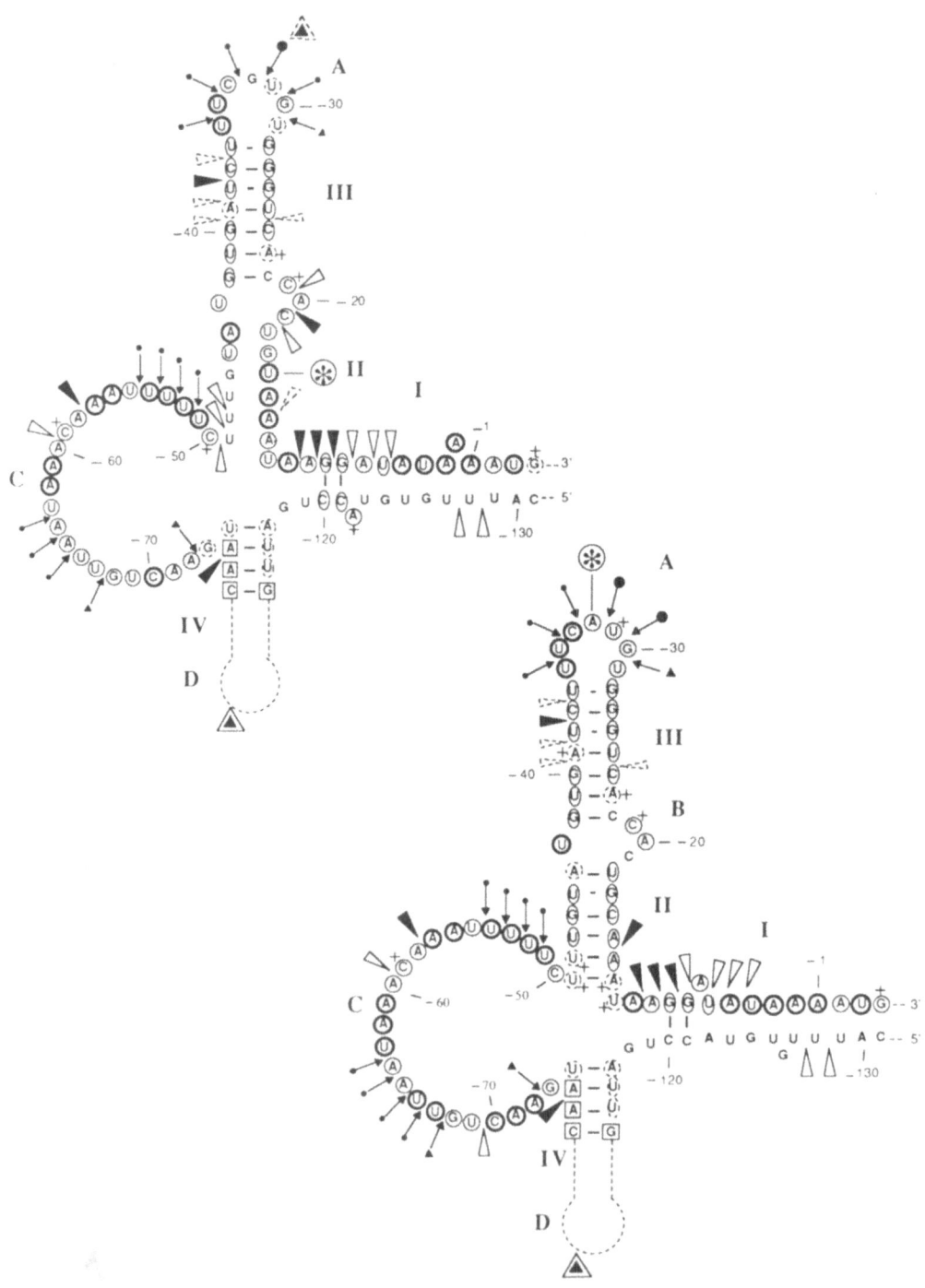

**Fig. 2. Reactivity of nucleotides +3 to −131 of mutants L6−8 (bottom) and M4−11 (top).** Same legend as Fig. 1. Helix IV and loop D are represented by a dotted line since their reactivity remains unchanged.

are summarized in Fig. 1-3. In the case of the wild type mRNA, strong protection could be detected at $G_{-30}$ and $G_{-32}$ in the anticodon-like region (site 1) and at $G_{-95}$ in the external loop of domain IV (site 2). Unexpectedly, in the case of the three constitutive mutants, $G_{-95}$ remained protected, suggesting that even in the absence of the regulatory control, the synthetase binds to the mRNA in a region (site 2) which is not affected by the different point mutations. When $G_{-32}$ is replaced by an adenine (mutant L6-8), no protection could be found at $G_{-30}$. This strongly suggests that though ThrRS still binds to the mRNA, the anticodon-like loop is no longer in close contact with the protein. In mutant M4-11 ($C_{-16} \rightarrow U$), destabilization of helix II has no effect on the structure of the anticodon-like arm. In that case, $G_{-30}$ is not protected whereas $G_{-32}$ in the anticodon-like sequence is weakly protected. It was shown that the destabilization of helix II leads to a large disordered region in the central area of the RNA. Therefore, a dynamic equilibrium between several conformations may exist, resulting in different orientations of the anticodon-like arm. The synthetase might be in contact to some extent with one of these forms, thus resulting in a weak and partial protection of $G_{-32}$. In mutant M6-1 ($G_{-40} \rightarrow A$), the large conformational change of the anticodon-like region led to a complete rearrangement of the anticodon arm. In the resulting structure, $G_{-30}$ and $G_{-32}$ are base paired, thus leading to the loss of RNase T1 cleavage sites in this region. Nevertheless, as in the other mutants, loop D is still protected by the protein but no additional protection could be observed.

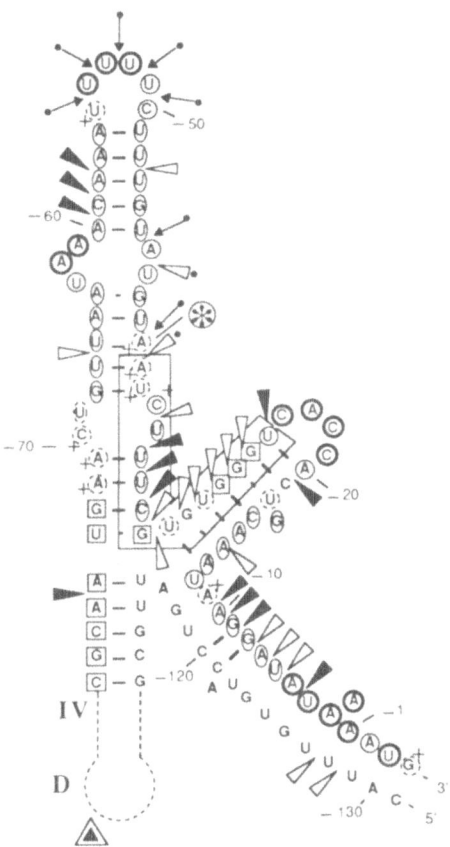

Fig. 3. Reactivity of nucleotides +3 to -131 of mutant M6-1. *Same legend as Fig. 1. Nucleotides corresponding to the anticodon-like arm in the wild type mRNA are boxed.*

Filter binding assays performed under conditions used to measure the specific complex formation between ThrRS and tRNA$^{Thr}$, confirm that the synthetase binds to the thrS mRNAs. These experiments together with the structure probing experiments suggest that site 2 is required for stabilizing of the ThrRS-mRNA complex, whereas interaction with site 1 containing the anticodon-like structure is required for the repression process.

At least two likely mechanisms can be imagined, which would explain translation repression : either the repressor blocks the ribosome on the translational initiation site or it prevents the binding of the ribosome on the messenger. In order to obtain more information on how ThrRS regulates the expression of its own messenger, "toeprinting" experiments have been performed as described by Hartz et al. (1988). In these experiments, a translational initiation complex can be formed with initiation factor-free ribosomal 30S subunits, deacylated tRNA and mRNA. Specific binding of the 30 S subunit to the mRNA stops elongation with reverse transcriptase of a DNA primer hybridized to a region of the mRNA 3' to the initiation codon. The addition of ThrRS prevents binding of the ribosomal subunit to the wild type mRNA leading to the loss of the above extension stop. In the three mutant mRNAs however, primer extension inhibition still occurs. This suggests : i) In the case of the wild type mRNA, ThrRS competes with the ribosome for binding to the translational initiation site. ii) The interaction of ThrRS with the tRNA-like region is responsible for this competition.


TRANSACTING REGULATORY MUTANTS

The results reported above indicate that ThrRS recognizes two different RNA substrates : threonine specific tRNAs and its specific mRNA. In order to find out whether the gene thrS is divided in separate domains for the recognition of these two RNAs (one domain corresponding to regulation and the other to the enzymatic activity), we selected mutants with an altered regulatory behaviour and then analysed kinetic properties of these mutant enzymes (Springer et al., 1988). Two different selections yielded two different types of regulatory mutants, the first consists of ThrRS mutants that are unable to repress thrS mRNA translation (R$^-$ mutants) and the second are ThrRS mutants that superrepress thrS mRNA translation (R$^{++}$ mutants).

ThrRS mutants unable to repress thrS mRNA translation were selected using the same strategy which enabled us to obtain the operator mutants. The presence of a thrS carrying plasmid makes it impossible for a strain carrying a thrS-lacZ fusion to grow on lactose. After mutagenesis of the thrS carrying plasmid, strains are selected that grow on lactose and that still overproduce ThrRS normally and which complement chromosomal thrS mutants. All such the mutants have a low specific activity for aminoacylation under normal conditions. Three mutants giving extracts with the highest threonine aminoacylation activities were analyzed in details.

The three mutants were shown to map at the 3' end of the gene. Nucleotide sequencing of the mutants demonstrate that in two cases the Arg$_{583}$ residue is changed to an histidine and in the third case Arg$_{583}$ and Arg$_{612}$ are both changed to cysteine. It is remarkable that in all three cases the basic aminoacid is replaced by a more acidic one.

Synthetases catalyze in addition to the aminoacylation of tRNA a partial reaction (exchange reaction) which concerns only two substrates ATP and the aminoacid :

ATP + aa + Enz $\rightleftharpoons$ [AMPaa Enz.] + PP

The mutant enzymes have a wild type Vmax and a wild type Km for threonine and ATP in this exchange reaction. In the overall reaction which also concerns the third substrate of the enzyme tRNA$^{Thr}$, the $V_{max}$ is unchanged but the Km for tRNA$^{Thr}$ is about 10 times higher than for the wild-type enzyme. Therefore the mutated enzymes have a low affinity for tRNA$^{Thr}$ but behave normally with the two other substrates ATP and threonine.

The ThrRS mutants that super-repress thrS mRNA translation were found using a slightly different isolation method. A strain containing an operator mutant of a thrS-lacZ fusion grows in lactose as it is not repressed by a multicopy plasmid carrying thrS. The ThrRS is incapable of regulating mRNA translation most probably because of a low affinity for mutant mRNA. Mutant plasmids in thrS that repress operator constitutive mutants (compensatory mutants) for the defect of the operator were selected. It was possible to isolate these super repressor ThrRS mutants only with the two operator mutants M6-1 and M4-11, both being located in the stem of the operator. All the mutant plasmids behave as superrepressors with a wild type thrS-lacZ fusion, i. e. they repress β-galactosidase synthesized from such a fusion more strongly than the wild type thrS gene on the same multicopy plasmid. The overproduction of the R$^{++}$ ThrRS mutant causes a repression of β-galactosidase synthesis that varies between 1 and 3 000 fold depending on the operator mutant (Table 1). Constitutive mutants located in the loop of the operator site I are not repressed at all by R$^{++}$ 4-11-8. Some other operator mutants such as the one located in the stem of the operator are repressed to an intermediary extent which varies between 3 and 30 fold depending on the mutant. These results indicate that the level of repression caused by R$^{++}$4-11-8 is allele specific, i. e., it depends on the nature of the operator allele.

Besides R$^{++}$4-11-8 only one other mutant represses β-galactosidase synthesis from a wild type thrS-lacZ fusion in such an extreme way : the others have a weaker effect. The two extreme mutants isolated independently are identical, they map in the middle of the gene and both carry a double change in which Glu$_{258}$ and Glu$_{259}$ residues are changed to Lys residue. It should be noticed that in contrast to R$^{-}$ mutants acidic residues are changed to basic ones. This indicates that electrostatic interactions play a role, but allele specificity indicates that the interaction is not only due unspecific charge effects.

Enzymatic tests on crude extracts of the strain carrying the R$^{++}$ 4-11-8 overproducing plasmid indicate that all the kinetic parameters for ThrRS activities are approximately wild type in the exchange reaction. In the aminoacylation reaction the Km for the tRNA was approximately 5 times lower than with the wild type enzyme. Therefore the thrS mutant which could be considered to have a higher affinity for mRNA (higher repressor activity) also has a higher affinity for threonine specific tRNAs.

CONCLUSION

Our studies show that the mRNA region upstream of the thrS initiation codon folds in several distinct structural domains. One of these ressembles the anticodon arm of tRNA $^{Thr}$ isoacceptors : 1) It consists of a seven base loop containing at the right place the two nucleotides shared by the anticodon sequence of all known isoacceptors. 2) It also displays a stem with several base pairs identical to those of the anticodon stem of different tRNA$^{Thr}$ isoacceptors.

Genetic selection permitted the identification of a site in the mRNA called the "operator" which is essential to the translational negative autocontrol of thrS. This site is localised precisely in the region of the tRNA anticodon-like arm. This suggested that threonine-tRNA synthetase controls its own synthesis by interacting with the region that shares structural features with its natural substrate. This is now known to be true since the synthetase was shown 1) to protect the operator region of the mRNA from enzymatical and chemical reagents 2) to bind the mRNA on nitrocellulose filters with a low but measurable affinity. Studies with mutant mRNAs indicate that the control is lost either by a conformational change of the anticodon-like arm or by a more specific change of one in the two bases shared by all tRNA$^{Thr}$ anticodons and which do not alter the mRNA conformation, suggesting that those two bases are, per se, essential to the control.

However, biochemical as well as genetic evidence indicates that the mRNA domain implicated in regulation extends beyond the region which resembles tRNA$^{Thr}$. Several other domains are also implicated in the level and regulation of the expression of threonyl-tRNA synthetase. In particular, in addition to the binding site of the synthetase to the anticodon-like arm, a second binding site was identified in vitro. This site still exists in mutants which lose control. The precise function of this "protein anchoring" site has yet to be determined.

At least two likely mechanisms can be imagined which would explain translation repression. The binding of the synthetase traps the ribosome on the translational initiation site or it prevents the binding of the ribosome to the messenger. We have shown that the second mechanism is the most likely, i. e. , that an excess of ThrRS prevents the binding of the ribosomal small subunit to the wild type mRNA. The synthetase does not prevent ribosome binding to this site in constitutive mutant mRNAs. This suggests a competition between ThrRS and the ribosome for binding to the translation initiation site and the interaction of ThrRS with the tRNA-like region is responsible for the competition.

The selection of ThrRS mutants with altered control properties (either incapacity or supercapacity to repress thrS mRNA translation) and the study of their enzymatic properties suggest that the binding domains of the mRNA and the tRNA on the synthetase are overlapping.

Other examples of tRNA-like structures are known and the best documented case is the presence of such structures at the 3' end of several plant viral RNAs (see J. P. Ebel et al. , this volume). Recently, two mitochondrial aminoacyl-tRNA synthetases, the tyrosine-tRNA synthetase of Neurospora crassa (Akins and Lambowitz, 1987) and the leucine-tRNA synthetase of Saccharomyces cerevisiae (Herbert et al. , 1988) were shown to be directly involved in mRNA splicing. This new property of aminoacyl-tRNA synthetases could be explained by an interaction between the synthetases and tRNA-like structures in the introns. A tRNA-like structure exists also in the leader mRNA of the histidine attenuator region of S. typhimurium (Ames et al. , 1983). These tRNA-like structures could in fact be molecular fossils of the original RNA world. They were suggested to be used by primitive living systems as tags on RNA genomes to identify them as substrates for replicases (Weiner and Maizels, 1987). Another possibility is that these tRNA-like structures arose through duplication of a tRNA gene and transposition in front of another gene whose expression would then become sensitive to macromolecules binding the original tRNA (Ames et al. , 1983).

ACKNOWLEDGEMENTS

We are grateful to J. W. B. Hershey for critically reading the manuscript and to L. Paineau for typing it. This work was supported by grants from the Centre National de la Recherche Scientifique (LP 6201 and UA1139), from the Ligue Nationale Française contre le Cancer, from the Association pour la Recherche contre le Cancer, from the Fondation pour la Recherche Médicale and from E. I. du Pont de Nemours (to M. G.-M.) and from the Ministère de la Recherche et de l'Enseignement Supérieur (87. C. 0562) (to B. E. and M. S.).

26

REFERENCES

Akins R. A. and Lambowitz A. M. , 1987, A protein required for splicing group I introns in Neurospora Mitochondria is mitochondrial tyrosyl-tRNA synthetase or a derivative thereof, Cell, 50: 331-345.

Amos B. N. , Tsang T. H. , Buck M and Christman M. F. , 1983, The leader mRNA of the histidine attenuator region ressembles tRNA$^{His}$. Possible general regulatory implications, Proc. Natl. Acad. Sci. USA, 80: 52.

Butler, J. S. , Springer, M. , Dondon, J. , Grunberg-Manago, M. , 1986, Posttranscriptional autoregulation of E. coli threonyl-tRNA synthetase expression in vivo, J. Bacteriol. , 165: 198-203.

Cech, T. R. and Bass B. L. , 1986, Biological catalysis by RNA, Ann. Rev. Biochem. , 55: 599-629.

Ehresmann, C. , Baudin, F. , Mougel, M. , Romby, P. , Ebel, J. -P. and Ehresmann, B. , 1987, Probing the structure of RNAs in solution", Nucleic Acids Res. , 15: 9109-9128.

Hartz, D. , McPheeters, D. S. , Traut, R. and Gold, L. , 1988, Extension inhibition analysis of translation initiation complexes, Methods in Enzymology, in press.

Herbert C. J. , Labouesse M. , Dujardin G. and Slonimski P. P. , 1988, The NAM2 proteins from S. cerevisiae and S. douglasii are mitochondrial leucyl-tRNA synthetases, and are involved in mRNA splicing", EMBO J. , 7: 473 -483.

Kikuchi, Y. , Ando, Y. and Shiba, T. , 1986, Unusual priming mechanism of RNA-directed DNA synthesis in copia retrovirus-like particles of Drosophila, Nature 323: 824-826.

Lestienne, P. , Plumbridge, J. A. , Grunberg-Manago, M. and Blanquet, S. , 1984, Autogenous repression of E. coli threonyl-tRNA synthetase expression in vitro, J. Biol. Chem. 259: 5232-5237.

Moine, H. , Romby, P. , Springer, M. , Grunberg-Manago, M. , Ebel, J. -P. , Ehresmann, C. and Ehresmann, B. , 1988, Messenger RNA structure and gene regulation at the translational level in Escherichia coli : the case of threonine tRNA$^{Thr}$ ligase, Proc. Natl. Acad. Sci USA, 85: 7892-7897.

Springer, M. , Plumbridge, J. A. , Butler, J. S. , Graffe, M. , Dondon, J. , Mayaux, J. -F. , Fayat, G. , Lestienne, P. , Blanquet, S. and Grunberg-Manago, M. , 1985, Autogenous control of E. coli threonyl-tRNA synthetase expression in vivo, J. Mol. Biol. 185: 93-104.

Springer M. , Graffe M. , Butler J. S. and Grunberg-Manago M. , 1986, Genetic definition of the translational operator of the threonine-tRNA ligase gene in Escherichia coli, Proc. Natl. Acad. Sci. USA, 83: 4384-4388.

Springer, M. , Graffe, M. , Dondon, J. , Grunberg-Manago, M. , Romby, P. , Ehresmann, B. , Ehresmann, C. and Ebel, J. -P. , 1988, Translational control in E. coli : the case of threonyl-tRNA synthetase, in : "Genetics of Translation", M. F. Tuite et al. , eds. , NATO ASI Series, Vol. H14, Springer Verlag, pp. 463-478

Theobald, A., Springer, M., Grunberg-Manago, M., Ebel, J.-P. and Giégé, R., 1988, Tertiary structure of <u>Escherichia coli</u> tRNA$_3^{Thr}$ in solution and interaction of this tRNA with the cognate threonyl-tRNA synthetase, <u>Eur. J. Biochem.</u>, 175:511-524.

Weiner A.M. and Maizels N., 1987, tRNA-like structures tag the 3' end of genomic RNA molecules for replication : implications for the origine of protein synthesis, <u>Proc. Natl. Acad. Sci. USA</u>, 84:7383-7387.

# STRUCTURAL AND FUNCTIONAL tRNA MIMICRY OF THE 3'-END OF TURNIP YELLOW MOSAIC VIRUS RNA

Jean-Pierre Ebel, Richard Giegé and Catherine Florentz

Institut de Biologie Moléculaire et Cellulaire du
Centre National de la Recherche Scientifique,
15, rue R. Descartes, 67084 Strasbourg-Cedex, France

## INTRODUCTION

The 3'-ends of several plant viral RNAs show a number of functional characteristics of tRNAs[1,2]; they are recognized by a set of tRNA-specific proteins, including aminoacyl-tRNA synthetases. So, the RNA of turnip yellow mosaic virus can be valylated by yeast valyl-tRNA synthetase[3,4] with kinetic constants very close to those of the aminoacylation of yeast tRNA$^{Val}$ by this enzyme[5]. The 3'-ends of these viral RNAs however, are lacking several characteristic primary structural features of tRNAs such as strategic D- or T-loop sequences and modified bases. Moreover, they cannot be folded *a priori* into a canonical tRNA cloverleaf. In the case of TYMV RNA, the question arose as to "how do two structures as different as tRNA$^{Val}$ and the 3'-region of the viral RNA behave in such a similar fashion in the presence of valyl-tRNA synthetase ?". This question might be answered if one supposes that similar structural domains are recognized by the synthetase and exist at the level of the three dimensional structures and even at the level of their secondary structures. In this view, the secondary structure of the 3'-end of TYMV RNA was established by enzymatic footprinting methodologies[6]. A three-dimensional L-shaped conformation mimicking tRNA, but involving a new RNA folding principle, the pseudoknot, was proposed by the Leiden group[7,8] for the 86 last nucleotides of this RNA. A rigourous graphical modelling allowed to assess the reality of this pseudoknotted folding[9]. To understand the involvement of the 3'-end of the TYMV RNA in aminoacylation, the direct contact points between this part of the RNA and yeast valyl-tRNA synthetase were determined[10]. Moreover, the study of the valylation of tRNA-like transcripts from cloned cDNA of TYMV RNA permitted to determine the minimal length of the RNA necessary for optimal valylation[11].

## STRUCTURAL ORGANIZATION OF THE tRNA-LIKE END OF TYMV RNA

### A Peculiar Secondary Structure

When the sequence of the tRNA-like 3'-end from TYMV was established[11,12], it could not be unambiguously folded into a secondary structure mimicking the tRNA cloverleaf. Thus, the secondary structure of the 3'-end of TYMV RNA, fully active in valylation[14], was established from partial nuclease digestion patterns[6]. This approach has already been used to establish the secondary structures of several small RNAs and to check structural features in tRNAs and rRNAs[15]. The method depends on the differential accessibility of residues or domains to nucleases and uses rapid RNA sequencing methods and analysis of results after autoradiography of the sequencing gels. Four enzymes were used : the base-specific RNAse T1 and RNAse CL3 and the structure specific nuclease S1 and cobra venom ribonuclease. In such a way results obtained with one nuclease could be corroborated with those obtained with the others. Experiments were conducted on both 5' and 3' end-labelled molecules to distinguish the artefactual secondary cuts in the mapping experiments (for a discussion see 16).

Figure 1A shows the secondary structure of the tRNA-like fragment of TYMV RNA deduced from the nuclease accessibility studies. This model displays six main loop and stem regions. When compared to the cloverleaf structure of yeast tRNA[Val] (Fig. 1B) striking similarities as well as significant differences appear. The anticodon arm of tRNA[Val] resembles arm III in the tRNA-like fragment. Analogies appear between the T- and D-arms of tRNA[Val] and arms II and IV of the viral RNA, respectively. Some structural constants found in tRNA are conserved in the tRNA-like fragment, i.e. the CCA-end, the G-C base pair at the top of the T-stem and the length of the T- and anticodon-loops. Furthermore, the model shows a single stranded region which is homologous to the extra-loop of a tRNA. The main difference between the two structures concerns the amino acid accepting region. Whereas the tRNA possesses a 7 base pair stem, the viral tRNA-fragment has a much more complex structure composed of the two extra-arms I and V. Arm VI does not have any resemblance with a tRNA element ; since this region contains the last 50 nucleotides of the coat protein cistron it is probably not involved in the tRNA-like folding.

An L-Shaped Three-Dimensional Model

Three-dimensional similarities with parts of the classical L-shape can be found within the remaining 110 nucleotides of the TYMV RNA fragment corresponding to arms I to V. The behaviour of loops II, III and IV, in terms of accessibility towards the enzymatic probes and in terms of sequence and size, resembles that observed in several tRNAs for T-, anticodon- and D-loops, respectively, suggesting that the three-dimensional structure of arms II, III and IV is similar to that of the corresponding parts in tRNA. More direct evidence was brougth by an alkylation experiment of the phosphates in these regions with the chemical probe ethylnitrosourea. In all tRNAs tested so far, two phosphates of the T-loop are inaccessible to this chemical probe. In the tRNA-like structure, two inaccessible phosphates are found in loop II at topologically similar positions (see Fig. 1A) as in all other tRNAs tested so far[17]. These two phosphates are located in the hinge region of tRNA where the two helical domains form the characteristic L-shaped structure, more precisely in the cleft where the D-loop overlaps the T-loop. The ethylnitrosourea experiment shows that a similar structure exists in the viral RNA : arms III and IV would form the anticodon limb of a L-shaped structure and arm II the begining of the amino acid accepting limb of the L. Precise analysis of

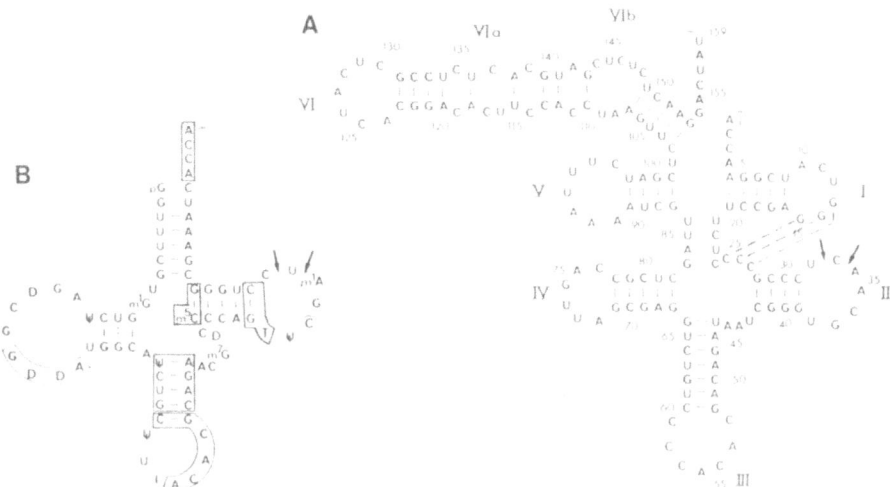

Figure 1. Secondary structure of the 3'-end tRNA-like fragment (n=159) of TYMV RNA. (A) Secondary structure of the viral RNA fragment and (B) of yeast tRNA[Val]. The nucleotides of tRNA[Val] common to the viral RNA are boxed. The arrows indicate phosphates of the T-loop or T-like loop which are inaccessible to alkylation by ethylnitrosourea. Dashed lines show the Watson-Crick pairings needed for the pseudoknotted folding of the amino acid accepting arm (see Fig. 2 and 3).

structural and chemical accessibilities of the nucleotides of the region comprising arm I to arm II, as well as sequence comparisons, leed Pleij and coworkers[7,8] to the proposal of an overall L-shaped conformation for the tRNA-like fragment, where the amino acid accepting limb arises from a peculiar pseudoknotted folding of the region 1 to 42. Stem and loop V are not required in this folding. For the pseudoknot, data pointed to base pairing involving the triple G sequence G13 to G15 of loop I with the triple C sequence C25 to C27 of the single stranded domain separating hairpins I and II (Fig. 1). The coaxial staking of stem I, stem II and of the previously mentioned three G-C base pairs defines an amino acid acceptor arm of twelve stacked base pairs as found in classical tRNAs (Fig. 2). Two short single-stranded stretches, comprising 3 (A10 to U12) and 4 (U21 to C24) nucleotides, respectively, cross the helix from one strand to the other and ensure the chain continuity. Calculations of the minimal number of nucleotides needed to cross either the shallow or the deep groove showed that the proposed model should be feasible[18]. Although the resulting wire model (Fig. 2) was very attractive, it was important to test the proposal by submitting it to rigourous graphical modelling[9].

The graphical modelling was done using the program FRODO[19] on the PS300 graphics system of Evans and Sutherland. Building of the pseudoknot was performed as follows : examination of the wire model previously proposed for the amino acid acceptor arm suggests a perfect helical continuity between residues U9 and C25 as well as between G13 and C42 (Fig. 2). An A-type helix was thus built comprising all the 12 stacked bases of the amino acid arm, ignoring for that purpose the true sequence continuity. The backbone was then restored by chain cuts and addition of the missing stretches A10, C11, U12 and U21, C22, U23, C24 (named

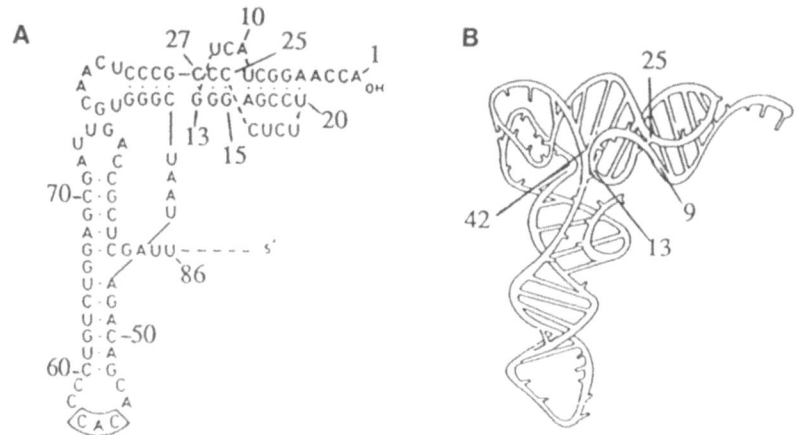

Figure 2. Folding of the minimal RNA sequence required to model an L-shaped conformation of the TYMV tRNA-like sequence. In (A) the sequence is drawn such as to emphazise the wire three dimensional model shown in (B). (Adapted from 8).

respectively loops L2 and L1). In fact, concerning the pseudoknot, given its overall folding, the major problem to be solved was fixing the conformation for these two short stretches. This remark is of importance since it shows that no much hypothesis was needed for the building of the pseudoknot itself, which adds to the confidence of its model. In fact the conformation of the two loops L1 and L2, allowing the chain to come back twice in the helical part of the pseudoknot, seems to be accurately determined since their is a very small number of residues comprised in each loop and since the position and orientation of the ends of both loops was perfectly known.

The construction of the other parts of the molecule, i.e. T-stem, anticodon arm and D-stem analogues, was achieved by following as a guide as much as possible the structure of classical tRNA. Yeast tRNA[Asp] was taken for that purpose. The building of the interactions between the two loops corresponding to the D- and T-loops in canonical tRNAs, was rather dependent on hypothesis made during the course of modelling. Consequently, this part of the proposed model can only be considered as a putative conformation.

The overall TYMV tRNA-like structure (86 nucleotides) and for comparison, the well known structure of tRNA[Asp] [20,21] are shown in Figure 3. The resemblance is striking, even when considering the *a priori* poblematic CCA-stem. Indeed, the pseudoknot is well integrated in a confomation giving the amino acid accepting stem the appearance of a double-stranded RNA helix. However, the presence of the two loops, L1 and L2 which are integral parts of the pseudoknot confers a protrusion to one side of the tRNA-like structure[7,9]. An important point can readily be stressed as a direct consequence of the different chain folding in the two structures. While in classical tRNA the CCA-stem is covalently anchored at two different points far from each other, in the tRNA-like structure the same stem has only one single attachment point with the rest of the molcule, i.e. the phosphodiester bond between C27 and G28. Thus, comparison of the two molecules allows to infer much more independence of the two limbs of the L-shaped tRNA-like structure. This could give

A

B

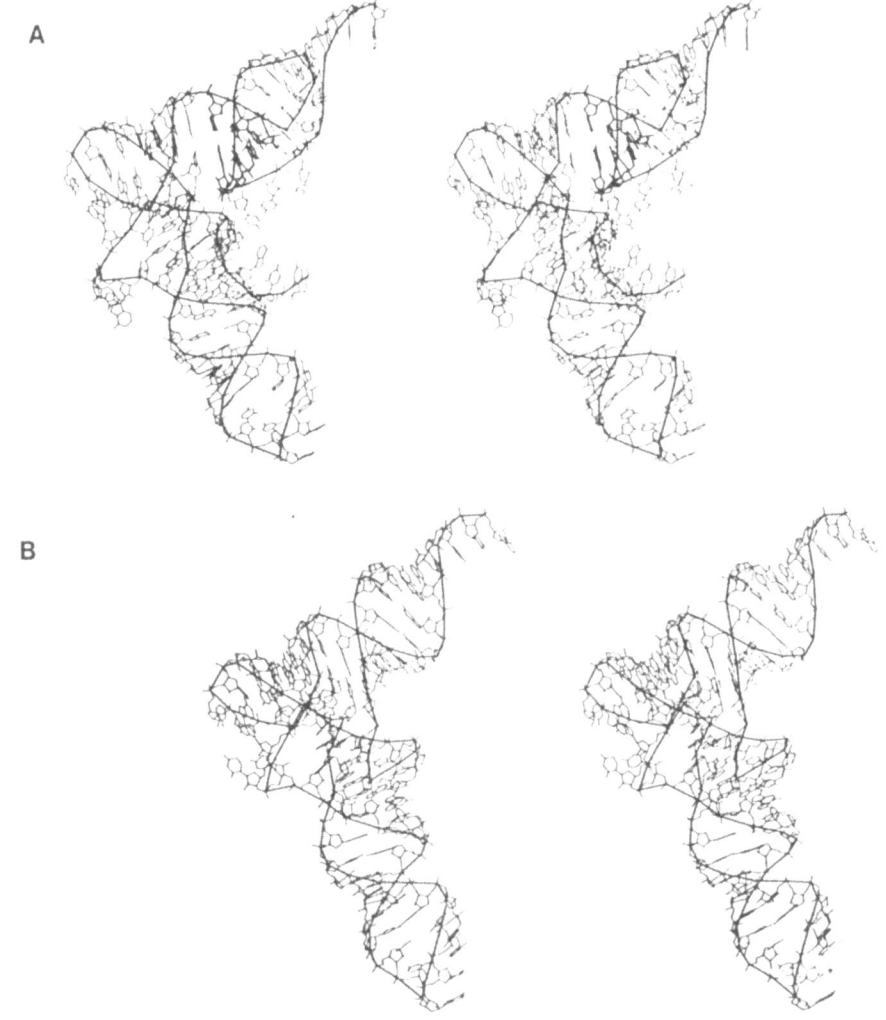

Figure 3. Stereoscopic views of the modellized TYMV tRNA-like fragment (nucleotides A1 to U86) (A) and of the crystallographic structure of yeast tRNA[Asp] (B).

to the latter the ability to modulate the recognition of tRNA- or RNA-specific proteins.

This work represents the first attempt for a precise graphical modelling of a large RNA structure presenting very novel features. Beside large extra-loop tRNA[Ser 22,23] and D-loop deprived mitochondrial tRNA[Ser 24], the present model is the most detailed structure for a (t)RNA molecule. Meanwhile, this approach combining chemical mapping and computer modelling, was employed to model chloroplastic 5S RNA[25]. Though the TYMV tRNA-like model represents only a first step in determining its exact three-dimensional structure, it can provide usefull information for the design of new experiments in order to further improve the model. For instance, the selection of positions for site directed mutagenensis can be done in a more sophisticated manner. Also the design of new chemical mapping experiments can benefit from such a model. Thus one can imagine a procedure which essentially cycles from the model to new experiments and back to the model again. This approach should be especially benefic to test and refine the present conformation in the D- and T-loop regions for which hypothesis were necessary for building. Such a feedback procedure has to be envisaged as the approach of choice for structural determination of RNA molecules or domains as long as crystals, suitable for X-ray analysis, are not available.

## DOMAINS OF THE tRNA-LIKE FRAGMENT RELEVANT TO VALYLATION

Many methods, including genetics, photo-crosslinking, nuclease digestion and chemical accessibility, have been used to study the interaction domains between tRNA and their cognate aminoacyl-tRNA synthetases[26,27]. With tRNA-like structures, in addition to the recent tyrosylation asssays on mutant brome mosaic virus RNAs synthesized *in vitro*[28,29], only sequence comparisons with homologous tRNAs have provided structural information. For instance the importance of the anticodon arm in the 159 nucleotide long fragment from TYMV has already been stressed for the valylation reaction of the viral RNA[6,10], as it was for that of canonical tRNA$^{Val}$ [30]. As seen in Figure 1, yeast tRNA$^{Val}$ and the TYMV tRNA-like structure present indeed their highest sequence homology in the anticodon stem and loop. Such data, however, are of indirect nature and do not allow to ascertain the degree of involvement of this region in a recognition site. Therefore two series of experiments were designed to address this question.

### Contact Points of the Viral Fragment with Valyl-tRNA Synthetase

Because aminoacylation of TYMV RNA by valyl-tRNA synthetase is particularly efficient, it was important to know the direct contact points of the RNA interacting with the synthetase[10] and to compare them with those found for yeast tRNA$^{Val}$ [31]. Experiments consisted in footprinting of the complexed viral RNA fragment with S1 nuclease and cobra venom ribonuclease, structural probes for single stranded and higher order conformations in RNAs, as well as with ethylnitrosourea, the alkylating reagent specific for phosphate groups. The interaction areas were deduced from comparison of hydrolysis products of free RNA with those of molecules complexed with valyl-tRNA synthetase.

With S1 nuclease, clear protections were found in the anticodon loop. Due to technical difficulties, it was not possible to probe the CCA-end, which should obviously interact with the synthetase. With cobra venom ribonuclease, the main protections were observed on both sides of the anticodon arm and at the 5'-ends of loops I and II. Additionally, weak protections were observed between arms I and II and near stem V. It is interesting to note that in other tRNA/synthetase complexes the anticodon stem of the tRNA is also protected against cobra venom ribonuclease hydrolysis[16,32]. In contrast to tRNAs, the viral fragment possesses several single-stranded regions sensitive to the cobra enzyme that are protected in the synthetase complexed fragment. But, as seen in the tertiary structure model, they are all involved in higher-order structures (Fig. 4). Summarizing nuclease experiments, it appears that residues located in both the anticodon and amino acid accepting limbs of the three-dimensional model of the viral structure are protected by valyl-tRNA synthetase.

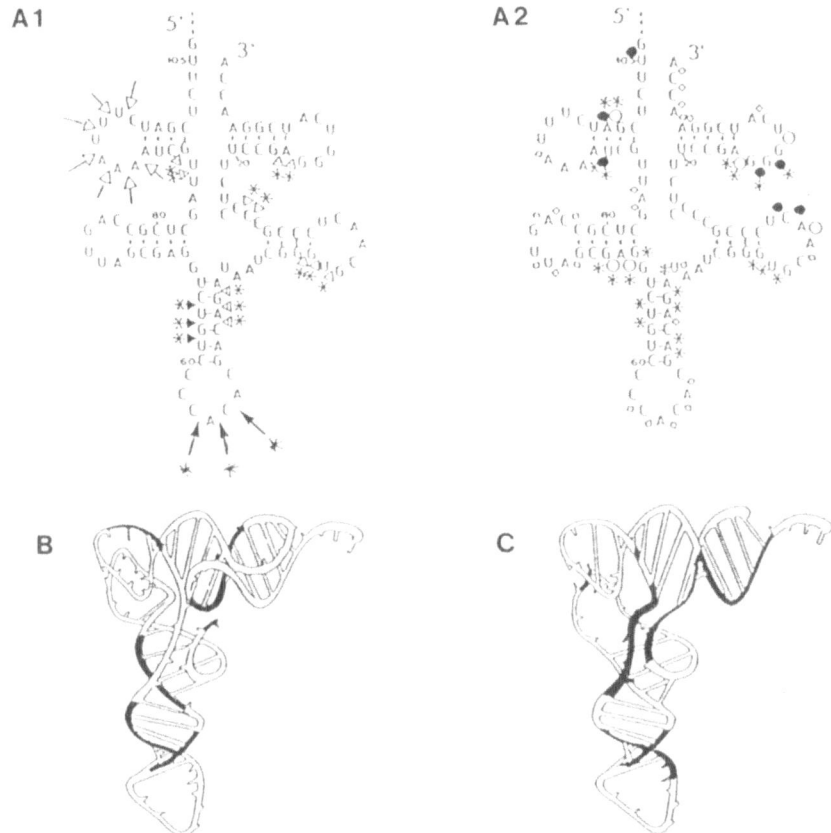

Figure 4. Contacts of the TYMV tRNA-like structure (A and B) and of yeast tRNA$^{Val}$ (C) with yeast valyl-tRNA synthetase. Contacts obtained by enzymatic (A1) or ethylnitrosourea (A2) mapping are shown in the secondary structure of the tRNA-like molecule. Triangles show the nucleotides in the free viral RNA, accessible to cobra venom digestion [(▶) strong, (◗) medium, (▷) weak cuts], and arrows those accessible to S1 nuclease. Circles represent protected phosphates in the native tRNA-like fragment [(●) strong, (o) weak protections]. The stars indicate residues protected by the synthetase. Diamonds represent non-testable phosphate groups. These results are displayed in a synthetic way (heavy lines) on the 3-D model of the tRNA-like model (B). For comparison, contacts in yeast tRNA$^{Val}$ with valyl-tRNA synthetase are given in (C) on a 3-D model of the tRNA.

To overcome steric hindrance problems which could occur with enzymatic probes (for a discussion see 10) further experiments were performed using ethylnitrosourea. This small probe is suitable for such studies because it does not inactivate synthetases[31-33]. Results summarized in Figure 4, show protected phosphates grouped in several stretches scattered all over the secondary structure of the RNA. In agreement with the nuclease experiments, some of them are located in arms I, II and III. To mention is the similar location of phosphates in the anticodon stem of both complexed yeast tRNA$^{Val}$ [31] and viral RNA.

In fact, considering the similar kinetic behaviour of tRNA[Val] and the tRNA-like fragment of TYMV in the valylation reaction, one expects both RNAs to be recognized by the synthetase in a similar fashion. Indeed most of the protected regions in both types of RNA are localized at topologically similar positions in their three-dimensional structures (that of tRNA[Val], probably similar to the crystallographic structures of elongator tRNA[Phe] and the non-crystallographic model of the viral fragment, see Fig. 4). Taken together, these experiments show an interaction of valyl-tRNA synthetase with the two branches of the L-shaped viral RNA so that the nucleic acid is surrounded by the protein. This agrees with earlier results demonstrating contacts of the T-, D- and anticodon arms of yeast tRNA[Val] with its cognate synthetase[31,34] and showing that the RNA is buried in the protein[35]. The fact that the synthetase is able to recognize two RNA molecules exhibiting a structural variability in the amino acid accepting stem, leads to the conclusion that a perfectly defined conformation in this part of the molecule is not required for tRNA/synthetase recognition. Hence, rather than identical conformations, the size of the L-structures in tRNA and tRNA-like molecules is important for recognition. As found by the Leiden group, the amino acid accepting branch of the L is formed by a stack of 12 base-pairs in both RNAs[7]. Thus, the correct positioning of the CCA-end in the catalytic site of the synthetase can be ensured. The specificity of the interaction will then be mostly governed by the anticodon stem and loop structure.

## The L-Shaped tRNA-like Conformer is not the Most Efficient Structure for Aminoacylation

Aminoacylation of partially digested TYMV RNA has shown that the minimal structure acylatable by valyl-tRNA synthetase is 84 ± 2 nucleotides long[37]. This sequence, corresponding to the L-shaped folding, does not contain arm V. Mapping experiments with ethylnitrosourea and cobra venom ribonuclease, however, show that the synthetase protects both sides of stem V as well as the intervening single-stranded sequence between arms IV and V against these probes (Fig. 4). Although the presence of arm V does not seem to be an absolute structural prerequisite for aminoacylation, it is likely that it contributes to a better interaction with the synthetase, thus favouring a more efficient enzymatic reaction. Valylation of *in vitro* synthesised tRNA-like transcripts of cloned TYMV cDNA, allowed us to ascertain this proposal[11].

A family of four clones permitting transcription of various lengths of 3'-coterminal TYMV RNA fragments was constructed. The shortest resulting transcript contains 82 nucleotides of the TYMV sequence and corresponds to the tRNA-like L-configuration. The other transcripts are progressively longer, containing 93, 109 and 258 nucleotides of TYMV sequence. Results of comparative aminoacylation by wheat germ valyl-tRNA synthetase of the different transcripts (Table 1) indicate that, while the L-shaped tRNA-like structure is the essential core required for activity, upstream nucleotides are important in achieving maximal rates of valylation. The main effect of these nucleotides in kinetic terms is in achieveing high Vmax values ; Vmax increases progressively with length, with a 14 fold increase observed between the longest and the shortest transcripts. On the contrary, the Km values look rather similar for all transcripts.

Table 1. Kinetic parameters for the valylation of 3'-terminal TYMV RNAs of different lengths.

| RNA Length | TYSma 258[a] | TYDde 109[a] | TYDra 93[a] | TYAlu 82[a] |
|---|---|---|---|---|
| Km (μM) | 0.55±0.08 | 0.33±0.05 | 0.82±0.07 | 0.42±0.08 |
| Vmax[b] | 87.9±8.4 | 34.3±2.9 | 29.4±2.1 | 6.3±0.8 |
| Vmax/Km | 159.8 | 103.9 | 35.8 | 15.0 |

(a) Length of TYMV RNA segment tested ; (b) pmol/min/ml.

In conclusion, the results reported here confirm the identity of the 82 nucleotide long L-shaped tRNA-like domain at the 3'-end of TYMV RNA as the core feature determining valylation by a plant valyl-tRNA synthetase. Maximally efficient valylation, however, depends on upstream sequences, which may facilitate tighter binding of the synthetase to the RNA. These interactions may compensate the effects of the non-canonical structure of the L-shaped core, and of the intrinsic flexibility between the two arms of the L, both effects that would probably be unfavourable for optimal binding by the synthetase.

ACKNOWLEDGEMENTS

We thank our colleagues, J.P. Briand, P. Dumas, L. Hirth, R. Mengual, D. Moras and P. Romby (Strasbourg), P. Verlaan, A. Van Belkum and C.W.A. Pleij (Leiden, The Netherlands) and T. Dreher (Corvallis, OR, USA), who participated in the project on the tRNA-like structure of TYMV.

REFERENCES

1. Hall, T.C., 1979, Transfer RNA-like structures in viral genomes, Int. Rev. Cytol., 60:1.
2. Haenni, A.L., Joshi, S. and Chapeville, F., 1982, tRNA-like structures in the genomes of RNA viruses, Prog. Nucl. Acid Res. Mol. Biol., 27:85.
3. Pink, M., Yot, P., Chapeville, F. and Duranton, H., 1970, Enzymatic binding of valine to the 3'-end of TYMV RNA, Nature, 226:954.
4. Yot, P., Pinck, M., Haenni, A.L., Duranton, H. and Chapeville, F., 1970, Valine specific tRNA-like structure in turnip yellow mosaic virus RNA, Proc. Natl. Acad. Sci. USA, 67:1345.
5. Giegé, R., Briand, J.P., Mengual, R., Ebel, J.P. and Hirth, L., 1978, Valylation of two RNA components of turnip yellow mosaic virus and specificity of the aminoacylation reaction, Eur. J. Biochem., 84:251.
6. Florentz, C., Briand, J.P., Romby, P., Hirth, L., Ebel, J.P. and Giegé, R., 1982, The tRNA-like structure of turnip yellow mosaic virus RNA : structural organization of the last 159 nucleotides from the 3'-OH terminus, The EMBO J., 1:269.
7. Rietveld, K., Van Poelgeest, R., Pleij, C.W.A., Van Boom, J.H. and Bosch, L., 1982, The tRNA-like structure at the 3'

terminus of turnip yellow mosaic virus RNA. Differences and similarities with canonical tRNA, <u>Nucleic Acids Res.</u>, 10:1929.

8. Rietveld, K., Pleij, C.W.A. and Bosch, L., 1983, Three-dimensional models of the tRNA-like 3'-termini of some plant viral RNAs, <u>The EMBO J.</u>, 2:1079.

9. Dumas, P., Moras, D., Florentz, C., Giegé, R., Verlaan, P., Van Belkum, A. and Pleij, C.W.A., 1987, 3-D graphics modelling of the tRNA-like 3'-end of turnip yellow mosaic virus RNA : structural and functional implications, <u>J. Biomol.Struct. Dyn.</u>, 4:707.

10. Florentz, C. and Giegé, R., 1986, Contact areas of the Turnip Yellow Mosaic Virus tRNA-like structure interacting with yeast valyl-tRNA synthetase, <u>J. Mol. Biol.</u>, 191:117.

11. Dreher, T.C., Florentz, C. and Giegé, R., 1988, Valylation of tRNA-like transcripts of turnip yellow mosaic virus demonstrate that the L-shaped region at the 3' end of the viral RNA is not sufficient for optimal valylation, <u>Biochimie</u>, in press.

12. Briand, J.P., Jonard, G., Guilley, H., Richards, K.E. and Hirth, L., 1977, Nucleotide sequence (n=159) of the amino acid accepting 3'-OH extremity of turnip yellow mosaic virus RNA and the last portion of its coat protein, <u>Eur. J. Biochem.</u>, 72:453.

13. Silberklang, M., Prochiantz, A, Haenni, A.L. and RajBhandary, U.L., 1977, Studies on the sequence of the 3' terminal region of turnip yellow mosaic virus RNA, <u>Eur. J. Biochem.</u>, 72:465.

14. Florentz, C., Mengual, R., Briand, J.P. and Giegé, R., 1982, Large scale purification of the 3'-OH-terminal tRNA-like sequence (n=159) of turnip yellow mosaic virus RNA, <u>Eur. J. Biochem.</u>, 123:89.

15. Branlant, C., Krol, A., Ebel, J.P., Gallinaro, H., Lazar, E. and Jacob, M., 1981, The conformation of chicken, rat and human U1A RNAs in solution. <u>Nucleic Acids Res.</u> 9:841.

16. Favorova, O.O., Fasiolo, F., Keith, G., Vassilenko, S.K. and Ebel, J.P., 1981, Partial digestion of tRNA-Aminoacyl-tRNA synthetase complexes with cobra venom ribonuclease, <u>Biochemistry</u>, 20:1006.

17. Romby, P., Carbon, P., Westhof, E., Ehresmann, C., Ebel, J.P., Ehresmann, B. and Giegé, R., 1987, Importance of conserved residues for the conformation fo the T-loop in tRNAs, <u>J. Biomol. Struct. Dyn.</u>, 5:669.

18. Pleij, C.W.A., Rietveld, K. and Bosch, L., 1985, A new principle of RNA folding based on pseudoknotting, <u>Nucleic Acids Res.</u>, 13:1717.

19. Jones, T.A., 1978, A graphic model building and refinement system for macromolecules, <u>J. Appl. Crys.</u>, 11:268.

20. Moras, D., Thierry, J.C., Comarmond, M.B., Fischer, J., Weiss, R., Ebel, J.P. and Giegé, R., 1980, Three-dimensional structure of yeast tRNA[Asp], <u>Nature</u>, 288, 669.

21. Westhof, E., Dumas, P. and Moras, D., 1985, Crystallographic refinement of yeast aspartic transfer RNA, <u>J. Mol. Biol.</u>, 184:119.

22. Brennan, R. and Sundaralingam, M., 1976, Structure of transfer RNA molecules containing the long variable loop, <u>Nucleic Acids Res.</u>, 3:3235.

23. Dock-Bregeon, A.C., Westhof, E., Giegé, R. and Moras, D., 1988, Solution structure of a tRNA with a large variable region : yeast tRNA[Ser], submitted for publication.

24. De Bruijn, M.H.L. and Klug, A., 1983, A model for the tertiary structure of mammalian mitochondrial transfer RNAs lacking the entire dihydrouridine loop and stem, The EMBO J., 2:1309.
25. Romby, P., Westhof, E., Toukifimpa, R., Mache, R., Ebel, J.P., Ehresmann, C. and Ehresmann, B., 1988, Higher order structure of chloroplastic 5S ribosomal RNA from spinach, Biochemistry, 27, 4721.
26. Ebel, J.P., Renaud, M., Dietrich, A., Fasiolo, F., Keith, G., Baltzinger, M., Remy, P., Bonnet, J. and Giegé, R., 1979, Interaction between tRNA and aminoacyl-tRNA synthetase in the valine and phenylalanine systems from yeast, In "Transfer RNA : Structure, Properties and Recognition", Cold Spring Harbor Lab. (Söll, D., Abelson, J.N. & Schimmel, P. eds) 235.
27. Schimmel, P.A., 1987, Aminoacyl-tRNA synthetases : general scheme of structure function relationships in the polypeptides and recognition of RNAs, Ann. Rev. Biochem., 56:125
28. Dreher, T.W., Bujarski, J.J. and Hall, T.C., 1984, Mutant viral RNAs synthesized in vitro show altered aminoacylation and replicase activities, Nature, 311:171.
29. Dreher, T.W. and Hall, T.C., 1988, Mutational analysis of the tRNA mimicry of brome mosaic virus RNA, sequence and structural requirements for aminoacylation and 3'-adenylation, J. Mol. Biol., 201:41.
30. Kisselev, L.L., 1985, The role of anticodon in recognition of tRNA by aminoacyl-tRNA synthetase, Prog. Nucl. Acid Res. Mol. Biol., 32:237.
31. Vlassov, V.V., Kern, D., Romby, P., Giegé, R. and Ebel, J.P., 1983, Interaction of tRNA[Phe] and tRNA[Val] with aminoacytl-tRNA synthetases: a chemical modification study, Eur. J. Biochem., 132:537.
32. Théobald, A., Springer, M., Grunberg-Manago, M., Ebel, J.P. and Giegé, R., 1988, Tertiary structure of Escherichia coli tRNA[Thr3] in solution and interaction of this tRNA with the cognate threonyl-tRNA synthetase, Eur. J; Biochem. 175:511.
33. Romby, P., Moras, D., Bergdoll, M., Dumas, P., Vlassov, V.V., Westhof, E., Ebel, J.P. and Giegé, R., 1985, Yeast tRNA[Asp] tertiary structure in solution and areas of interaction of the tRNA with aspartyl-tRNA synthetase, J. Mol. Biol., 184:455.
34. Renaud, M., Dietrich, A., Giegé, R., Remy, P. and Ebel, J.P, 1979, Interaction between yeast tRNA[Val] and yeast valyl-tRNA syntehtase studied by monochromatic ultraviolet light incued cross-linking, Eur. J. Biochem., 101:475. 1979
35. Zaccaï, G., Morin, P; Jacrot, B., Moras, D., Thierry, J.C. and Giegé, R., 1979, Interactions of yeast valyl-tRNA synthetase with RNAs and conformational changes of the enzyme, J. Mol. Biol., 129:483.
36. Rietveld, K., 1984, Three-dimensional folding of the tRNA-like structures of some plant viral RNAs. A new principle in the folding of RNA. PhD University of Leiden.
37. Joshi, S., Chapeville, F. and Haenni, A.L., 1982, Length requirements for tRNA specific enzymes and cleavage specificity at the 3'-end of turnip yellow mosaic virus RNA, Nucleic Acids Res., 10:1947.

# PROTEIN ENGINEERING OF ELONGATION FACTOR Tu

M. Jensen, K.K. Mortensen, H.U. Petersen, T.F.M. la Cour, J. Nyborg, M. Kjeldgaard, S. Thirup and B.F.C. Clark

Division of Biostructural Chemistry, Department of Chemistry, Aarhus University, 8000 Aarhus C, Denmark

## Summary

The general goal of our programme is to study how alterations in a protein's structure can modify its biological function. The protein selected for study is the elongation factor EF-Tu. This was chosen because of its own fundamental biological importance, because of its recently-recognized significance as a member of a group of GTP-binding proteins of medical relevance, and finally because of existing experience and expertise with this protein in our laboratory. The long-term aim of our research programme is to study this factor with special reference to modification of its substrate specificity, its sensitivity towards antibiotics and its thermal stability. In this contribution, we shall look at the important property of EF-Tu that we have just alluded to, namely, its ability to bind specifically to the energy-rich substrate molecule GTP.

## Elongation Factor Tu

EF-Tu plays a central role in the translation of messenger RNA into protein, and therefore it is ubiquitous in biology. In eukaryotic cells, it is called EF-1α, for historical reasons, while the name EF-Tu is reserved for the protein from prokaryotes.

EF-Tu and EF-1α, which have been sequenced for diverse organisms, have molecular weights around 50 000. They are water-soluble and are largely monomeric in aqueous solution. There is a high degree of sequence homology (70-80%) within the prokaryotic and within the eukaryotic groups, and a somewhat lower degree (~40%) between these groups. The latter figure may be regarded as rather high, in view of the long time since the divergence of the groups and especially in view of the eukaryotic cell's more complex machinery of protein synthesis. Not surprisingly, most of the research effort on this factor has so far been concentrated upon EF-Tu from *E. coli*. The functions of EF-Tu may conservatively be listed as: (1) binding and protection of aa-tRNA; (2) helping to attach the aa-tRNA to the ribosome; (3) providing the energy-rich GTP and catalysing its hydrolysis to GDP at the right moment; (4) contributing to the accuracy of translation. EF-Tu also interacts with the companion elongation factor EF-Ts, with important antibiotics, such as kirromycin and possibly with a second tRNA molecule. None of these processes has been elucidated in molecular detail; the single exception is the site of binding of GDP, which has been determined by X-ray crystallography.

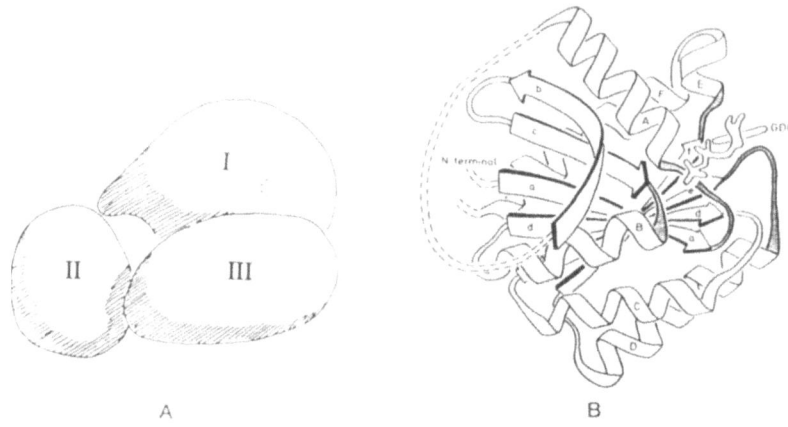

Figure 1. A: the domain structure of EF-Tu [1]. The location of the GDP molecule in the G-binding domain I is outlined. B: a structure cartoon of domain I [2].

In spite of its small size and globular structure, EF-Tu has proved to be surprisingly intractable towards structural studies; although the first low-resolution electron density maps were produced over ten years ago, the determination of the structure at atomic resolution is still incomplete. The reason for this appears to be the existence of domains with different degrees of structural order, as shown in Figure 1 (left), and known as the

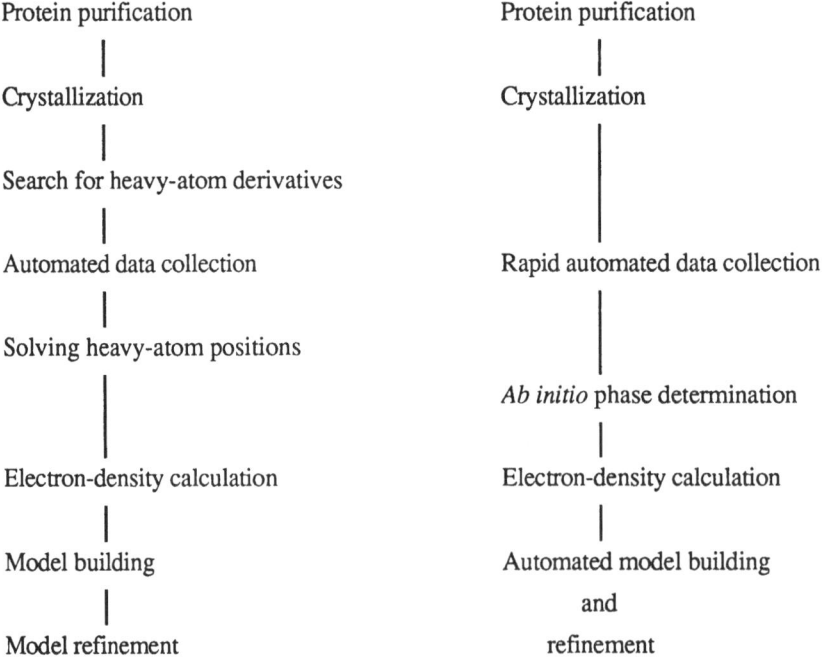

Figure 2. Major steps at present (left) and expected in the mid-nineties (right), in a protein structure determination *de novo*. After J. Drenth [3].

"tight", "loose" and "floppy" domains (nowadays labelled as domains I, III and II, respectively); these contain amino acids 1-200, 300-393 and 201-299, respectively. The disposition of these three domains is shown in Figure 1A; the binding site for GDP is the "tight" domain I, which has therefore been re-named the G-domain. A structure model for this domain based upon our X-ray studies has been proposed by la Cour *et al.* [2] and is also shown in Figure 1. This model shows us which residues of EF-Tu are in contact with GDP - which means that we are beginning to study the behaviour of this protein on an atomic level. The details of this we reserve for the next section, in order here to comment on some recent developments which may cause a revolution in protein crystallography within the next decade. As mentioned, the structure determination of EF-Tu is barely complete after more than ten years of slow, uphill progress. The introduction of large-scale automated data collection and model-building, together with the new possibility of phase determination *ab initio*, makes it likely that such a determination will soon be performed in a fraction of the time now taken. Figure 2 outlines the strategy for protein structure determination by X-ray crystallography as practised in 1988 and in 1998.

## Related Proteins

An aspect of EF-Tu of recent topical interest is its relationship with other proteins that bind and hydrolyse GTP and related nucleotides. Extensive comparison of sequences has shown that certain elements of these sequences are retained across the board - irrespective of whether the protein is an elongation factor, an initiation factor, an oncogene product, a transducing protein, a kinase, an ATPase, a nitrogenase, a phosphotransferase or a contractile or DNA-processing protein. These common elements of sequence are of two types. The first type embraces regions of general homology, stretching over one hundred amino acids or more. For example, EF-Tu's from two bacterial species have a homology of 70% over the whole length of the molecule (here we are comparing EF-Tu from *E. coli* and from *Th. thermophilus* [4]. If we include conservative substitutions, their homology is 85%. Therefore, these two proteins will probably have very similar structures, which is not surprising, as they come from closely-related organisms and have a similar, perhaps identical, function.

More interesting are molecules whose functions are dissimilar but related. A region of the initiation factor IF2 from *E. coli* is homologous with the G-domain of EF-Tu. We therefore guessed that this region of IF2 must be the molecule's G-domain and, using a computer to smooth out the small structural differences arising from the amino-acid substitutions present, made a prediction (model) for its tertiary structure [5]. A less closely-related protein is the *ras* p21 oncogene product, whose function is not thought to be related to those of EF-Tu or IF-2. However, it binds and hydrolyses GTP, and this functional resemblance leads to homology of their G-domain sequences. A computer prediction of the structure of *ras* p21 was likewise made; it also turned out to be very similar to the G-domain of EF-Tu [6]. The structure of the *ras* p21 protein has now been determined by X-ray crystallography [7]; the GDP binding site in the crystal bears a remarkably high structural resemblance to that in EF-Tu. This confirms the usefulness of our model-building strategy and encourages us to extend it to other G-binding proteins of as yet unknown three-dimensional structure.

The second type of common elements of sequence is the recurrence, throughout a large and diverse group of proteins, of particular groupings of amino-acid residues. Looking for common elements of this kind, W.C. Merrick and co-workers [8] have completed an analysis of all currently known protein sequences and have shown that certain amino-acid sequence elements are common to all proteins that bind GTP/GDP and probably only to these. Merrick's consensus elements all occur in loops of EF-Tu that we showed to be in contact with GDP in the X-ray crystal structure. This is shown in Table 1. In the Table, the loops connecting α-helix and β−sheet are defined by appropriate upper- and lower-case letters; for example, aA is the loop running from the C-terminal end of strand **a** in the β−sheet to the N-terminal end of α-helix A. It is seen that the residues found to be

Table 1. Comparison of consensus sequences of G-binding proteins with sequences of loops in EF-Tu that are close to bound GDP. The conserved residues (left) are emphasized.

| Consensus sequence (Merrick *et al.*) | Loop sequence (Aarhus) |
|---|---|
| G X X X X G K | $aA = G_{18}$ H V D H G $K_{24}$ |
| D X X G | $cB = D_{80}$ C P $G_{83}$ |
| N K X D | $eD = N_{135}$ K C $D_{138}$ |
| | $fE = S_{173}$ A L $K_{176}$ |

conserved (left column) reflect well the residues found to be in contact with the GTP (right column, large capitals).

## Protein Engineering

The general approach and particular methods of protein engineering cannot be treated in detail here. We should like to summarise the former in an unusual way (Figure 3)

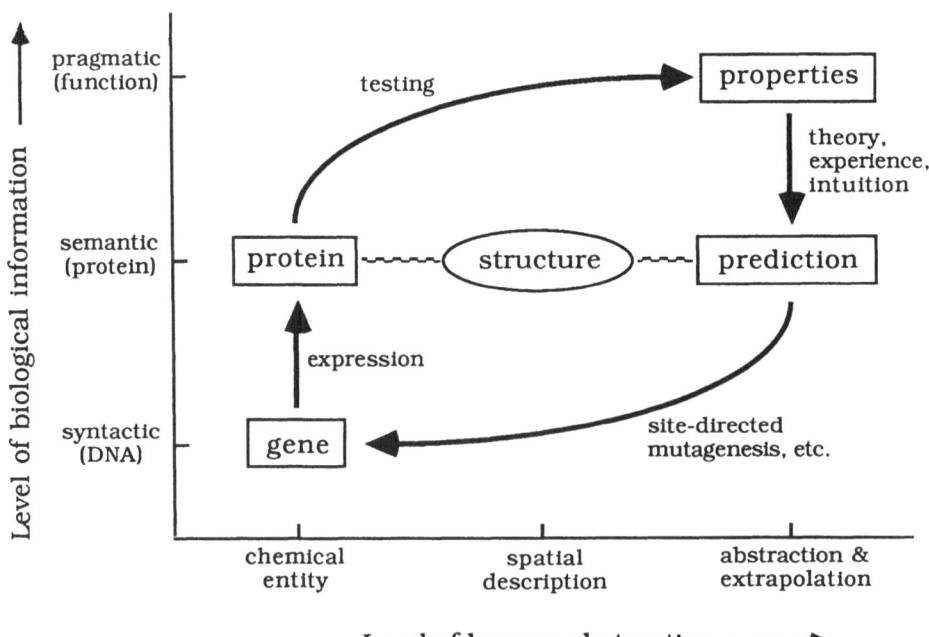

Figure 3. The protein engineering cycle. This representation emphasizes the central position of structure-determination in protein engineering.

Figure 3 emphasizes the complementary out-into-the-lab and back-to-the-drawing-board aspects of the protein engineering cycle. We start usually with a given protein and, by analysis of its structure and function, predict new sequences that we expect to have a new, interesting or at least different set of properties. The altered protein is prepared by site-directed mutagenesis and expression techniques, and the process starts again. But this is a very idealised scheme. Although many proteins have now been modified in this way, only in a very few cases have the products been refined by further passages around the protein engineering cycle.

For site-directed mutagenesis of EF-Tu we used as template the single-stranded form of the plasmid pEMBL, in which a 2-kb chromosomal fragment containing the entire *tuf*A gene had been cloned. Mutation was performed by the gapped-duplex method. Problems with the expression of the mutated gene products were avoided by cloning after mutation into a plasmid with temperature-regulated runaway replication. All of the results to be described here were obtained in an EEC-supported collaboration with the laboratories of Professors A. Parmeggiani (Palaiseau) and L. Bosch (Leiden), to whose members we are greatly indebted.

## Mutagenesis of EF-Tu

In initial mutagenesis experiments on EF-Tu, we have aimed at various putative functional amino-acid residues in the G-domain, such as those involved in guanine base recognition and specificity, Asn-135, Lys-136 and Asp-138; those involved in binding and hydrolysis of the phosphates, Val-20, Asp-80, Pro-82 and His-84; structurally interesting residues such as Glu-117; and finally one residue believed to be involved in tRNA binding, Cys-81.

In order to change the nucleotide specificity, we replaced Asn-135 with aspartic acid. Model studies predicted a decreased affinity for guanine, as Asn-135 hydrogen-bonds to the keto group of the base. On the other hand, Asp at that position could hydrogen-bond to the amino group of adenine. Preliminary results [9] show that the mutant factor binds GDP/GTP; however, a quantitative determination of nucleotide specificity has still to be made. Homology studies by Lebermann et al. [10] indicated the necessity of a second mutation (Lys-136 → Ile) in order to obtain increased affinity to ADP/ATP. This double mutation has been made and is being characterised. It is becoming clear from these studies that a simple hydrogen-bonding pattern does not define completely the specificity of the G-binding site for guanine, since the base interacts both with the polypeptide backbone and with another stretch of amino acids around positions 174 and 175.

Mutations around the phosphate binding site of EF-Tu obtained so far include the mutation Val-20 → Gly. This position is particularly interesting because the homologous position in the proto-oncogene product p21 is glycine, whereas the corresponding oncogenic protein contains valine at this position. The mutant led to an unexpected result. As mentioned, the normal p21 protein contains a glycine residue at position 12 in the sequence. Mutation of this to a valine causes oncogenicity. The same mutation lowers the GTPase activity of the p21 protein to a level around one-tenth of that of the wild-type, non-oncogenic protein (literature in reference 6). In EF-Tu, the corresponding position (position 20) is normally occupied by a valine, and this was replaced with a glycine by site-directed mutagenesis and expression. It would be expected that the mutation should be accompanied by a rise in the GTPase activity, since the mutation is the reverse of that in p21 protein. But, on the contrary, the replacement of Val-20 with Gly in EF-Tu led to a decrease of some 75% in GTPase activity. Although we cannot yet interpret this result in terms of structure, it makes it quite clear that the effects even of single, relatively straightforward-looking mutations can be very complex, and one cannot in general speak of simple additivity of effects.

We interpret the difficulties frequently encountered in producing mutants in the phosphate binding site by assuming that such mutant proteins are deleterious and that their expression is fatal for the host cell. This is in itself interesting, since most of the attempts involved the replacement of an amino-acid residue with glycine - that is, a side-chain was merely pruned away; this would not normally be regarded as a very harmful mutation.

## Expression and Characterisation of the G-domain alone

The G-domain of EF-Tu consists, as mentioned, of the N-terminal half of EF-Tu. Using standard methods of site-directed mutation, we produced a "mutant" in which the C-terminal half of EF-Tu was deleted. This mutant is 203 amino acids long, containing residues 1-202 and, for technical reasons, residue 393. The molecular weight of the purified fragment is 21 kDal as measured by SDS-polyacrylamide gel electrophoresis, the same value as calculated from the amino acid sequence. Functional tests of the fragment have shown that this isolated G-domain binds GTP and GDP with constants shown in Table 2.

Table 2. Apparent dissociation constants for the binary complexes between guanosine nucleotides and native and modified EF-Tu. The values (from M. Jensen, dissertation; to be published elsewhere) were obtained by nitro-cellulose filter tests [11].

| Ligand | $K'_d$ for EF-Tu ($\mu$M) | $K'_d$ for EF-Tu G-domain |
|--------|---------------------------|----------------------------|
| GTP | 0.6 | 4 |
| GDP | 0.006 | 3 |

At 30°C the addition of nucleotide-free EF-Tu to GTP causes an initial, brief hydrolysis of the GTP. However, the GDP formed binds strongly to the factor, thereby inhibiting further hydrolysis. Adding the G-domain to GTP results in continuous hydrolysis. In this case, in agreement with the measured dissociation constants, no strong binding of GDP inhibits the reaction.

Site-specific mutations of the EF-Tu G-domain will be a useful tool in the investigation of the relationships between structure and function in the class of G-binding proteins. Therefore, it is clear that the use of EF-Tu as a model protein will lead to valuable, new insight, in particular into the mechanism of action of the factor and in general into the problems associated with production of mutated proteins, such as harmful effects on the host, unpredictable levels of expression, or the denaturation and/or precipitation of the desired and overproduced product. The isolated G-domain is currently being prepared in large quantities for exhaustive and detailed structural investigation including X-ray crystallography.

## G-specificity of other G-proteins

Our location of the binding site for GDP in EF-Tu [2] led to the conclusion that certain residues were important for the binding of GDP. This implied that mutation of these side chains would affect the performance of EF-Tu in binding guanosine nucleotides, as has now been confirmed in several ways by independent work.

First, as already mentioned, exhaustive sequence searching by W. Merrick et al. [8] has revealed that the residues common to all G-proteins are some of those that we have found in contact with GDP.

A second source of confirmation is provided by recent work by Wagner et al. [12] with the ras-related yeast protein YPT. This protein binds and hydrolyses GTP. If the lysine or the asparagine corresponding to the K (top line in the Table) or the N (third line) is mutated, the protein cannot bind GTP and the mutation is lethal. However, mutations of two of the variable residues influenced GTP binding and GTPase activity only to a relatively slight extent.

Thirdly, we have collaborated with the group of Dr. M. Grunberg-Manago (IMBC, Paris) in a functional investigation of initiation factor IF2. It was shown [5] that removal of a 289-amino-acid-long N-terminal fragment from IF2 leaves most of the functions of IF2, in particular the binding or hydrolysis of GTP, unchanged. However, the retention of GTP binding and hydrolysis is in agreement with our postulate that the G-domain of IF2 is analogous with that of EF-Tu, since our postulated G-domain begins at residue 393, so it is not removed by the cleavage at residue 289. Potentially interesting positions for site-directed mutation of EF-Tu can therefore be predicted, on the basis of the above model, with some confidence.

Fourthly, D. Miller and co-workers [13] have recently prepared a mutant of EF-Tu in which the residue Asp-138 was replaced by asparagine. The crystal structure [2] had suggested that this residue is important in the recognition of the G base in GDP. Indeed, the change from Asp to Asn at this position suppresses the binding of guanosine phosphates; instead, xanthine phosphates are bound, precisely as the hydrogen-bonding pattern would lead us to expect. The mutant factor complexed with xanthosine triphosphate possesses all of the functional properties of the wild-type EF-Tu.GTP complex. Another substitution at this position is Asp-138 → Trp, found naturally in certain enzymes. Here, the hydrogen-bonding pattern is destroyed, so that the nucleotide specificity is lost to some extent, and these enzymes can use either guanosine or inosine triphosphate [8].

To conclude, we feel that our preliminary results have provided sound evidence that one can indeed predict structural/functional changes and realise them by protein engineering. This gives us confidence for future developments and commercial applications in this field.

## Acknowledgements

We thank Paul Woolley for making valuable suggestions for this manuscript and for consenting to our use of Figure 3. We also gratefully acknowledge essential financial support from the Danish Natural Science Research Council (J. no. 4.17.41.04), the Aarhus University Bioregulation Research Centre and the European Community (BAP 0058 DK).

## References

1. K. Morikawa, T.F.M. la Cour, J. Nyborg, K.M. Rasmussen, D.L. Miller & B.F.C. Clark, *J. Mol. Biol.* **125**, 325-338 (1978).

2. T.F.M. la Cour, J. Nyborg, S. Thirup & B.F.C. Clark, *EMBO J.* **4**, 2385-2388.

3. J. Drenth, personal communication.

4. A. Kushiro, M. Shimizu & K.-I. Tomita, *Eur. J. Biochem.* **170**, 93-98 (1987); L. Seidler, M. Peter, F. Meissner & M. Sprinzl, *Nucleic Acids Res.* **15**, 9263-9277 (1987).

5. Y. Cenatiempo, F. Deville, J. Dondon, M. Grunberg-Manago, C. Sacerdot, J.W.B. Hershey, H.F. Hansen, H.U. Petersen, B.F.C. Clark, M. Kjeldgaard, T.F.M. la Cour, K.K. Mortensen & J. Nyborg, Biochemistry **26**, 5070-5076 (1987).

6. F. McCormick, B.F.C. Clark, T.F.M. la Cour, M. Kjeldgaard, L. Nørskov-Lauritsen & J. Nyborg, *Science* **230**, 78-82 (1985).

7. A.M. de Vos, L. Tong, M.V. Milburn, P.M. Matias, J. Jancarik, S. Noguchi, S. Nishimura, K. Miura, E. Ohtsuka & S.-H. Kim, *Science* **239**, 888-893 (1988).

8. T.E. Dever, M.L. Glynias & W.C. Merrick, *Proc. Natl. Acad. Sci. U.S.A.* **84**, 1814-1818 (1987).

9.  A. Parmeggiani, P.H. Anborgh, D. Canceill, E. Jacquet, J. Jonak, M. Merola, K.K. Mortensen & G.W.M. Swart, in *"Structure, Function and Genetics of Ribosomes"* (eds. B. Hardesty & G. Kramer), Springer Verlag, New York, 1986, pp. 672-685.

10. R. Lebermann & U. Egner, *EMBO J.* **3**, 339-341 (1984).

11. A. Parmeggiani, G.W.M. Swart, K.K. Mortensen, J. Jensen, B.F.C. Clark, L. Dente & R. Cortese, *Proc. Natl. Acad. Sci. U.S.A.* **84**, 3141-3145 (1987).

12. P. Wagner, C.M.T. Molenar, A.J.G. Rauh, R. Brökel, H.D. Schmitt & D. Gallwitz, *EMBO J.* **6**, 2373-2379 (1987).

13. Y.-W. Hwang & D.L. Miller, *J. Biol. Chem.* **262**, 13081-13085 (1987).

TRANSLATIONAL CONTROL BY PHOSPHORYLATION OF

MAMMALIAN INITIATION FACTORS

John W. B. Hershey, Roger F. Duncan, Susan C.
Milburn, Vinay K. Pathak, Sang Y. Choi and
Randal J. Kaufman[+]


Dept. Biological Chemistry, School of Medicine
University of California, Davis, CA  95616, USA
+Genetics Institute
Cambridge, MA  02140, USA

## INTRODUCTION

Translational control is defined as a change in the efficiency
of translation of mRNAs.  This may involve a quantitative change in
the overall amount of protein synthesized, called global control.
Alternatively, the change may affect the synthesis of specific
proteins, called mRNA-specific control.  In either case, efficiency of
translation is usually measured by radiolabeled amino acid
incorporation into protein, or by the number and activity of
ribosomes.  Since the rate limiting step in protein synthesis appears
to be at the initiation phase, most control mechanisms are thought to
regulate this step.  We therefore focus on the pathway of initiation
and address studies of how global rates of protein synthesis may be
controlled.

## THE INITIATION PATHWAY

Initiation of protein synthesis involves the binding of the
initiator methionyl-tRNA and a mRNA to the 80S ribosome to form a
complex capable of entering the elongation cycle.  The non-covalent
binding interactions between these macromolecules are promoted by a
large number of proteins, called initiation factors.  A list of
mammalian initiation factors is provided in Table 1.  These proteins
have been purified and their physical and functional characteristics
have been elucidated (reviewed by Moldave, 1985; Pain, 1986).  Their
involvement in the pathway of initiation is depicted in Figure 1.  Two
steps in this pathway are critical control points, and have received
intensive study:  1) the binding of methionyl-tRNA to the 40S
ribosome; and 2) the binding of mRNA to the 40S ribosome.

The binding of methionyl-tRNA to ribosomes is catalyzed by eIF2.
This initiation factor comprises three non-identical subunits, $\alpha$, $\beta$
and $\gamma$.  The cDNAs for the first two subunits have been cloned and
sequenced, providing insight into their structures and function (Ernst
et al., 1987; Pathak et al., 1988a).  eIF2 first forms a binary

49

**Table 1.**  Mammalian Initiation Factors

| Name | mass (kDa) | phospho-rylated? | function |
|---|---|---|---|
| eIF-1 | 15 | | stabilizes 40S·Met |
| eIF-2 | 126 | | binds Met-tRNA to 40S |
| α | 36 | Y | site of control |
| β | 38 | Y | GTP, RNA binding? |
| γ | 52 | N | GTP phosphoryl binding |
| eIF-2B | 300 | | GTP exchange factor |
| α | 26 | | |
| β | 39 | | |
| γ | 58 | | |
| ∂ | 67 | Y | |
| ε | 82 | Y | confers regulation |
| eIF-3 | 550 | | stabilizes 40S·Met; |
| p35 | 35 | | binds RNA? |
| p36 | 36 | | |
| p40 | 40 | | |
| p44 | 44 | | |
| p47 | 47 | | |
| p66 | 66 | Y | |
| p120 | 120 | Y | |
| p170 | 170 | | |
| eIF-4A | 46 | N | binds mRNA; ATPase. |
| eIF-4B | 80 | Y | binds mRNA; |
| eIF-4C | 18 | | Rb dissociation? |
| eIF-4D | 16 | N | contains hypusine; i80S |
| eIF-4E | 25 | Y | cap binding protein |
| eIF-4F | 290 | | cap binding protein complex |
| p25 | 25 | Y | same as eIF-4E |
| p46 | 46 | N | same as eIF-4A |
| p220 | 220 | Y | cleaved by poliovirus |
| eIF-5 | 150 | Y | GTPase; Rb junction step |
| eIF-6 | 25 | | Rb dissoc.; binds 60S |

complex with GTP, then a ternary complex with methionyl-tRNA.  The ternary complex is an obligate intermediate in the binding of the tRNA to the ribosome.  The ternary complex with all of its components remains bound to the ribosome until eIF5 catalyzes the 60S junction step, which results in the hydrolysis of GTP to GDP and Pi, and the release of eIF2 as a binary complex with GDP.  In order for eIF2 to recycle,

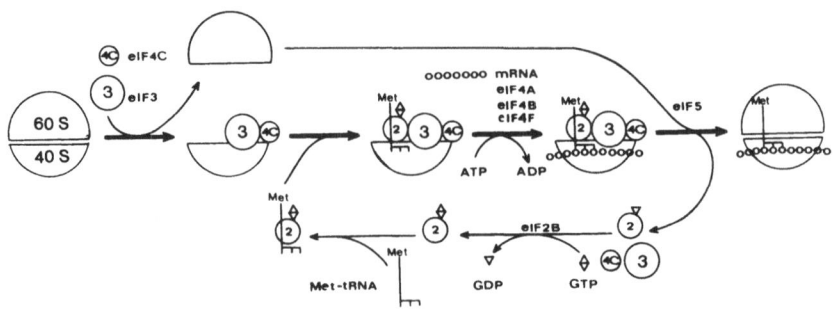

Figure 1.  Proposed pathway of initiation.

the GDP must be exchanged for GTP, a reaction that requires catalysis by another initiation factor, eIF2B (aka GEF). Control of the activity of eIF2 is expected to regulate global rates of protein synthesis, since its major function, to bring methionyl-tRNA to the 40S ribosome, occurs before the mRNA is selected.

The mRNA-binding step is considerably more complicated, and is less well characterized. The current working hypothesis is as follows. Three initiation factors, eIF4A, eIF4B and eIF4F, interact with the 5'-terminal m$^7$G cap structure, resulting in the melting of possible secondary structure in the 5'untranslated leader region of the mRNA. The 40S ribosome, with its bound methionyl-tRNA, then joins to the 5'-end of the mRNA and begins to "scan" the mRNA for a suitable AUG initiator codon. The factor binding and ribosome scanning require the hydrolysis of ATP, but molecular details of how these processes occur are lacking. A stable complex forms, involving at least in part the anticodon of the initiator tRNA and the initiator codon and its flanking sequences. The consensus sequence for a strong initiator codon has been deduced from analyses of many mRNAs and from mutational studies (Kozak, 1987): ACCACCAUGG. Recent studies suggest that an alternate pathway of mRNA binding to 40S ribosomes exists. This pathway bypasses the m$^7$G cap structure and involves a direct binding of the ribosome to internal regions of mRNAs. The components promoting internal initiation have not yet been identified. Once the 40S preinitiation complex is formed, the subsequent 60S junction reaction occurs rapidly and completed 80S initiation complexes result. The interaction of eIF4A, eIF4B and eIF4F with mRNAs offers another opportunity to control translation, either globally or in a mRNA-specific fashion.

## PHOSPHORYLATION OF INITIATION FACTORS

Analyses of purified initiation factor proteins indicate that many are phosphoproteins. Since many translational control phenomena are both rapid and reversible, it is attractive to postulate that the activities of these proteins might be modulated by protein kinases and phosphoprotein phosphatases. Nearly half of the 25 initiation factor polypeptides have been shown to be phosphorylated in vivo or in vitro (see Table 1). Each of these phosphoproteins therefore is a candidate for a control point.

The phosphorylation of the α-subunit of eIF2 has been studied extensively in the rabbit reticulocyte (reviewed by Pain, 1986; Proud, 1986; Ochoa, 1983). When reticulocyte lysates are incubated in the absence of heme, an inhibition of initiation of protein synthesis sets in after a short lag. A dominant inhibitor, called the heme-controlled repressor (HCR), is activated which is a protein kinase specific for eIF2α. Another protein kinase, called DAI, is regulated by dsRNA, induced by interferon, phosphorylates the same serine site in eIF2α, and causes an inhibition of protein synthesis. The simple view that phosphorylated eIF2 is inactive is not correct, however, since P-eIF2 is capable of forming ternary complexes and promoting the binding of methionyl-tRNA to ribosomes. Furthermore, strong inhibition occurs when only about 30% of the eIF2 molecules are phosphorylated. The mechanism of the inhibition became clear when it was shown that P-eIF2 does not undergo the guanine nucleotide exchange reaction catalyzed by eIF2B (Figure 1). The current view is that P-eIF2 has a high affinity for eIF2B and in effect ties up this less abundant initiation factor so that it is not available to catalyze the GTP exchange reaction on non-phosphorylated eIF2 (Rowlands et al.,

1988). There is also controversial evidence that P-eIF2 function may be altered at other points in the initiation pathway.

A strategy for elucidating possible mechanisms of translational control involving phosphorylation is to phosphorylate purified initiation factor proteins *in vitro*, then test for differences in specific activity in the various assays available. This approach is highly risky, since such phosphorylations might not occur *in vivo*, or assays may not detect functional changes. An alternate approach, which we have selected, is first to establish that phosphorylation occurs *in vivo*, then to correlate the change in extent of phosphorylation with a change in initiation rates. Once correlative

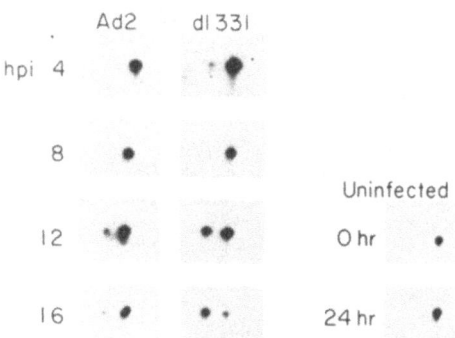

Figure 2. Immunoblot analysis of adenovirus-infected cell lysates. Suspension cultures of HeLa cells were infected with wild-type adenovirus (Ad2) or with the VA RNAI-negative mutant Ad5dl331 (dl331). Lysates were prepared at different hours post-infection (hpi), as indicated in the figure , or from uninfected cells. Proteins were fractionated by IEF/SDS-PAGE, and transferred to nitrocellulose. eIF-2α forms were visualized by immunoblotting with an antibody specific for eIF-2α, followed by [125]I-labeled goat anti-rabbit IgG. The figure shows an autoradiogram, with the more acidic (phosphorylated) form of eIF-2α to the left of the more basic (non-phosphorylated) form.

information is obtained, the phosphorylation event can be studied in greater detail *in vitro*. In order to detect phosphorylation events and measure the extents of phosphorylation occuring *in vivo*, we developed antisera specific for a number of initiation factor proteins and employed IEF/SDS-PAGE and immunoblotting. Cells are lysed directly into denaturing urea buffer to preserve the phosphorylation status of the proteins, and isoelectric variants of the factors are readily detected in the 2D gel blots. The method has been used to demonstrate initiation factor modification changes following heat shock (Duncan and Hershey, 1984), serum deprivation (Duncan and Hershey, 1985) and chemical stresses (Duncan and Hershey, 1987).

An example of this approach is a collaborative study of mammalian cells infected with adenovirus, done with Mike Mathews and Bob Malley at the Cold Spring Harbor Laboratory. When adenovirus infects a cell, it generates a small amount of double-stranded RNA which activates an eIF2α kinase called DAI. DAI activity results in an inhibition of protein synthesis and a decrease in virus yield. However, adenoviruses have developed a mechanism to prevent the activation of DAI by expressing large amounts of viral-associated RNAs (VA-I and VA-II). Mutant viruses lacking the gene for VA-I fail to prevent DAI activation and produce low viral yields (Schneider and Shenk, 1987). It has been shown that such infected cells contain activated DAI which can phosphorylate exogenous eIF2, but a direct demonstration that endogenous eIF2 is phosphorylated was lacking. Using immunoblotting (Figure 2), we show that endogenous eIF2 phosphorylation occurs to a great extent in the mutant virus-infected cell, whereas little phosphorylation occurs in cells infected by the wild-type virus.

Table 2. Changes in Phosphorylation of Initiation Factors

| Conditions | eIF2α | eIF4B | eIF4E |
|---|---|---|---|
| **INHIBITORY** | | | |
| severe heat shock | ↑ | ↓ | ↓ |
| serum deprivation | ↑ | ↓ | ↓ |
| amino acid starvation | ↑ | ↘ | |
| extremes of pH | ↑ | ↘ | |
| interferon treatment | ↑ | → | |
| hypertonic shock | → | → | |
| **STIMULATORY** | | | |
| serum stimulation | ↓ | ↑ | |
| sea urchin fertilization | ↓ | | ↑ |
| **VIRUS INFECTION** | | | |
| EMC | → | → | |
| polio | → | → | |
| VSV | ↗ | → | |
| Reo | ↗ | → | |
| vaccinia | ↗ | ↘ | |
| adeno | ↗ | ↗ | |

Such analyses have been carried out on mammalian cells grown under a large variety of physiological conditions that cause changes in global translation rates. We find that in general, inhibition of initiation of protein synthesis correlates with the phosphorylation of eIF2α and a dephosphorylation of eIF4B and eIF4E. A summary of the cell types analyzed and the results obtained are provided in Table 2. Given our knowledge about the effects of phosphorylation of eIF2α, it is reasonable to propose that this modification may cause or contribute to the inhibitory phenomena listed in the table. However, little or no *in vitro* evidence is yet available that demonstrates changes in the specific activities of eIF4B or eIF4E upon phosphorylation.

STUDIES OF eIF2a PHOSPHORYLATION *IN VIVO*

The studies cited above demonstrate that eIF2α phosphorylation may be a general method for inhibiting protein synthesis. However, such correlative evidence does not prove that eIF2α phosphorylation is the primary cause of the inhibition, or is a sufficient condition for inhibition. To address these issues, we have begun *in vivo* studies with transfected cells in which the eIF2α cDNA is expressed. Such experiments are based on our previous cloning of the eIF2α cDNA from rat and human libraries (Ernst et al., 1987), and identification of the site of phosphorylation by HCR and DAI (Pathak et al., 1988b). The strategy was to mutate the cDNA such that the protein product would contain alanine instead of serine at the phosphorylation site. We hoped that the mutated protein might inhibit cellular eIF2α kinases and thereby prevent the phosphorylation of endogenous eIF2, with a concommitant effect on translation rates.

To follow this strategy, two test systems were devised. The first involves the translation of mRNAs from transfected plasmids, where it has been observed that such mRNAs are translated poorly unless adenovirus VA-I RNA is present in the cells (Kaufman and Murtha, 1987). An explanation is that bidirectional transcription of such plasmids results in partially double-stranded RNA structures which locally activate DAI and inhibit protein synthesis through eIF2α phosphorylation nearby. Such local inactivation of mRNAs by apparent activation of DAI has been observed *in vitro* by de Benedetti and Baglioni (1984). We asked whether or not the presence of mutant eIF2α forms might substitute for VA-I RNA and thereby prevent the action of DAI. In the second test system, we proposed to construct long-term transfected cell lines expressing eIF2α forms, then test for adenovirus yields following infection with wild-type and mutant viruses unable to express VA-I RNA. Again, we asked whether or not the eIF2α forms might interfere with eIF2α phosphorylation in the mutant virus infections (as shown above in Figure 2) and thereby increase viral protein synthesis and yield. Towards these ends, the eIF2α cDNA was mutated to code for Ala instead of Ser at positions 48 (Ser-48-Ala) and at the phosphorylation site, 51 (Ser-51-Ala). The cDNAs were inserted into the expression vector shown in Figure 3, where eIF2α mRNA is transcribed by the adenovirus major late promoter.

In the first test system, expression plasmids for DHFR (dihydrofolate reductase) and ADA (adenine deaminase), similar to that shown in Figure 3, were co-transfected into COS-1 cells. Translation of there mRNAs is low (Figure 4, lane 2) unless VA-I RNA also is expressed (lane 1). When DHFR and wild-type eIF2α plasmids are co-

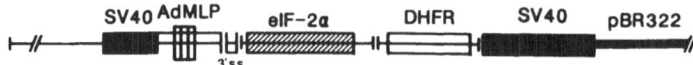

Figure 3. Expression vector for eIF-2α. The basic vector system is pBR322-based, and carries the SV-40 enhancer and origin. The cDNA is inserted downstream from the major late promoter and most of the tripartite leader of adenovirus, and is followed by the DHFR gene and SV-40 polyadenylation signal.

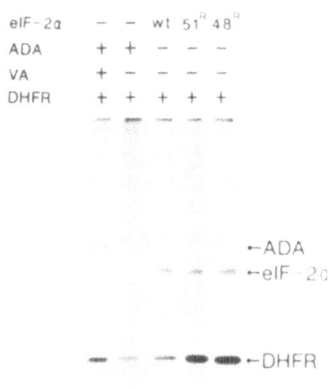

Figure 4. Effect of eIF-2α expression on DHFR synthesis in transfected cells. COS-1 cells were co-transfected with vectors expressing DHFR, ADA and/or eIF-2a forms as indicated at the top of the figure. Three forms of eIF-2a were expressed: wild-type (wt); the Ser-51-Ala mutant ($51^R$); and the Ser-48-Ala mutant ($48^R$). Transfected cells were sorted from non-transfected cells with a FACS (Kaufman and Murtha, 1987), pulse-labeled with [$^{35}$S]methionine, and lysate proteins were fractionated by SDS-PAGE. The figure shows an autoradiogram of the gel.

transfected in the absence of VA-I RNA, DHFR synthesis is low. In contrast, eIF2α synthesis is relatively high and is not stimulated by the presence of VA-I RNA (results not shown). When either the Ser-51-Ala or Ser-48-Ala mutant forms of eIF2α are expressed along with DHFR, DHFR synthesis is very strong (lanes 4 and 5). Figure 4 shows the protein labeling patterns of a fairly pure population of transfected cells. This is achieved by staining the cells with fluorocein-methotrexate followed by sorting by a FACS; transiently transfected cells make DHFR, bind the fluorescent probe, and can be distinguished from non-transfected cells. It is therefore possible to conclude that VA-I RNA and mutant eIF2α proteins affect only the translation of plasmid-derived mRNAs and do not stimulate global rates of cellular protein synthesis. The replacement of VA-I RNA by the eIF2α mutant proteins indicates that DAI kinase activity has been inhibited. Since both the Ser-48-Ala and Ser-51-Ala mutants inhibit, it is not clear by what mechanism the DAI activity is reduced. It also is not yet clear whether the active form of these mutant eIF2α proteins is the free subunit or a trimeric eIF2 factor formed by exchange with the overexpressed subunit.

In preliminary experiments, we also have constructed a long-term cell line that expresses the Ser-48-Ala mutant, and have infected the cells with adenovirus and with the dl331 mutant that fails to synthesize VA-I RNA. The mutant virus grows better on transfected than on non-transfected cells, suggesting that eIF2α phosphorylation is the primary cause of low virus yields in this system. This cell line should be very useful for determining whether or not eIF2α phosphorylation is the primary cause of translational inhibition in those cases listed in Table 2.

Further evidence that eIF2α phosphorylation affects protein synthesis *in vivo* was obtained by transiently transfecting cells with mutant eIF2α cDNAs where Ser at positions 48 and 51 was replaced by Asp. The aspartic acid residue, carrying a negative charge, might mimic phospho-serine and thereby interfere with initiation. Cells transfected with the Ser-48-Asp mutant exhibited normal rates of protein synthesis, and overproduced the mutant eIF2α protein which was phosphorylated to a high extent. This argues that phosphorylation at Ser-48, if it occurs, would not affect translation rates. In contrast, transfection with the Ser-51-Asp mutant resulted in barely detectable synthesis of the mutant eIF2α protein and a strong inhibition of global rates of protein synthesis. This is the expected result if the Ser-51-Asp protein functions like P-eIF2α. Only a small amount of the protein (ie., ca. 0.3 equivalents of the endogenous level of eIF2) may be sufficient to inactivate the low levels of eIF2B in the cell, and such amounts are nearly two orders of magnitude below the amounts normally produced transiently in tranfected cells. The results suggest that eIF2α phosphorylation at Ser-51 is a sufficient condition for inhibiting protein synthesis in intact cells. Such cells may be useful in identifying mRNAs that resist the effects of eIF2α phosphorylation.

## CONCLUSIONS

Phosphorylation likely plays an important role in regulating initiation factor function. This is now rather well established for eIF2α both *in vitro* and *in vivo*, but is not yet certain for the other initiation factors. However, the techniques are available to establish the roles of other factor modifications, and should be adequate to enlarge the list of firmly established examples. In addition to initiation factor phosphorylation, it is known that phosphorylation of ribosomal protein S6 and elongation factors EF-2 and EF-1α also occurs and likely influences translational efficiency. A major current research interest is to identify the protein kinases involved in phosphorylating the translational machinery and to elucidate how these kinases and their associated phosphoprotein phosphatases are regulated.

## REFERENCES

De Benedetti, A., and Baglioni, C., 1984, Inhibition of mRNA binding to ribosomes by localized activation of dsRNA-dependent protein kinase. Nature, 311:79-81.

Duncan, R., and Hershey, J.W.B., 1984, Heat-shock-induced translational alteractions in HeLa cells. J. Biol.Chem., 259:11882-11889.

Duncan, R., and Hershey, J.W.B., 1985, Regulation of initiation factors during translational repression caused by serum deprivation. J. Biol. Chem., 260:5493-5497.

Duncan, R.F., and Hershey, J.W.B., 1987, Translational repression by chemical inducers of the stress response occurs by different pathways. Arch. Biochem. Biophys., 256:651-661.

Ernst, H., Duncan, R.F., and Hershey, J.W.B., 1987, Cloning and sequencing of complementary DNAs encoding the α-subunit of translational initiation factor eIF-2. J. Biol. Chem., 77:1286-1290.

Kaufman, R.J., and Murtha, P., 1987, Translational control mediated by eukaryotic initiation factor-2 is restricted to specific mRNAs in transfected cells. Molec. Cell. Biol. 7:1568-1571.

Moldave, K., 1985, Eukaryotic protein synthesis. Annu. Rev. Biochem. 54:1109-1149.

Ochoa, S., 1983, Regulation of protein synthesis in eukaryotes. Arch. Biochem. Biophys., 223:325-349.

Pain, V.M., 1986, Initiation of protein synthesis in mammalian cells. Biochem. J., 235:625-637.

Pathak, V.K., Nielsen, P.J., Trachsel, H., and Hershey, J.W.B., 1988a, Structure of the B subunit of translational initiation factor eIF-2. Cell, in press.

Pathak, V.K., Schindler, D., and Hershey, J.W.B., 1988b, Generation of a mutant form of protein synthesis initiation factor eIF-2 lacking the site of phosphorylation by eIF-2 kinases. Molec. Cell. Biol., 8:993-995.

Proud, C.G., 1986, Guanine nucleotides, protein phosphorylation and the control of translation. Trends Biochem., 11:73-77.

Rowlands, A.G., Panniers, R., and Henshaw, E.C., 1988, The catalytic mechanism of guanine nucleotide exchange factor action and competitive inhibition by phosphorylated eukaryotic initiation factor 2. J. Biol. Chem., 263:5526-5533.

Schneider, R.J., and Shenk, T., 1987, Impact of virus infection on host cell protein synthesis. Annu. Rev. Biochem., 56:317-332.

# Ca²⁺-CALMODULIN-DEPENDENT PROTEIN KINASES AND PROTEIN KINASE C:

# FUNCTIONAL SIMILARITIES

Theodore G. Sotiroudis, Symeon M. Kyriakidis, Leonidas G. Baltas, Vasilis G. Zevgolis and Athanasios E. Evangelopoulos

The National Hellenic Research Foundation
48 Vassileos Constantinou Avenue
Athens 116 35 – Greece

Protein serine and threonine kinases can be classified into individual groups or subclasses on the basis of the type of regulation of their activities (Krebs, 1986). Two of the most intensively studied groups are Ca²⁺-regulated, i.e. the Ca²⁺/calmodulin (CaM)-dependent and the Ca²⁺-phospholipid (diacylglycerol)-dependent protein kinases. Of the enzymes belonging in the category of Ca²⁺/CaM-dependent kinases, myosin light chain kinases (MLCK) are distinguished by their high degree of substrate specificity and CaM dependency (Edelman et al, 1987). Phosphorylase kinase (PhK) another member of the same group is characterized by a broader substrate specificity. Its primary substrate is phosphorylase b but the enzyme may catalyze the phosphorylation of other proteins (Chan & Graves, 1984). In addition, a number of Ca²⁺/CaM-dependent multifunctional protein kinases (Ca²⁺/CaM PrK) identified in a variety of tissues shows a broad substrate specificity suggesting that such a group of CaM-dependent protein kinases may play important roles in the control of different cellular processes (Shenolikar et al, 1986). On the other hand, protein kinase C (PKC) is a multifunctional protein kinase identified by Nishizuka and co-workers as a Ca²⁺- and phospholipid-dependent protein kinase that plays a crucial role in the signal transduction for a variety of biologically active substances involved in cellular function and proliferation (Nishizuka, 1984). In the presence of limiting amounts of Ca²⁺ and phospholipids its activity is stimulated by sn-1,2-diacylglecerols or by phorbol esters (Nishizuka, 1984) and the kinase phosphorylates a broad range of cellular proteins (Kikkawa and Nishizuka, 1986).

During the last years a model has evolved which predicts a mechanism for enzyme activation by CaM (Gietzen et al, 1981; Kyriakidis et al, 1986).This model is based primarily on the basic similarity of CaM-dependent enzymes in their inhibition by hydrophobic drugs and their activation by acidic amphiphiles and limited proteolysis, three features that also characterize PKC, although this enzyme is known to be not responsive to CaM (Takai et al, 1979). On the basis of this hypothetical model it would be reasonable to assume that PKC obeys to the same general mechanism of hydrophobic activation and inhibition of CaM-dependent enzymes.

Studies in our laboratory, were focused on the understanding of the hydrophobic properties of cytoplasmic PhK in an effort to test the hypothetical model by which most types of activation and inhibition phenomena of Ca$^{2+}$/CaM- and Ca$^{2+}$/phospholipid-regulated protein kinases can be ascribed to a common general mechanism involving similar hydrophobic and ionic interactions. This type of interactions may lead under certain intracellular conditions to the transformation of the above Ca$^{2+}$-dependent kinases from a "soluble" to a membrane-bound form. We describe below the experimental evidence supporting such a model.

## Limited proteolysis

PhK from rabbit skeletal muscle, a hexadecamer composed of four dissimilar subunits $\alpha_4\beta_4\gamma_4\delta_4$, has been shown to exist in a non-activated and an activated form at pH 6.8. The non-activated kinase can be converted to the activated form by covalent modification processes including protein phosphorylation and limited proteolysis (Chan and Graves, 1984; Pickett-Giese & Walsh, 1986). In most cases, initial proteolysis activation is concomitant with a marked degradation of the $\alpha$ and $\beta$ subunits, whereas the $\gamma$ and $\delta$ subunits remain intact. Limited proteolysis eliminates the absolute requirement of PhK for the binding by extrinsic CaM but increases sensitivity to Ca$^{2+}$ activation via the intrinsic $\delta$-subunit (Pikett-Giese & Walsh, 1986). PhK from chicken gizzard (Nikolaropoulos & Sotiroudis, 1985) and bovine stomach (Zevgolis et al, in prep.) smooth muscle, a type of Ca$^{2+}$-CaM protein kinase, lacking $\alpha$ and $\beta$ subunits (one low Mr protein band was shown upon SDS-PAGE), has also been found to be effectively activated after limited proteolysis by trypsin. Similarly, light tryptic proteolyisi of MLCK has been shown in certain cases to activate the enzyme and to remove the absolute dependence on the presence of Ca$^{2+}$/CaM (Stull et al, 1986). PKC is now known to be a large family of proteins with multiple subspecies that have subtle individual enzymological characteristics and at the last count seven subspecies have been identified ($\alpha, \beta_I, \beta_{II}, \gamma, \delta, \epsilon, \zeta$) (Nishizuka, 1988). PKC can also be irreversibly activated by proteolysis and the liberated kinase domain free of the regulatory domain is active in the absence of Ca$^{2+}$, phospholipid and diacylglycerol (Kishimoto et al, 1983). Concerning the other class of Ca$^{2+}$-dependent protein kinases, the multifunctional Ca$^{2+}$/CaM PrK (Stull et al, 1986), there is incomplete experimental knowledge on the relation between limited proteolysis and Ca$^{2+}$/CaM-dependent catalytic activity (Woodgett et al, 1983) and therefore we cannot draw decisive conclusions.

## Effect of naturally occuring lipids and other hydrophobic compounds

Another means by which Ca$^{2+}$ might alter the activity or subcellular distribution of a protein kinase is to facilitate the association of hydrophobic ligands with protein. One such enzyme is PKC, which depends on Ca$^{2+}$ as well as phospholipid for its activation (Nishizuka, 1984). Diacylglycerol, which is generated by hormone-induced turnover of phosphatidylinositol in the plasma membrane dramatically increases reaction velocity and decreases the activation constant of phospholipid as well as of Ca$^{2+}$ in such a way that the enzyme can be activated at physiological Ca$^{2+}$ concentrations (0.1 μM) (Kikkawa & Nishizuka, 1986; Castagna et al, 1985). Among various phospholipids tested the order of potency for phospholipid in supporting enzyme activity is as follows: phosphatidyl serine > phosphatidic acid > phosphatidylinositol > phosphatidylethanolamine > sphingomyelin > phosphatidylcholine (Castagna et al, 1985). Tumor-promoting phorbol esters, have also been shown to bind to and stimulate PKC, by substituting for diacylglycerol in the activation of the enzyme (Kikkaea & Nishizuka, 1986). Another type of PKC activation reported recently, involves the interaction of the kinase with cis-fatty acids, in the absence of $Ca^{2+}$ and phospholipids (Murakami et al, 1986), while in a recent paper (Sakai et al, 1987) the authors presented evidence suggesting that purified PKC phosphorylates clupein sulfate in a phosphatidyl serine-dependent but $Ca^{2+}$-independent process; in the later case, the effect of diolein was not significant. Moreover, PKC-dependent phosphorylation of profillin has been shown to be specifically stimulated by phosphatidylinositol biphosphate, whereas phosphatidylserine was ineffective (Hansson et al, 1988). In general, the dependence of PKC on phospholipid and diacylglycerol varies greatly with the phosphate acceptor proteins and the known members of PKC are fully active to phosphorylate some substrate proteins in absence of phospholipid and diacylglycerol (Ono et al, 1988). Other hydrophobic activators of PKC which are known to substitute for phosphatidylserine include E.coli lipid X (Wightman & Raetz, 1984), the lipophilic muranyltripeptide MTP-PE (Meyer et al, 1986) and the synthetic naphthalenesulfonamide derivative SC-9 (Ito et al, 1986). It has been also observed that sphingosine (Hannun et al, 1986), lysosphingolipids (Hannun & Bell, 1987) and gangliosides (Kreutter et al, 1987), a specific type of lipids ubiquitous in eukaryotes, potently and reversibly inhibited PKC activity (in presence of phosphatidylserine and discylglycerol) while, glaglioside GM3 (Mamoi, 1986) and cerebroside sulfate (Fujiki et al, 1986) were able to activate PKC as substitutes for phosphatidylserine. It was proposed that sphingosine inhibition of PKC may have physiological significance acting as a negative effector of the enzyme (Hannun et al, 1986) and that lysosphingolipids represent the functional missing link between the accumulation of sphingolipids and the pathogenesis of sphingolipidoses (Hannun & Bell, 1987). More interestingly, a recent communication suggests that phosphatidylinositol biphosphate may effectively antecede diacylglycerol as activator of PKC, reducing the $Ca^{2+}$-requirement of this kinase, and that phosphatidylinositol biphosphate rather diacylglycerol is the primary pre-activator of PKC (Chauhan &Brockerhoff, 1988). It is also important to emphasize that the newly found PKC subspecies with δ- and ε-sequences do not show an absolute requirement for phospholipid and diacylglycerol with H1 histone as substrate (Ono et al, 1988).

Although initial studies in other laboratories have shown that hydrophobic organic solvents and unsaturated fatty acids stimulated PhK activity at neutral pH, they failed to find any significant effect of phospholipids on phoaphorylase b to a conversion (Singh and Wang, 1979; Negami et al, 1986). In this respect, we were able to demonstrate that the activity of non-activated PhK at pH 6.8 can be stimulated by a variety of phospholipids and certain anionic amphiphiles, the activating effect being largely dependent on the size of lipid vesicles which in turn is connected with the procedure of vesicle preparation (Kyriakidis et al, 1986a). In parallel, both acidic phospholipids and SDS drastically increase autophosphorylation of PhK (Negami et al, 1986; Kyriakidis et al, 1986a) while in presence of mixed acidic phospholipids, the autoactivation of PhK, at pH 6.8, is enhanced (Negami et al, 1986). Diacylglycerol was also found to slightly stimulate phospholipid activation of phosphorylase b to a conversion while PhK was unable to bind phorbol esters (unpublished results).

In our effort to gain better insight into the hydrophobic behaviour of PhK and to compare its lipid-binding properties with those of PKC we have tested a number of lysosphingolipids and gangliosides on several reactions catalyzed by PhK (Baltas et al, in prep.). We found that sphingosine and galactosylsphingosine (phsychosine) potently inhibited phosphorylase b to a conversion. Psychosine was also able to stimulate the autophosphorylation of PhK and this stimulation was accompanied with a sharp decrease of the rate of autoactivation at pH 6.8. Moreover the activity of nonactivated PhK at pH 6.8 could be stimulated by a number of brain gangliosides. Among the individual gangliosides tested the activation potency of DG1$_a$ and GT1$_b$ was higher than that of GM$_1$. Most important, GD1$_a$ dramatically increaes the affinity of the enzyme for Ca$^{2+}$.

Furthermore, evidence was obtained that smooth muscle MLCK can also be activated by naturally occuring lipids (phosphatidylinositol, phosphatidylserine, arachidonic acid) and SDS (Tanaka and Hidaka, 1980). As far as the effect of phospholipids on the activity of the multifunctional Ca$^{2+}$/CaM PrK is concerned, there is a recent report (Baudier and Cole, 1987) showing that phosphorylation of tau proteins by this broad substrate specificity kinase is stimulated by a number of phospholipids, although in this case the authors suggest that the effect of phospholipids is on substrate protein rather on activation of the kinase. However, further studies are needed with a number of other physiological substrates and phospholipids in order to clarify if this type of Ca$^{2+}$-dependent kinase is able to interact with lipid molecules. In this respect it is interesting to note that smooth muscle PhK form bovine stomach, which has been proved by our group to be Ca$^{2+}$/calmodulin dependent and to possess similar structural characteristics with those of multifunctional Ca$^{2+}$/CaM kinase (Zevgolis et al, in prep.), was drastically inhibited by psychosin and activated by ganglioside GD1$_a$ (Baltas et al, in prep.); At the same time the action of phosphatidylinositol was inhibitory in contrast ot the activating effect observed with skeletal PhK (Kyriakidis et al, 1986a). The above results suggest that each Ca$^{2+}$-dependent kinase may have its own specific structural requirements for effective interaction with lipid molecules.

## Effect of other hydrophobic compounds

Several hydrophobic compounds interacting with phospholipids were found to inhibit PKC to various extents. These include psychotic drugs, local anesthetics, W-7, polyamines, melittin, heparin, polymyxin B and some other drughs (Kikkawa & Nishizuka, 1986). In addition, the hydrophobic flavonoid drug quercetin, known to affect membrane-linked activities and to inhibit cAMP-independent kinases have also been shown to inhibit Ca$^{2+}$/phospholipid dependent activity (Gschwendt et al, 1983). Most of the above compounds have been shown to inhibit a number of Ca$^{2+}$/CaM dependent regulatory systems and therefore they have become known as anti-calmodulin agents or CaM inhibitors.

Concerning PhK, a number of studies in this and other laboratories have shown that many anticalmodulin agents effectively alter the catalytic and structural properties of the enzyme. Thus, we found that polymyxin B inhibited both cytoplasmic (Ktenas et al, 1985) and sarcoplasmic reticulum (SR)-bound PhK (Ktenas et al, in prep.). In the first case, we observed also that this cyclic polycationic peptide greatly stimulated autophosphorylation and autoactivation of the kinase at pH 6.8. Moreover, when SR membranes were $^{32}$P labelled by endogenous kinase(s), in presence of exogenous calmodulin, it was revealed that polymyxin B increased $^{32}$P incorporation into a protein band with a similar Mr to that of $\alpha$ subunit of PhK (ktenas et al, in prep.), suggesting that the drug is able to stimulate also autophosphorylation of membrane-bound kinase. In another series of experiments we have examined the interaction of flavonoids with skeletal muscle PhK (Kyriakidis et al, 1986b). From 14 flavonoids tested the flavones quercetin and fisetin were proved to be efficient inhibitors while the flavanone hesperetin stimulated PhK activity. Quercetin inhibited also the atuophosphorylation of the kinase while this flavonoid was found to be a competitive inhibitor of ATP for the phosphorylation of phosphorylase b. Shenolikar et al (1979) studying the role of CaM in the structure and regulation of PhK have observed that the antipsychotic drug trifluoperazine completely abolished CaM-stimulated PhK activity. On the other hand Hessova et al (1985) have shown that heparin reversibly stimulated $Ca^{2+}$-independent pH 7.0 activity with concomitant inhibition of $Ca^{2+}$-dependent,pH 8.2 activity. At the same time the aggregation state of the enzyme was drastically changed.

As in the case of PKC and PhK, it is well known that a number of the anticalmodulin agents effectively inhibited both MLCK (Stull et al, 1986) and multifunctional $Ca^{2+}$/CaM kinase (Schulman, 1984) competing with CaM. In contrast cyclic-nucleotide dependent protein kinases are not susceptible to this type of drugs (Kikkawa and Nishizuka, 1986).


## Association with membranes

The most important property of PKC concerning signal transuction mechaisms, is its ability to be rapidly and reversibly distributed between soluble and membrane-bound forms (Kraft & Anderson, 1983). The translocation process is closely related to the activation of the enzyme itself, since its physiological regulation requires, as a prerequisite, a specific interaction with membrane phospholipids (Nishizuka, 1986). To this point, PKC represents the best known example of an "amphitropic" protein kinase, the term "amphitropic" being recently introduced by Burn (1988) for the proteins which can exist both in a cytoplasmic soluble form as well as in a membrane embedded form. It is thus important to emphasize that a prerequisite for a $Ca^{2+}$/CaM-dependent protein kinase to be characterized as a functional equivalent of PKC must be the capacity of the kinase for a reversible distribution between soluble and membrane-bound forms. In this respect it is known that several microsomal preparations from brain (Nairn et al, 1985), liver (Famulski and Carafoli, 1984) and cardiac (Lindemann and Watanabe, 1985) or skeletal muscle (Tuana and MacLennan, 1984) contain membrane bound CaM-dependent protein kinase activities which control $Ca^{2+}$/ATPase function (Famulski and Carafoli, 1984; Tuana and MacLennan, 1984), the phosphorylation of the cardiac sarcolemmal protein phospholamban (Lindemann and Watanabe, 1985) and the phosphorylation of the neuronal protein synapsin I (Nairn et al, 1985). The structural relationship between these membrane-bound CaM-dependent protein kinases and their soluble counterparts is not well understood. However, recently it has been indicated that the membrane-bound CaM-dependent protein kinase that phosphorylates phospholamban in cardiac membranes resembles the multifunctional $Ca^{2+}$/CaM kinase II and that this kinase may be present in both the particular and soluble heart fractions (Jett et al, 1987). In additon, a multifunctional $Ca^{2+}$/CaM protein kinase purified from cytosol of rabbit skeletal muscle was considered to be indentical with a 58 kDa protein in SR and sactolemma of the same tissue (Sato et al, 1988).

To our knowledge, there is not any experimental data in literature on the interaction of MLCK with cellular membranes, although myosin light chain served as good substrate for the cytosolic multifunctional $Ca^{2+}$/CaM kinase which is also present in SR and sarcolemma of skeletal muscle (Sato et al, 1988).

Concerning PhK, the other broad substrate specificity $Ca^{2+}$/CaM dependent kinase it is known that although cytosolic enzyme, this kinase is also associated with SR or plasma membranes (Hörl et al, 1978; Dombradi et al, 1984). Recently Thieleczek et al (1987) have localized molecular structures related to PhK at the SR of rabbit skeletal muscle employing polyclonal antibodies against the holenzyme as well as monoclonal antibodies specific for its α-, β-, or γ-subunits. In our effort to further investigate the relation between cytoplasmic and membrane bound

PhK we have chosen as a membrane system the inside-out human erythrocyte vesicles a system also used for the study of the mechanism of PKC-membrane interaction (Wolf et al, 1985). We found (Kyriakidis et al, 1988) that at pH 7.0 PhK binds to the inner face of the erythrocyte membrane in a $Ca^{2+}$ and $Mg^{2+}$ dependent manner and the sharpest increase of this association occurs between 70 and 550 nM free $Ca^{2+}$ when 3 mM $Mg^{2+}$ is present. CaM-decreases the original binding capacity about 50% suggesting that this $Ca^{2+}$ binding protein may block some sites on the kinase molecule responsible for enzyme-membrane association. These sites are possibly located on exposed areas of $\alpha$ or/ and $\beta$ subunits which are susceptible to proteolysis, a hypothesis compatible with a highly reduced capacity of trypsinolyzed PhK for erythrocyte membrane binding (Kyriakidis et al, 1988). It was found also that several proteins of the membrane can be labelled with $^{32}P$ by PhK, a 93 KD band being the most prominent phosphorylated protein and possibly represents band 3 polypeptide of red blood cell membrane. In contrast, we could not find any significant $Ca^{2+}$-dependent binding of PhK to SR either in presence or in absence of glycogen, suggesting that endogenous PhK molecules associated very tightly with SR (Thieleczek et al, 1987) do not permit exogenous PhK to interact with specific SR membrane binding sites.

## Activation by lanthanide ions and $Cd^{2+}$

Although both $Ca^{2+}$/CaM-and $Ca^{2+}$/phospholipid-dependent kinase activation seem to be specific for $Ca^{2+}$ (Pickett-Giese & Walsh, 1986; Stull et al, 1986; Kikkawa & Nishizuka, 1986), experimental evidence suggests that trivalent lanthanide ions ($Ln^{3+}$) and $Cd^{2+}$ are able to substitute for $Ca^{2+}$ in regulating the activity of the above kinase systems. Thus, it has been shown that $Ln^{3+}$ were effective $Ca^{2+}$ substitutes for both MLCK (Mazzei et al, 1983) and multifunctional $Ca^{2+}$/CaM PrK (Kuret and Schulman, 1984). On the other hand, PKC although less sensitive to $Ca^{2+}$ substitution by $Ln^{3+}$ shows an enhanced ability to be activated by suboptimal $Ca^{2+}$, when these metal ions are present (Mazzei et al, 1983). Similarly, we obserbed that $Ln^{3+}$ effectively mimic the stimulatory action of $Ca^{2+}$ on PhK (Sotiroudis, 1986) but in contrast to the effect on MLCK and PKC which is biphasic (stimulation followed by inhibition with increasing metal cation concentration), $Ln^{3+}$ exhibited only a stimulatory action on PhK. $Cd^{2+}$ was also found to mimic effectively, potentiate and antagonize the stimulatory actions of $Ca^{2+}$ on MLCK and PKC in a biphasic manner (Mazzei et al, 1984). We obtained similar results when $Cd^{2+}$ was substituted for $Ca^{2+}$ in PhK activity assays both at pH 6.8 and 8.2 (Sotiroudis, 1986). It must also be noted that both $Ln^{3+}$ and $Cd^{2+}$ were able to replace $Ca^{2+}$ required for the stimulation of PhK by exogenous CaM (Sotiroudis, 1986).

According to the hypothesis of the existence of two classes of sites for divalent and trivalent cations in CaM (capital and auxiliary) (Cox, 1988), $Ca^{2+}$ binds only to capital sites but $Ln^{3+}$ and $Cd^{2+}$ are able to occupy both type of sites. Concerning PKC, the location of $Ca^{2+}$ binding site(s) is unclear and there is no typical E-F hand structure present that would provide a CaM like $Ca^{2+}$-binding site. Moreover there is experimental evidence that diacylglycerol and phospholipid binding sites are located within this regulatory domain (Parker & Ullrich, 1987). Recently, evidence has been presented (Murakami e al, 1987) that PKC possesses high and low affinity $Ca^{2+}$-binding sites and that at least one $Zn^{2+}$-binding site (auxiliary site?) exists which is dinstict from $Ca^{2+}$-binding sites.

## Conclusions

Many hormones and extracellular messengers regulate cell function, in part at least, by inducing an increase in $Ca^{2+}$ concentration of the cell cytosol. $Ca^{2+}$/CaM and $Ca^{2+}$/phospholipid-dependent protein kinases represent a very important group of $Ca^{2+}$-receptor proteins which is responsible for the propagation and amplification of the signal formed by a variety of physiolocigal stimuli.

An essential phase in the mechanism of activation of enzymes by CaM is a conformational change induced in CaM by binding of $Ca^{2+}$ and as a consequence a hydrophobic site becomes exposed on the surface of the molecule (Cox, 1988). Peptides modeled on the CaM-binding domains of several enzymes have demonstrated both $Ca^{2+}$-dependent CaM-binding and the potential to form amphiphilic helices. Overall, studies with MLCK, PhK (γ subunit) (Lucas et al, 1986) and $Ca^{2+}$/CaM-dependent protein kinase (Hanley et al, 1988) provide a model for CaM binding domains in structurally diverse CaM binding proteins, that contain clusters of basic residues within potential amphiphilic α-helical structures. On the other hand from the amino acid sequence of bovine PKC one can predict amphiphatic helices that could provide a site for hydrophobic interaction, although no single hydrophobic stretch was identified within the regulatory domain (Parker and Ullrich, 1987). In this context, experimental evidence from this and other laboratories permits us to develop a hypothesis according to which the hydrophobic activation and inhibition of both $Ca^{2+}$/CaM and $Ca^{2+}$/phospholipid-dependent protein kinases can be described by the assumption of similar hydrophobic domains, which possibly determine in a similar way the responsiveness of these kinases to transmembrane signalling. Moreover, the data available argue against the existence of specific (e.g. phosphatidylserine) phospholipid-activated kinases indicating instead that the $Ca^{2+}$-dependent stimulation of the above protein kinases reflects the more general actions of CaM and phospholipids as hydrophobic probes (Juskevich et al, 1983). As far as the inability of different research groups to identify a CaM-induced activation of PKC there are several explanations: Thus, CaM may substitute for phospholipid for the phosphorylation of specific substrates (Hansson et al, 1988) or that PKC may lose its sensitivity to be activated by CaM, during purification (Juskevich et al, 1983). Accordingly, the CaM-induced inhibition of phosphatidylserine-dependent PKC activity (Albert et al, 1984) may be due to a direct competition between CaM and phospholipid for the same hydrophobic site on the kinase.

Finally we suggest, that under mobilizing $Ca^{2+}$ conditions $Ca^{2+}$/CaM kinases and PKC may translocate to membrane compartments where they serve for a specialized function not necessarily connected with the corresponding cytosolic catalytic action. This translocation may lead under certain conditions to an insertion of the kinase into the membrane (Bazzi & Nelsestuen, 1988) and thus the integral membrane form of the kinase may function as a long-term cell regulator.

## References

Albert,K.A., Wu,W.C-S., Nairn, A.C. and Greengard,P.,1984, Inhibition of calcium/phospholipid-dependent protein phosphorylation,Proc.Natl. Acad.Sci. USA 81:3622.

Baltas,L.G., Zevgolis,V.G., Kyriakidis,S.M., Sotiroudis,T.G. and Evangelopoulos,A.E. in preparation

Baudier,J. and Cole,R.D.,1987, Phosphorylation of tau proteins to a state like that in Alzheimer's brain is catalyzed by a calcium/calmodulin-dependent kinase and modulated by phospholipids, J.Biol.Chem., 262:17577.

Bazzi,M.D. and Nelsestuen,G.L.,1988, Constitutive activity of membrane-
    inserted protein kinase C, Biochem.Biophys.Res.Commun.,152:336.
Burn,P.,1988, Amphitropic proteins: A new class of membrane proteins,
    Treds Biochem.Sci. 13:79.
Castagna,M., Pavone,C., Bazgar,S., Couturier,A., Chevalier,M. and
    Fiszman,M.,1985, Phospholipid/$Ca^{2+}$-dependent protein kinase,
    cell differentiation and tumor promotion, in: "Hormones and Cell
    Regulation" J.E. Dumont et al, eds, Elsevier Science Publishers BV
Chan,K.-F.J. & Graves,D.J.,1984, Molecular properties of phosphorylase
    kinase, in:"Calcium & Cell Function" W.Y. Cheung, ed., Academic
    Press, New York
Chauhan,V.P.S. and Brockerhoff,H.,1988, Phosphatidylinositol, -4-5 biphos-
    phate antecede diacylglycerol as activator for protein kinase C,
    FASEB J.2:A349
Cox,J.A., 1988, Interactive properties of calmodulin, Biochem.J., 249:621
Dombradi,V.K., Silberman,S.R., Lee,E.Y.C., Caswell,A.H. & Brandt,N.R.,1984,
    The association of phosphorylase kinase with rabbit muscle T-tubules,
    Arch.Biochem.Biophys., 230:615
Edelman,A.M., Blumenthal,D.K. and Krebs,E.G.,1987, Protein serine-threonine
    kinases,Ann.Rev.Biochem.,56:567
Famulski,K.S. and Carafoli,E.,1984, Calmodulin-dependent protein
    phosphorylation and calcium uptake in rat liver microsomes, Eur.
    J.Biochem., 141:15
Fujiki,H., Yamashita,K., Suganuma,M., Horiuchi,T., Taniguchi,N. and
    Makita,A.,1986, Involvement of sulfatide in activation of protein
    kinase C by tumor promoters,Biochem.Biophys.Res.Commun.,138:153
Gietzen,K., Sadorf,I. and Bader,H.,1981, A model for the regulation of the
    calmodulin-dependent enzymes  erythrocyte $Ca^{2+}$-transport ATPase
    and brain  phosphodiesterase by activators and inhibitors,
    Biochem.J., 207:541
Gschwendt,M., Horn,F., Kittstein, W. and Marks,F., 1983, Inhibition of
    the calcium- and phospholipid-dependent protein kinase activity
    from mouse brain cytosol by quercetin,Biochem.Biophys.Res.Commun.,
    117:444
Hanley,R.M., Means,A.R., Kemp,B.E. and Shenolikar,S., 1988, Mapping of
    calmodulin-binding domain of $Ca^{2+}$/calmodulin-dependent protein
    kinase II from rat brain,Biochem.Biophys.Res.Commun.,152:122
Hannun,Y.A., Loomis,C.R., Merill,A.H.Jr and Bell,R.M.,1986, Sphingosine
    inhibition of protein kinase C activity and of phorbol dibutyrate
    binding in vitro and in human platelets,J.Biol.Chem., 261:12604
Hannun,Y.A. and Bell,R.M.,1987, Lysosphingolipids inhibit protein kinase C:
    Implications for the sphingolipidoses,Science, 235:670
Hansson,A., Skoglund,G., Lassing,I., Lindberg,U. and Ingelman-Sundberg,M.,
    1988, Protein kinase C-dependent phosphorylation of profilin is
    specifically stimulated by phosphatidylinositol biphosphate ($PIP_2$),
    Biochem.Biophys.Res.Commun.,150:526.
Hessova,Z., Varsanyi,M. & Heilmeyer,L.M.G.,Jr.,1985, Dual function of
    calmodulin (δ) in phosphorylase kinase,Eur.J.Biochem., 146:107
Hörl,W.H., Jennissen,H.B. and Heilmeyer,L.M.G.,Jr.,1978, Evidence for the
    participation of a $Ca^{2+}$-dependent protein kinase and a protein
    phosphatase in the regulation of the $Ca^{2+}$-transport ATPase of
    the sarcoplasmic reticulum. 1.Effect of inhibitors of the $Ca^{2+}$-
    dependent protein kinase and protein phosphatase, Biochemistry,
    17:759
Ito,M., Tanaka,T., Inagaki,M., Nakanishi,K. and Hidaka,H.,1986,
    N-(6-Phenylhexyl)-5-chloro-1-Naphthalenesulfonamide. A novel
    activator of protein kinase C,Biochemistry, 25:4179
Jett,M.-F., Schworer,C.M., Bass,M. and Soderling,T.R.,1987, Identification

66

of membrane-bound calcium, calmodulin-dependent protein kinase II in canine heart,Arch.Biochem.Biophys., 255:354

Juckevich,J.C., Kuhn,D.M. and Lovenberg,W.,1983, Phosphorylation of brain cytosol proteins. Effects of phospholipids and calmodulin,J.Biol. Chem., 258:1950

Kikkawa,V. and Nishizuka,Y.,1986, Protein kinase C, in: "The Enzymes", P. Boyer amd E.G. Krebs, eds, Academic Press, New York

Kishimoto,A., Kajikawa,N., Siota,M. and Nishizuka,Y. ,1983, Proteolytic activation of calmodulin-activated, phospholipid-dependent protein kinase by calcium-dependent neutral protease, J.Biol.Chem., 258:1156

Kraft,A.S. and Anderson,W.B. ,1983, Phorbol esters increase the amount of $Ca^{2+}$ phospholipid-dependent protein kinase associated with plasma membrane, Nature, (London) 301:621

Krebs,E.G., 1986, The Enzymology of control by phosphorylation, in:"The Enzymes", P. Boyer and E.G. Krebs, eds, Academic Press New York

Kreutter,D., Kim,J.Y.H., Goldenring,J.R., Rasmussen,H., Ukomadu, C., DeLorenzo,R.J. and Yu,R.K., 1987, Regulation of protein kinase C activity by gangliosides, J.Biol.Chem., 262:1633

Ktenas,T.B., Sotiroudis,T.G., Nikolaropoulos,S. and Evangelopoulos,A.E., 1985, Interaction of phosphorylase kinase with polymixins, Biochem.Biophys.Res.Commun.,133:891

Ktenas,T.B., Sotiroudis,T.G. and Evangelopoulos,A.E. in preparation

Kuret,J. and Schulman,H., 1984, Purification and characterization of a $Ca^{2+}$/calmodulin-dependent protein kinase from rat brain Biochemistry, 23:5495

Kyriakidis,S.M., Sotiroudis,T.G. and Evangelopoulos,A.E., 1986a, Stimulation of glycogen phosphorylase kinase with phospholipids, Biochem. Inter., 13:853

Kyriakidis,S.M., Sotiroudis,T.G. and Evangelopoulos,A.E., 1986b, Interaction of flavonoids with rabbit muscle phosphorylase kinase, Biochim. Biophys.Acta, 871:121

Kyriakidis,S.M., Sotiroudis,T.G. and Evangelopoulos,A.E., 1988, $Ca^{2+}$ and $Mg^{2+}$-dependent association of phosphorylase kinase with human erythrocyte membranes, submitted for publication

Lindemann,J.P. and Watanabe,A.M., 1985, Phosphorylation of phospholamban in intact myocardium. Role of $Ca^{2+}$-calmodulin-dependent mechanisms,J.Biol.Chem., 260:4516

Lucas,T.J., Burgess,W.H., Prendergast,F.G., Lau,W. and Watterson,D.M., 1986, Calmodulin binding domains:Characterizarion of a phospho-rylating and calmodulin binding site from myosin light chain kinase, Biochemistry, 25:1458

Mamoi,T., 1986, Activaton of protein kinase C by ganglioside GM3 in the presence of calcium and 12-o-tetradecanoylphorbol-13-acetate, Biochem.Biophys.Res.Commun., 138:865

Mazzei,G.J., Qi,D.-F., Schatzman,R.C., Raynor,R.L., Turner,R.S. and Kuo,J.F., 1983, Comparative abilities of lanthanide ions $La^{3+}$ and $Tb^{3+}$ to substitute for $Ca^{2+}$ in regulating phospholipid-sensi-tive $Ca^{2+}$-dependent kinase and myosin light chain kinase, Life Sci., 33:119

Mazzei,G.J., Girrard,P. and Kuo,J.F., 1984, Environmental pollutant $Cd^{2+}$ biphasically and differentially regulates myosin light chain kinase and phospholipid/$Ca^{2+}$-dependent protein kinase, FEBS Lett., 173:124

Meyer,T., Fabro,D., Eppenberger,U. and Matter,A., 1986, The lipohilic muramyltripeptide MTP-PE, a biological response modifier, is an activator of protein kinase C, Biochem.Biophys.Res.Commun., 140: 1043

Murakami,K., Chan,S.Y. and Routtenberg.A., 1986, Protein kinase C activation
        by cis-fatty acid in the absence of Ca²⁺ and phospholipids,
        J.Biol.Chem., 261:15424
Murakami,K., Whitley,M.K. and Routtenberg,A., 1987, Regulation of protein
        kinase C activity by cooperative interaction of Zn²⁺ and Ca²⁺,
        J.Biol.Chem., 262:13902
Nairn,A.c., Hemmings,H.C.,Jr. and Greengard,P., 1985, Protein kinases in the
        brain, Ann.Rev.Biochem., 54:931
Negami,A.I., Sasaki,H. and Yamamura,H., 1986, Activation of phosphorylase
        kinase through autophosphorylation by membrane component
        phospholipids, Eur.J.Biochem., 157:597
Nikolaropoulos,S. and Sotiroudis,T.G., 1985, Phosphorylase kinase from
        chicken gizzard. Partial purification and characterization, Eur.J.
        Biochem., 151:467
Nishizuka,Y., 1984, The role of protein kinase C in cell-surface signal
        transduction and tumor promotion, Nature, 308:693
Nishizuka,Y., 1986, Studies and perspectives of protein kinase C, Science
        233:305
Nishizuka,Y., 1988, The molecular heterogeneity of protein kinase C and
        its implications for cellular regulation, Nature, 334:661
Ono,Y., Fujii,T. Ogita,K., Kikkawa,U., Igarashi,K. and Nishizuka,Y., 1988,
        The structure, expression and properties of additional members of
        the protein kinase C family, J.Biol.Chem., 263:6927
Parker,P.J. and Ullrich,A., 1987, Protein kinase C, J.Cell.Physiol.Suppl.,
        5:53
Pickett-Giese,C.A. & Walsh,D.A., 1986, Phosphorylase kinase,in:"The
        Enzymes", P. Boyer & E.G. Krebs, eds, Academic Press, New York
Sakai, K., Kobayashi,T., Komuvo,T., Nakamura,S., Mizuta,K., Sakanoue,Y.,
        Hashimoto,E. and Yamamura,H., 1987, Non-requirement of calcium on
        protamine phosphorylation by calcium-activated, phospholipid
        dependent protein kinase, Biochem.Inter., 14:63
Sato,H., Fukunaga,K., Araki,S., Ohtsuki,I. and Miyamoto,E., 1988,
        Identification of the multifunctional calmodulin-dependent protein
        kinase in the cytosol, sarcoplasmic reticulum and sarcolemma of
        rabbit skeletal muscle, Arch.Biochem.Biophys., 260:443
Schulman,H., 1984, Calcium-dependent protein kinases and neuronal function,
        Trends Pharmacol.Sci., 5:188
Shenolikar,S., Cohen,P.T.W., Cohen, P., Nairn,A.C. and Peryy,S.V., 1979,
        Role of calmodulin in the structure and regulation of phosphorylase
        kinase from rabbit skeletal muscle, Eur.J.Biochem., 100:329
Shenolikar,S., Lickteig,R., Hardie,D.G., Soderling,T.R., Hanley,R.M. and
        Kelly,P.T., 1986, Calmodulin-dependent multifunctional protein
        kinases. Evidence for isoenzyme forms in mammalian tissues, Eur.
        J.Biochem., 161:739
Singh,T.. & Wang,J.H., 1979, Stimulation of glycogen phosphorylase kinase
        from rabbit skeletal muscle by organic solvents, J.Biol.Chem.,
        254:8466
Sotiroudis,T.G., 1986, Lanthanide ions and Cd²⁺ are able to substitute
        for Ca²⁺ in regulating phosphorylase kinase, Biochem.Inter.,
        13:59
Stull,J.T., Nunnally,M.H. and Michnoff,C.H., 1986, Calmodulin-dependent
        protein kinases,in:"The Enzymes", P.Boyer and E.G. Krebs, eds,
        Academic Press, New York
Takai,Y., Kishimoto, A., Iwasa,Y., Kawahara,Y., Mori,T. and Nishizuka,Y.,
        1979, Calcium-dependent activation of a multifunctional protein
        kinase by membrane phospholipids, J.Biol.Chem., 254:3692
Tanaka,J. and Hidaka,H., 1980, Hydrophobic regions function in calmodulin
        enzyme(s) interactions, J.Biol.Chem., 255:11078
Thieleczek,R., Behle,G., Behle,G., Messer,A., Varsanyi, M., Heilmeyer,L.M.G.,
        Jr & Drenckhahn,D., 1987, Localization of phosphorylase kinase

68

subunits at the sarcoplasmic reticulum of rabbit skeletal muscle by monoclonal and polyclonal antibodies, Eur.J.Cell Biol., 44:333

Tuana,B.S. and MacLennan,D.H., 1984, Calmidazolium and compound 48/80 inhibit calmodulin-dependent $Ca^{2+}$ uptake but not $Ca^{2+}$-ATPase activity in skeletal muscle sarcoplasmic reticulum, J.Biol.Chem., 259:6979

Wightman,P.D. and Raetz, C.R.H., 1984, The activation of protein kinase C by biologically active lipid moieties of lipopolysaccharide, J.Biol. Chem., 259:10048

Wolf,M., LeVine III,H., May,S.,Jr, Cuatrecasas,P. and Sahyoun,N., 1985, A model for intracellular translocation of protein kinase C involving synergism between $Ca^{2+}$ and phosrbol esters, Nature, 317:546

Woodgett,J.R., Davison,M.T. and Cohen, P., 1983, The calmodulin-dependent glycogen synthase kinase from rabbit skeletal muscle. Purification subunit structure and substrate specificity, Eur.J.Biochem., 136: 481

Zevgolis,V.G., Sotiroudis,T.G. and Evangelopoulos,A.E. in preparation

# TRANSLOCATION OF PROTEINS INTO MITOCHONDRIA

Gottfried Schatz

Biocenter
University of Basel
CH-4056 Basel, Switzerland

## THE PROBLEM

Mitochondria, the eukaryotic organelles of oxidative phosphorylation, contain hundreds of different polypeptides. Only a small fraction of these (13 in humans) are encoded by mitochondrial DNA and made inside the mitochondria. All the others are encoded by nuclear genes, synthesized in the extramitochondrial cytoplasm, and then imported into one of the four major intramitochondrial compartments: outer membrane, intermembrane space, inner membrane, and matrix (Attardi and Schatz, 1988). How is this import achieved?

## Requirements for protein import into mitochondria

Import of a protein into mitochondria depends on the following: (a) a signal on the protein itself; (b) a loose conformation of the protein; (c) energy (ATP and, in most cases, a potential across the inner membrane); and (d) a proteinaceous import machinery that involves components in the cytosol as well as in the mitochondria. These four requirements are shared by protein transport across all other "translocation-competent" membranes. This has led to the view that all of these membranes transport proteins by basically the same mechanism (Schatz, 1986).

## Major import steps

Fig. 1 summarizes the major steps during import of a precursor protein into the mitochondrial matrix. Cleavage of the presequence can occur at any stage following the potential-requiring step.

STEPS        INHIBITORS

| STEPS | INHIBITORS |
|---|---|
| 1. INSERTION OF PRESEQUENCE | UNCOUPLERS |
| 2. UNFOLDING AND BINDING OF UNFOLDED CONFORMERS | FOLATE ANALOGS; LOW TEMPERATURE |
| 3. RELEASE OF UNFOLDED CONFORMERS FROM BINDING SITE (S) OR FURTHER UNFOLDING | ATP-TRAPS |
| 4. TRANSLOCATION OF UNFOLDED POLYPEPTIDE | |
| 5. REFOLDING INSIDE | |
| 6. REMOVAL OF PRESEQUENCE | MEMBRANE-PERMEANT CHELATORS; mas1 MUTATION |

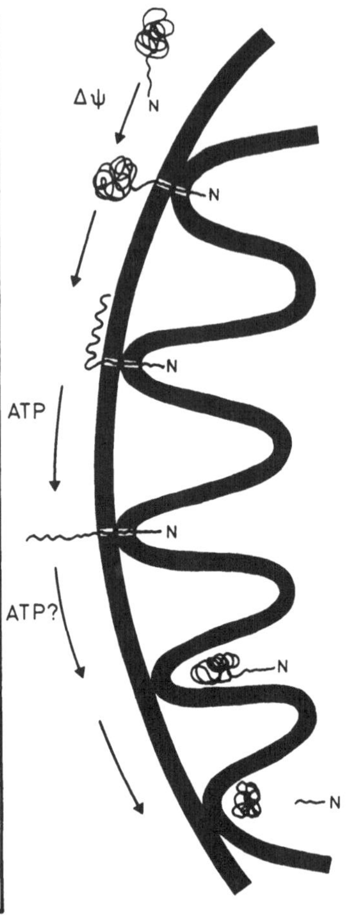

$\Delta\psi$    N

ATP

ATP?

Fig. 1 Major steps during import of a precursor protein into
the mitochondrial matrix. The precursor depicted in
this example is a fusion protein containing a mitochondrial presequence fused to mouse dihydrofolate reductase (Hurt et al., 1984).

## The mitochondrial import machinery: Receptor-like proteins

We have found that Fab fragments directed against a
major 45 kd protein of the yeast mitochondrial outer membrane blocks import of several precursors into mitochondria
(Ohba and Schatz, 1987). The sensitive step appears to be
the initial binding of precursors to the mitochondrial surface; once the precursors have been allowed to bind to mitochondria in the absence of antibody, subsequently-added antibody no longer inhibits import (A. Baker, in preparation;
Fig. 2).

We are now cloning the gene for this protein in order to clarify its role in the import process.

## The mitochondrial import machinery: Contact sites between the two membranes

Several observations suggest that import of proteins occurs at sites where the two mitochondrial membranes are in close apposition (Hackenbrock, 1968; Kellems et al., 1975; Ades and Butow, 1980; Suissa and Schatz, 1982; Schwaiger et al., 1987). Similar "contact sites" have been observed in chloroplasts (Pain et al., 1988).

| IgG | | | | | | | C | A | C | A |
|---|---|---|---|---|---|---|---|---|---|---|
| BINDING | 0°C | | + | + | + | + | + | + | + | + |
| CHASE | 0°C | | + | + | | | + | + | | |
| | 20°C | 20% | | | + | + | | | + | + |
| | ΔΨ | St | + | + | | | + | + | + | + |

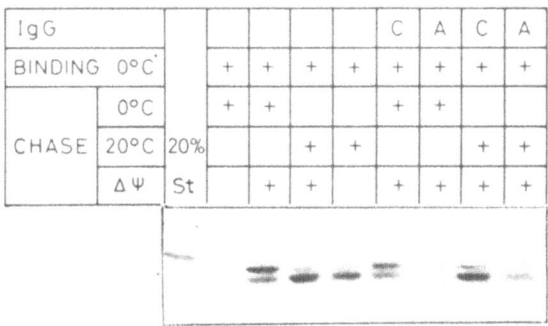

Fig. 2 Effect of antiserum against 45 kd outer membrane protein on import of cytochrome $b_2$ into yeast mitochondria. C and A, control- and antiserum, respectively. Pre-cytochrome $b_2$ was prebound to mitochondria at 0°C in the absence or presence of antiserum and then chased at 0°C or 20°C in the presence or absence of a membrane potential (A. Baker, in preparation).

In order to mark these contact sites, we constructed an artificial precursor protein that remains stuck in the mitochondrial protein import machinery. This precursor consists of three major parts (Fig. 3): (a) The first 22 residue of the yeast cytochrome oxidase subunit IV precursor (this part represents the mitochondrial targeting signal); (b) mouse dihydrofolate reductase (this 187-residue cytosolic protein is readily transported into mitochondria if fused to the mitochondrial targeting signal; Hurt et al., 1984, 1985); (c) bovine trypsin inhibitor attached via a bifunctional organic cross-linker to the C-terminus of the subunit IV-dihydrofolate reductase fusion protein (because of its three internal disulfide crosslinks, the trypsin inhibitor should not be able to unfold and should thus be unable to pass through the mitochondrial membranes; Eilers and Schatz, 1986). In order to permit the specific crosslink at the C-terminus of dihydrofolate reductase, cysteine residues in the targeting signal and within dihydrofolate reductase were removed and a cysteine was added to the C-terminus by site-directed mutagenesis (Vestweber and Schatz, 1988).

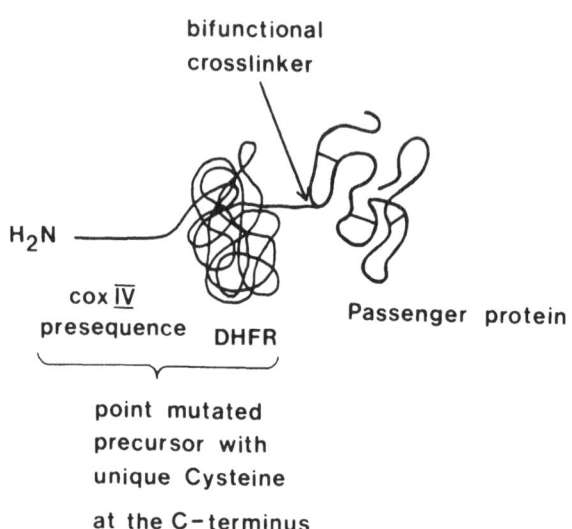

Fig. 3 A partly importable artificial precursor protein.
DHFR = dihydrofolate reductase
coxIV = cytochrome oxidase subunit IV precursor

When μg-amounts of this purified precursor protein were presented to isolated, energized yeast mitochondria, it was partly imported: its dihydrofolate reductase moiety was in the matrix, the attached targeting signal was cleaved off by the matrix-localized protease, and the trypsin-inhibitor moiety was exposed on the mitochondrial surface. Accumulation of this transmembraneous translocation intermediate blocked import of all authentic mitochondrial precursors tested (Vestweber and Schatz, 1988). Thus, the intermediate appeared be stuck within the import machinery, thereby "jamming" it. We calculated that each mitochondrial particle 1 mμ in diameter contained between 100 and 1000 import sites.

Subfractionation of the mitochondria containing the partly imported, radiolabeled precursor showed that virtually all of the precursor was present in a membrane fraction whose density was intermediate between that of inner and outer membrane. This "intermediate density fraction" also appeared to contain the binding sites for cytoplasmic ribosomes. Electron micrographs showed that this fraction was enriched in vesicles containing a closed, cytochrome oxidase-reactive space and an attached membrane that lacked cytochrome oxidase (L. Pon and B. Marshallsay, submitted). We conclude that this fraction contains the mitochondrial contact sites and that these sites are the entry point for proteins into mitochondria.

## The mitochondrial import machinery: Processing protease

In order to define components of the mitochondrial import system by genetic means, we screened several thousand temperature-sensitive yeast mutants for the accumulation of mitochondrial precursor proteins at the non-permissive temperature, 37°C (Yaffe and Schatz, 1984). Two mutants (mas1 and mas2, for mitochondrial assembly) were isolated. Each carried a mutation in a single nuclear gene; both genes were cloned and sequenced.

The two genes encode the two subunits of the matrix-localized processing protease that had been originally discovered by Böhni et al. (1980). The enzyme was recently purified from yeast and shown to contain a 51 kd subunit and a 48 kd subunit (Yang et al., 1988).

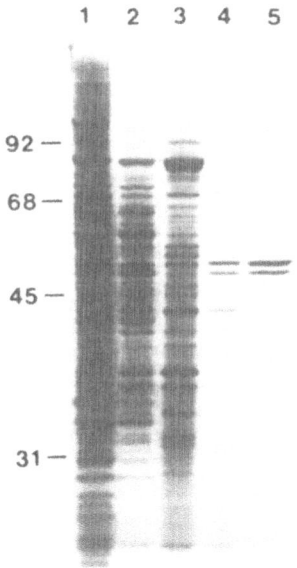

Fig. 4 Purification of the matrix-localized processing protease from yeast. A silver-stained gel is shown. Lanes: 1, total mitochondrial protein (6 µg); 2, total matrix (6 µg); 3-5, first, second, and final stage of purification (6, 0.2 and 0.2 µg protein, respectively)

The MAS1 gene encodes the smaller subunit (Witte et al., 1988) and the MAS2 gene encodes the larger subunit (Jensen et al., 1988). Both subunits are apparently made as larger precursors with N-terminal extensions which are cleaved upon import into mitochondria. Interestingly, the two subunits are homologous to each other (Fig. 5) and to the COR1 subunit of the cytochrome $bc_1$ complex from yeast mitochondria. Thus, we are beginning to define the components that mediate import of proteins into mitochondria.

REFERENCES

Ades, I.Z., and Butow, R.A., 1980, The products of mitochondria-bound cytoplasmic polysomes in yeast,
J. Biol. Chem., 255:9918-9924.
Attardi, G., and Schatz, G., 1988, Biogenesis of Mitochondria,
Annu. Rev. Cell Biol., in press.
Boehni, P., Gasser, S., Leaver, C., and Schatz, G., 1980, A matrix-localized mitochondrial protease processing cytoplasmically-made precursors to mitochondrial proteins, in: The organization and expression of the mitochondrial Genome,
Kroon, A.M. and Saccone, C., eds., Elsevier/North Holland, Amsterdam, 423-433.
Eilers, M., and Schatz, G., 1986, Binding of specific ligand inhibits import of a purified precursor protein into mitochondria,
Nature, 322:228-232.
Hackenbrock, C.R., 1968, Chemical and physical fixation of isolated mitochondria in low-energy and high-energy states,
Proc. Natl. Acad. Sci. USA, 61:598-605.
Hurt, E.C., Pesold-Hurt, B., and Schatz, G., 1984, The cleavable prepiece of an imported mitochondrial protein is sufficient to direct cytosolic dihydrofolate reductase into the mitochondrial matrix,
FEBS Letters, 178:306-310.
Hurt, E.C., Pesold-Hurt, B., Suda, K., Oppliger, W., and Schatz, G., 1985, The first twelve amino acids (less than half of the pre-sequence) of an imported mitochondrial protein can direct mouse cytosolic dihydrofolate reductase into the yeast mitochondrial matrix,
EMBO J., 4:2061-2068.
Jensen, R.O., Schatz, G., and Yaffe, M.P., 1988, MAS2, an essential gene required for yeast mitochondrial protein import, encodes a matrix-localized protein homologous to the MAS1 product,
EMBO J., submitted.
Kellems, R.E., Allison, V.F., and Butow, R.A., 1975, Cytoplasmic type 80S ribosomes associated with yeast mitochondria. IV. Attachment of ribosomes to the outer membrane of isolated mitochondria,
J. Cell Biol., 65:1-14.
Ohba, M., and Schatz, G., 1987, Protein import into yeast mitochondria is inhibited by antibodies raised against 45-kd proteins of the outer membrane,
EMBO J., 6:2109-2115.

Pain, D., Kanwar, Y.S., and Blobel, G., 1988, Identification of a receptor for protein import into chloroplasts and its localization to envelope contact zones,
Nature, 331:232-236.
Schatz, G., 1986, Protein translocation: A common mechanism for different membrane systems?,
Nature, 321:108-109.
Schwaiger, M., Herzog, V., and Neupert, W., 1987, Characterization of translocation contact sites involved in the import of mitochondrial proteins,
J. Cell Biol., 105:235-246.
Suissa, M., and Schatz, G., 1982, Import of proteins into mitochondria: translatable mRNAs for imported mitochondrial proteins are present in free as well as mitochondria-bound cytoplasmic polysomes,
J. Biol. Chem., 257:13048-13055.
Vestweber, D., and Schatz, G., 1988, A chimeric mitochondrial precursor protein with internal disulfide bridges blocks import of authentic precursors into mitochondria and allows quantitation of import sites,
J. Cell Biol. submitted.
Witte, C., Jensen, R.E., Yaffe, M.P., and Schatz, G., 1988, MAS 1, a gene essential for yeast mitochondrial assembly, encodes a subunit of the mitochondrial processing protease,
EMBO J., 7:1439-1447.
Yaffe, M.P., and Schatz, G., 1984, Two nuclear mutations that block mitochondrial protein import in yeast,
Proc. Natl. Acad. Sci. USA, 81:4819-4823.

# tRNA GENES - TINKERING IN ORGANIZATION AND EXPRESSION?

Horst Feldmann

Institut für Physiologische Chemie
der Universität München
Schillerstraße 44, D-8000 München 2, F.R.G.

## INTRODUCTION

tRNA genes and components of their basic transcription machinery provide well-documented examples for the conservation of cis-acting elements and trans-acting factors from lower to higher eukaryotes. Like other genes that are transcribed by RNA polymerase III, tRNA genes contain intragenic control regions that direct their transcription. For tRNA genes, these highly conserved internal control regions (the A and B boxes) coincide with sequences of the D-loop and the T-loop of the tRNAs (review, Sharp et al., 1985).

In addition to polymerase III, tRNA gene transcription is strictly dependent on (at least) two transcription factors, TFIIIC and TFIIIB. Two seperate domains exist in TFIIIC that govern the binding to the A and B blocks, whereby direct binding studies have shown a strong role for the B box and a subsidiary role for box A sequences in stabilizing the complex which remains bound through multiple rounds of gene transcription (e.g. Ruet et al., 1984). Neither the spacing nor the relative helical orientation of the A and B blocks are critical for efficient transcription (Baker et al., 1987), thus reflecting the natural occurrence of tRNA genes some of which carry introns of variable length. In this respect, the signal elements in tRNA genes have features common to upstream activating sequences.

While the transcription of eukaryotic tRNA genes is directed by the internal control regions, studies in various systems have shown that sequences flanking the tRNA genes (preferably 5' sequences) serve to modulate transcription (review, Sharp et al., 1985). However, with a few exceptions, sequence similarities which could be attributed to defining transcriptional control sequences are not observed in these regions. This is also obvious from the fact that, in the majority of the cases, even the flanking regions of the individual members within the same tRNA gene family are not conserved.

In the yeast, Saccharomyces cerevisiae, analysis of a large variety of tRNA gene loci has revealed a unique feature in their genomic organization: they are found associated with transposable elements such as Ty and delta (Eigel and Feldmann, 1982), sigma (Sandmeyer et al., 1982; 1988) or tau (Genbauffe et al., 1984). In several cases, rather complex patterns are formed between sequences from two or more of these elements. So far, sigma and tau have been encountered exclusively in regions flanking tRNA genes, whereby sigma is always located 16-18 bp upstream of the coding sequences. In a recent study of extended genomic regions we found (Feldmann et al., unpublished) that (i) nearly all of the members of the tRNA(Glu3) gene

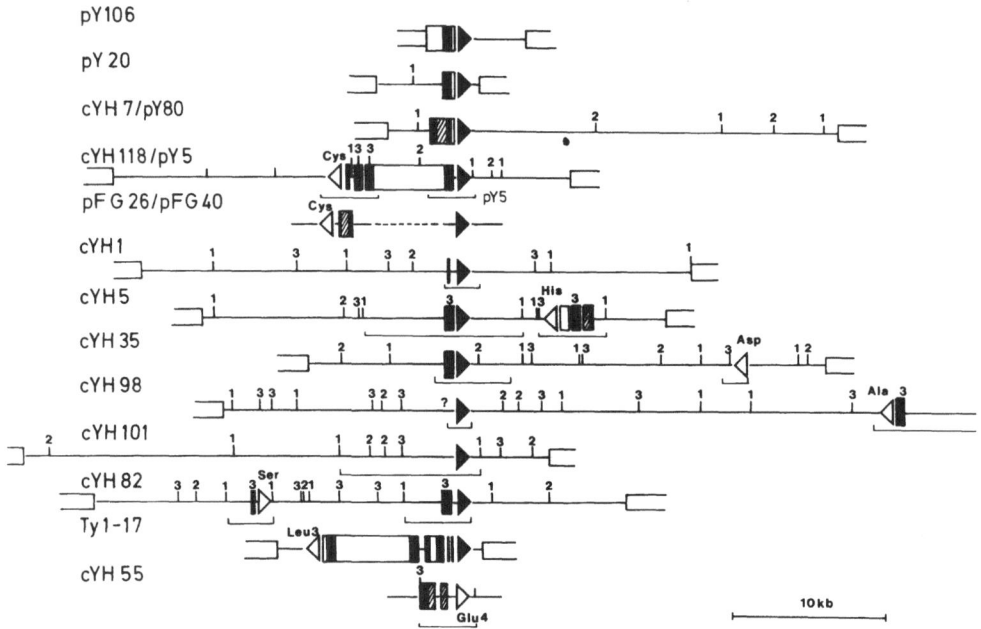

Fig.1. Schematic representation of genetic and restriction maps of clones carrying members of the yeast tRNA(Glu3) gene family, other tRNA genes, and transposable elements. Only the restriction sites for BamHI (=1), SalI (=2) and XhoI (=3) are listed. Filled triangles represent tRNA(Glu3) genes, indicating direction of transcription. Sequences from transposable elements: delta, filled bars; sigma, small open bars; tau, hatched bars; Ty, large open bars.

family are associated with such elements and (ii) that at these loci the Ty and delta sequences are exclusively located in regions flanking tRNA genes (Figure 1). Otherwise, no obvious sequence similarities could be detected in the gene flanking regions. Micro-injection into Xenopus oocyte nuclei of recombinants that carry different copies of the tRNA(Glu3) gene family showed that template activity varies probably depending on the type of element: expression was more efficient in those cases in which Ty, delta or tau sequences are affiliated to the genes.

SYN2, A TESTER tRNA GENE FOR IN VIVO STUDIES IN YEAST

Direct in vivo studies on tRNA gene transcription in the homologous system are hampered by the fact that normally multiple copies of a given tRNA gene are present in the genome. To be able to quantitate the in vivo expression of a particular tRNA gene flanked by different upstream sequences and simultaneously to examine the underlying chromatin structure, we started experiments by using the following approach (Krieg and Feldmann, unpublished). We synthesized a gene (SYN2) that in its body is identical to the native tRNA(Glu3) genes but contains a 30 bp intron-like sequence; however, this was designed in such a way that it cannot be spliced (Baldi et al., 1986; Greer et al., 1987). Therefore, SYN2 transcripts are distinguished from endogenous mature tRNAs by their different migration in gel electrophoresis. Furthermore, they can be monitored by hybridization with either tDNA(Glu3) or a DNA probe corresponding to the insert. For cloning, SYN2 was endowed with a 5' HindIII site and a 3' SalI site (Figure 2).

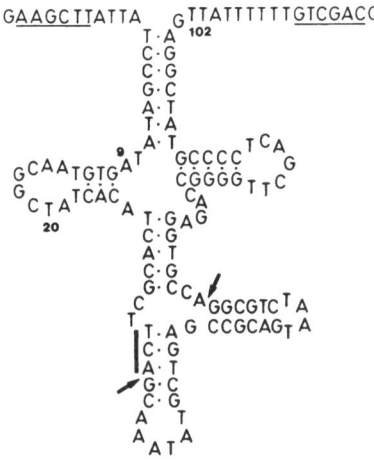

Fig.2. Clover-leaf structure of SYN2. The restriction sites for cloning are underlined; the position of the anticodon is indicated. The arrows point to the ends of the 'inserted' sequence.

Injection of SYN2 recombinants in Xenopus oocyte nuclei and subsequent analysis of the RNA showed that SYN2 is confidentially expressed but not spliced. Analyses of yeast SYN2 transformants confirmed that the SYN2 product is an RNA 102 nucleotides in length, not spliced but correctly processed at its 5' and 3' ends. Moreover, primer extention experiments revealed a "strong stop" at position 20 indicating that SYN2 RNA is a substrate for U → D modification. In other words, SYN2 behaves like a native tRNA gene.

Construction and transcriptional analysis of SYN2 variants

A set of SYN2 variants was constructed by employing the integrating yeast shuttle vector YIp5 and placing different yeast DNA segments upstream of SYN2. Yeast transformants were generated by integrating a single copy of the variant sequence at the URA3 locus.

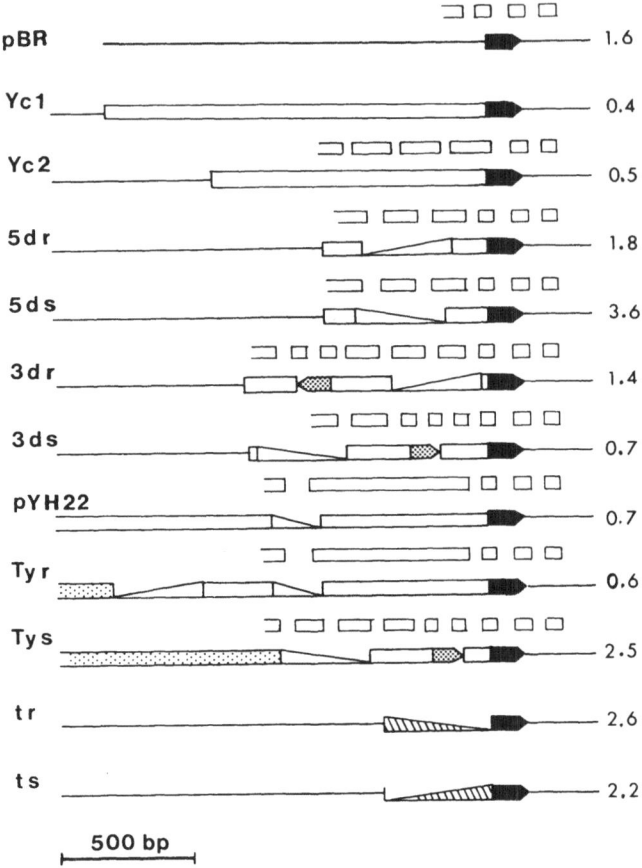

Fig.3. Schematic representation of constructs containing SYN2 and various
yeast DNA segments. Recombinants were obtained by simultaneous
integration of SYN2 and appropriate yeast restriction fragments
between the HindIII/SalI sites in YIp5. Plasmids linearized with
StuI were used to transform yeast cells. Black arrows, SYN2;
stippled arrows, tRNA(Trp) gene; triangles, delta sequences; hatched
triangles, tau sequences; open bars, yeast sequences; stippled bars,
Ty sequences; black lines, pBR sequences. Ty and delta segments (Ty,
5d ('promoter'delta), and 3d ('terminator' delta) are from pY109;
pYH22 is the Ty⁻-allele of pY109 (Hauber et al., 1985). The tau
sequence in tr and ts is a 400 bp TaqI fragment from pY80 (Nelböck
et al.,1985). "Control sequence" in Yc1 is a 1.5 kb segment upstream
from the yeast TYR1 gene (Stucka and Feldmann,unpublished); that in
Yc2 is a 1 kb fragment from the downstream flanking region of the
tRNA(Glu3) gene of pY80 (Nelböck et al.,1985); r and s refer to
inverted orientations of the inserted segments, respectively. The
boxes above the SYN2 variants indicate regions inaccessible to DNase
I. The numbers refer to relative transcriptional activities.

A comparison of the transcriptional activities of the SYN2 variants by the method outlined in Figure 4 showed that the levels of transcriptional activity vary by a factor up to about 9. The lowest activity is seen for SYN2 variants having upstream sequences not normally occurring in tRNA gene 5'-flanking regions (Yc1, Yc2, and pYH22) or carrying a Ty at a relatively large distance (Tyr). The highest template activities are found in those variants (d5r, d5s) that carry delta sequences in a distance that is similar to naturally occurring elements (90-230 bp). The tau variants (tr and ts) exhibit comparable high levels of transcription. If a delta is very close to SYN2, as in d3r, transcription is in an intermediate range. It remains at a low level, if a second tRNA gene (tRNA(Trp)) is placed between the delta and SYN2 (3ds). By contrast, transcriptional activity is fairly high in Tys, although the second tRNA gene is closer to SYN2 as in 3ds. Surprisingly, pBR322 sequences do not block transcription of SYN2.

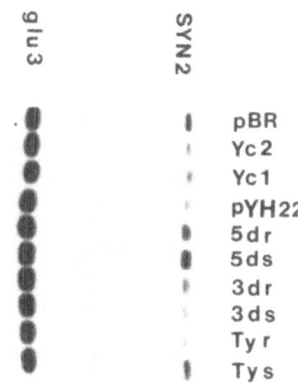

Fig.4. Transcriptional activities of SYN2 variants. Low molecular weight RNA was isolated from each of the transformants at early log-phase, separated on a 10% polyacrylamide gel and transferred to a nylon filter by electro-blotting. Subsequent hybridization with labeled tDNA(Glu3) allowed to measure the amount of SYN2 RNA synthesized in vivo with reference to bulk tRNA(Glu3) as an internal standard.


Analysis of the chromatin organization of the SYN2 variants

Several of the transformants were used to examine the DNase I sensitivity of the tRNA gene regions; the analyses of four examples are documented in Figure 5. The results obtained so far are schematically summarized in Figure 3. Size measurements map major cleavage sites to the tRNA gene regions in all of the variants. Within the structural parts of the tRNA genes, DNA segments are selectively exposed in chromatin, spanning a region that approximately extends from position 20 through 50 (in the standard nomenclature for tRNA). Additionally, in well or fairly well expressed SYN2 variants, these genes have chromatin-specific DNaseI sites mapping 30-40 bp upstream from the 5'-end, and 20-30 bp downstream from the 3'-end of the structural gene. The corresponding bands appear as a distinct triplet, with some variation in intensity. The spacing between these nuclease cleavage sites is very different from that expected for an evenly spaced arrangement of nucleosomes. Remarkably, the upstream cleavage site is nearly vanished in those SYN2 variants that exhibit low template activity (e.g. Yc2).

Products representing minor DNaseI cleavage sites are seen further upstream of the tRNA gene regions, at a spacing that could reflect a regular arrangement of nucleosomes (e.g., 5ds and 3dr, figures 3 and 5). However, in the variants with low template activity, the corresponding bands are rather faint (e.g. Yc2), reflecting a more 'condensed' chromatin structure than that seen in the variants with higher template activity. With all precautions, it appears that external sequences are of relevance to the chromatin organization of the tRNA genes and that, in turn, there is a correlation between a particular chromatin structure and transcriptional activity.

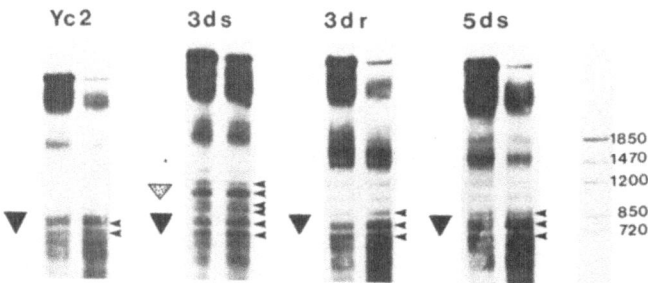

Fig.5. Mapping of DNase I sensitivity in SYN2 variants from Fig.3. Nuclei were prepared from the transformants (Almer and Hörz, 1986) and subjected to digestion with varying concentrations of DNase I. DNA was isolated, and after secondary cutting with StyI which cleaves downstream close to SYN2, run out on agarose gels. Sensitive sites were mapped by the indirect end-labelling technique (Nedopasov and Georgiev, 1980; Wu, 1980) using the StyI/NruI fragment of pBR322 as a probe. Control experiments were carried out with deproteinized DNA (not shown).

DISCUSSION

The studies presented here provide some insights into the genomic organization of tRNA genes in yeast and the chromatin structure of tRNA gene regions in vivo as examplified by the SYN2 variants; some findings may be of relevance to the transcriptional activation of tRNA genes. Our data are in agreement with earlier results obtained by in vitro or in vivo studies of tRNA gene transcription (review, Sharp et al.,1985; Raymond and Johnson,1987, and references therein) or formation of transcriptional complexes (e.g. Ruet et al.,1984; Huibregtse et al., 1987), and by studies on the chromatin structure of tRNA genes in Drosophila (DeLotto and Schedl, 1984). In the latter experiments (DeLotto and Schedl,1984), it has been shown by nuclease digestions that tRNA gene promoters are exposed in chromatin and may have an unusual conformation. Actually, the DNaseI cleavage patterns we obtained are very similar to the ones reported by these authors. The findings are in support of the view that the tRNA promoter elements would potentially be accessible in chromatin for interaction with the transcription factors and/or RNA polymerase III. In fact, the observed patterns could directly reflect the complex in which the transcription factors are tightly bound to the promoter elements. This interpretation is supported by DNaseI footprinting experiments of Huibregtse et al. (1987) showing that transcription complexes with TFIIIC formed in vitro accurately reflect aspects of the nucleoprotein structure of the genes in chromatin: in both instances, ca. 40 bp at the A block, and ca. 30 bp at the B block are protected from cleavage, whereas enhanced cleavage was observed between the protected regions.

The differential effects seen in transcriptional activity and chromatin organization of the SYN2 variants might suggest that the (upstream) flanking regions are directly involved in determining a more or less favorable conformation of the tRNA structural gene region for the interaction with the transcription apparatus. These effects need not be 'sequence-specific'. It is conceivable, however, that particular sequences, such as Ty, delta or tau, in certain constellations have a positive influence: this is reflected by the 'up-modulation' in transcription activity of some of our constructs.

Certainly, more detailed experiments are needed to precisely define the influence of flanking sequences towards the chromatin organization of tRNA gene regions and the correlation between nucleoprotein conformation and transcriptional activation of the tRNA genes. The picture that emerges, so far, would imply that particular promoter elements directing tRNA gene transcription are highly conserved during evolution but need not be precisely spaced within a tRNA gene. In addition, sequences in the environment of the tRNA genes might have evolved in a way as to permit the formation of transcriptionally active complexes via a favorable chromatin structure. In this regard, the accumulation of transposable elements in the vicinity of the yeast tRNA genes seems not only to be non-hazardous to the cell but even of some advantage.

REFERENCES

Almer, A. and Hörz, W., 1986, Nuclease hypersensitive regions with adjacent positioned nucleosomes mark the gene boundaries of the PHO5/PHO3 locus in yeast, EMBO J. 5:2681-2687.
Baker, R.E., Camier, S., Sentenac, A. and Hall, B.D., 1987, Gene size differentially affects the binding of yeast transcription factor tau to two intragenic regions, Proc. Natl. Acad. Sci. USA 84:8768-8772.
Baldi, M.I., Matoccia, E., Ciafrè, S., Attardi, D.G. and Tocchini-Valentini, G.P., 1986, Binding and cleavage of pre-tRNA by the Xenopus splicing endonuclease: two separable steps of the intron excision reaction, Cell 47:965-971.
DeLotto, R. and Schedl, P., 1984, Internal promoter elements of transfer RNA genes are preferentially exposed in chromatin, J. Mol. Biol. 179:607-628.
Eigel, A. and Feldmann, H., 1982, Ty1 and delta elements occur adjacent to several tRNA genes in yeast, EMBO J. 1:1245-1255.
Genbauffe, F.S., Chisholm, G.E. and Cooper, T.G., 1984, Tau, sigma, and delta, a family of repeated elements in yeast, J. Biol. Chem. 259:10518-10525.
Greer, C.L., Söll, D. and Willis, I., 1987, Substrate recognition and identification of splice sites by the tRNA-splicing endonuclease and ligase from Saccharomyces cerevisiae, Mol. Cell. Biol. 7:76-84.
Hauber, J. Nelböck, P. and Feldmann, H., 1985, Nucleotide sequence and characteristics of a Ty element from yeast, Nucleic Acids Res. 13:2745-2758.
Huibregtse, J.M., Evans, C.F. and Engelke, D.R., 1987, Comparison of tRNA gene transcription complexes formed in vitro and in nuclei, Mol. Cell. Biol. 7:3212-3220.

Nedopasov, S. and Georgiev, G., 1980, Non-random cleavage of SV40 DNA in the compact minichromosome and free in solution by micrococcal nuclease, Biophys.Biochem. Res. Comm. 92:532-539.

Nelböck, P., Stucka, R. and Feldmann, H., 1985, Different patterns of transposable elements in the vicinity of tRNA genes in yeast: a possible clue to transcriptional modulation, Biol. Chem. Hoppe-Seyler 366:1041-1051.

Raymond, G.J. and Johnson, J.D., 1987, The 5'flanking sequence of yeast tRNA(Leu3) genes enhances the rate of transcription from stable pre-initiation complexes, Nucleic Acids Res. 15:9881-9894.

Ruet, A., Camier, S., Smagowicz, W. Sentenac, A. and Fromageot, P., 1984, Isolation of a class C transcription factor which forms a stable complex with tRNA genes, EMBO J. 3:343-350.

Sandmeyer, S.B. and Olson, M.V., 1982, Insertion of a repetititve element at the same position in the 5'-flanking regions of two dissimilar yeast tRNA genes, Proc. Natl. Acad. Sci. USA 79:7674-7678.

Sandmeyer, S.B., Bilanchone, V.W., Clark, D.J. Morcos,P. Carle, G.F. and Brodeur, G.M., 1988, Sigma elements are position-specific for many different yeast tRNA genes, Nucleic Acids Res. 16:1499-1515.

Sharp, S.J., Schaack, J., Cooley, L. Burke, D.J. and Söll, D., 1985, Structure and transcription of eukaryotic tRNA genes, CRC Critical Rev. Biochem. 19:107-144.

Wu, C., 1980, The 5'ends of Drosophila heat shock genes in chromatin are hypersensitive to DNaseI, Nature 286:854-860.

TRANSCRIPTIONAL CONTROL BY RETROVIRAL LTR REGIONS

Niels Ole Kjeldgaard, Allan J. Bækgaard, Hong Yan Dai,
Michael Etzerodt, Poul Jørgensen, Steen Lovmand,
Henrik Steen Olsen and Finn Skou Pedersen

Institute of Molecular Biology and Plant Physiology
Aarhus University
DK-8000 Århus C, Denmark

INTRODUCTION

Murine leukemia viruses represent a polymorphic group of
retroviruses. Individual isolates differ in species and tissue
tropism and in their specificity and potency of pathogenic properties. Some viruses induce lymphomas with high incidence and a
latency period of a few months when injected into newborn mice
of various inbred strains, whereas others show a weaker and less
specific pathogenicity. The pathogenic effects of these weaker
viruses include various lymphomas and leukemias as well as neo-
plastic and non-neoplastic abnormalities in bone tissues. Recombination mapping between thymomagenic and weakly leukemogenic
viruses has localized a major determinant of the oncogenic potency to the long terminal repeat (LTR) region of the viruses,
(DesGroseillers et al., 1983; DesGroseiller and Jolicoeur, 1984;
Lenz et al., 1984). Recombination mapping between viruses that
yield thymic and erythroid leukemias has also localized the determinants for tissue specificity to the LTR region, (Chatis et
al., 1983,1984; Ishimoto et al., 1985; Vogt et al., 1985). The
LTR regions contain sequences necessary for the initiation  and
termination of retroviral transcription including promotor elements, termination signals and enhancers. Mapping studies of the
more potent MuLVs point to the transcriptional control region
containing an enhancer structure as a disease determinant,
(Chatis et al.,1984; Ishimoto et al.,1985; Bösze et al., 1986,
(MoMuLV, Friend MuLv); Des Groseillers and Jolicoeur, 1984,
(GrossA); Lenz et al., 1984; Celander and Haseltine, 1984,
(SL3-3)).
In an attempt to dissect the role of the U3 region in pathogenicity and tissue specificity we have concentrated our studies
on a family of retroviruses related to the Akv MuLV of AKR
mice.

VIRUSES

a) The Akv virus is the product of the endogenous, ecotro-
pic provirus, Emv11, of the inbred AKR strain of mice. It is
produced throughout the life of the animals, is weakly lymphoma-

genic and induce benign bone tumors in of some strains of mice, (Pedersen et al., 1987).

b) The SL3-2 and SL3-3 MuLVs are derived from cultures of a spontaneous T-cell lymphoma of AKR mice. They induce T-cell lymphomas with short latency periods in newborn mice of various inbred strains, (Pedersen et al., 1981).

c) The OA-I MuLV is derived from spontaneous bone tumors of strain 101/Nhg mice. The virus induces osteomas, osteopetrosis and malignant lymphomas with rather long latency periods in NMRI mice, (Schmidt et al., 1984; Leib-Mösch et al., 1986).


THE U3 REGION

The complete nucleotide sequence of a molecular clone of the Akv MuLV, (AKR λ623) has been determined (Etzerodt et al., 1984), as well as the sequences of the U3 regions of SL3-2 Dai et al., submitted), SL3-3 (Lenz et al., 1984) and OA-I (Leib-Mösch et al., 1984). The U3 regions of these viruses, contain the conventional CAAT and TATA promotor sequences at the 3' end and have tandem nucleotide repeat sequences located 85 basepairs upstream from the CCAAT site. Analysis of the nucleotide sequences of the U3 regions, have revealed a modular arrangement of the repeat structures involving five different sequence modules, (fig. 1). For each virus a characteristic array of modules is observed. It seems likely that it is this pattern which is a primary determinant for the phenotype of the virus, probably in combination with a few point mutations within certain of the modules.

Outside the repeat structure the nucleotide sequences of the U3 regions of the viruses above, are identical or almost identical. This is also true for several other MuLVs such as the FBJ virus. In the case of Moloney MuLV, the modular structure shows a few nucleotide differences from the Akv modules. The U3 nucleotide sequence upstream of the repeat structure has an extensive homology to the Akv sequence, whereas the downstream U3 sequence is very different.


TRANSCRIPTIONAL ENHANCERS

Enhancers, which are often composed of repeated sequence motifs, have been functionally defined. They often increase the rate of transcription from the promotor up to hundred folds, to a large extent independent of their localization and orientation relative to the promotor. The transposition of the Akv repeat structure to the *StuI* site distal to the *cat* gene in the plasmid pAkv8.3-cat (fig. 2) has confirmed that the modular repeat harbors a classical enhancer function. Furthermore enhancers contain multiple sequence recognition sites for cooperative protein factors, which by their interaction generate an active transcriptional initiation complex at the promotor, (Schaffner et al., 1988). An analysis of the protein binding sites in the repeat structure of Moloney MuLV has been done by Speck and Baltimore (1987).They recognized the binding sites for six distinct nuclear factors corresponding to the SV40 core motif, to the NF-⌐ binding site, to the glucocorticoid response element and to binding sites for three different leukemia virus factors (LVa, LVb,LVc). Recently Thornell et al., (1988) and Hallberg and Grundström, (1988) have described yet another nuclear protein, SEF1, which binds to a TGTGG core motif in retroviral enhancers.

A survey of a number of different core motifs along the arrays
of modules in all the retroviral repeat structures shows a clustering of the motifs on the plus strand and rather few on the
minus strand.

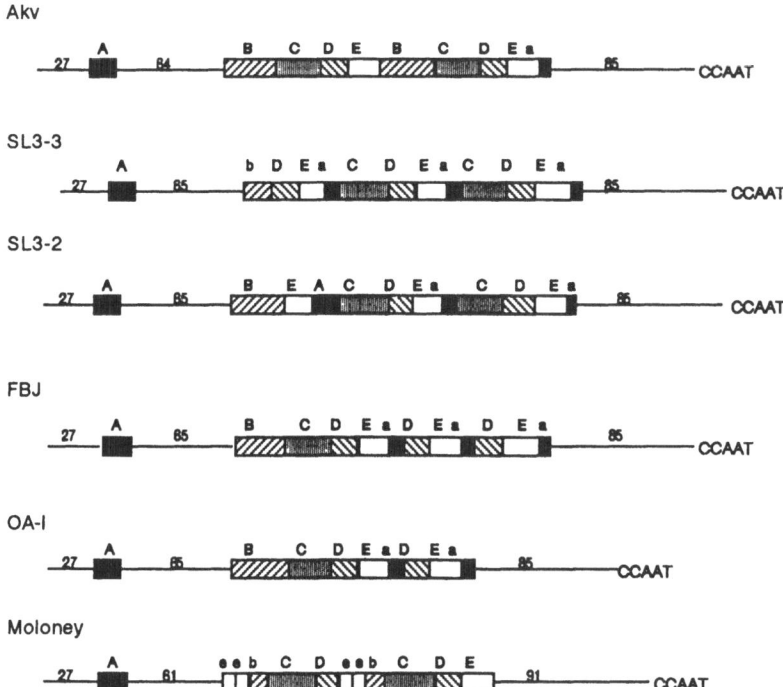

Fig.1. Modular arrays of the repeat structures of selected MuLV
       U3 regions. The sequences of the modules  and the point
       mutations relative to the Akv sequence are given in
       Pedersen et al. (1987). Modules indicated by small
       letters contain only part of the normal module
       sequence.The U3 sequences outside the modules are
       identical or nearly identical to Akv except for Moloney
       MuLV. Here the sequences upstream of the repeats are
       fairly similar to, whereas the sequences downstream of
       the repeats are very  different from the Akv sequence.

EXPRESSION VECTORS

     To dissect the role of the  U3 region in pathogenicity and
tissue specificity, we have generated a series of retroviral expression vectors, which allow an easy exchange and mutagenesis
of U3 sequences. Our vector design is aimed at studying viral
control functions in an immediate DNA sequence environment close
to that found in the provirus. We have normally retained a complete LTR sequence as well as a fairly long stretch of 5' untranslated sequences.
     The genomic map of the prototype expression vector pAkv6-
cat and the U3 replacement vector pL6-cat are shown in figure 2.

The transcription unit of pAkv6-cat contains the Akv LTR driving
the expression of a chloramphenicol acetyl transferase (CAT) reporter gene, a small intron and poly(A) signals from SV40. The
6382 bp. pAkv6-cat plasmid has been constructed through a large
number of separate cloning steps. When numbered from the unique
*Eco*RI site of the plasmid, the origin of the various nucleotide
segments are: 1-33, {pBR327, 1-33}; 34-1281, {pAKR59, 7512-456
(Lenz et al., 1982)}; 1282-1457, {pBR328, 149-321}; 1458-2943,
{pSV2-cat, 149-1634 (Gorman, 1985)}; 2944-4318, {pBR327, 3274-
1373}; 4319-5986, {pGEM1, 99-1766} and 5987-6382, {pBR327, 2879-
3274}.

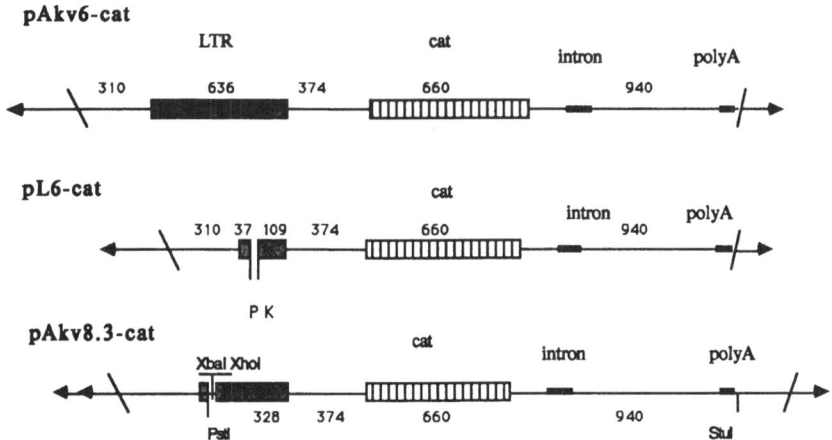

Fig.2. Genomic structure of expression vectors. The numbers indicate length of fragments in basepairs. The restriction
enzyme sites for *Pst*I (P) and *Kpn*I (K) in pL6-cat are
bordering a 23 bp. polylinker. The 310 bp.fragment to the
left of the LTR contains sequences from the *env* region of
Akv.The vectors contain an intron and poly(A) signals
from SV40.

The plasmid pL6-cat is derived from pAkv6-cat by deletion
of most of the LTR sequence from the *Pst*I site early in the U3
region to the *Kpn*I site in the center of the R region.
In the plasmid pAkv8.3-cat the *Pst*I - *Ava*II fragment of the
LTR in pAkv6-cat is replaced by a polylinker and a *Stu*I site is
introduced at position 2943 of the parent plasmid. This vector
is used for the insertion of DNA fragments of chemical or biological origin and to test fragments inserted in a position 3'
to the poly(A) site.
Transient expression assays with DNA from these plasmids
were performed for NIH 3T3 cells using the calcium phosphate
method (Graham and van der Eb, 1973) or for lymphoid cells using
the DEAE dextran method (Grosschedl and Baltimore, 1985). The
expression of chloramphenicol acetyl transferase was determined
36 hours after transfections (Gorman, 1985). With a number of
different plasmids primer extension analysis has confirmed that
the correct initiation site at the U3-R border is used in the
transfected cells.

DELETION ANALYSIS OF THE U3 REGION

## Akv MuLV

The U3 region of Akv contains a perfect tandem repeat structure with a unique restriction enzyme sites for *Apa*I in the 99
bp. repeat sequence. Digestion of the U3 DNA with this enzyme
followed by religation will create plasmids with only one 99 bp.
repeat(pAkv6-catΔ1-99). This plasmid will still confer a high
transient expression of cat in transfection experiments (fig.5
and table 2). Deletions introduced in the 99 bp. repeat structure by *Bal*31 digestion of two precursor plasmids to pAkv6-cat
(Pedersen et al., 1987), have produced a number of plasmids
where more than one repeat structure has been removed. The nucleotide sequences of the repeat region of some of these plasmids are shown in fig.3 together with the results from transient
expression experiments. Replacement of the E and D modules and a
few nucleotides of C with another sequence still yields a structure which enhance the promotor activity. Deletions of 18 nucle-

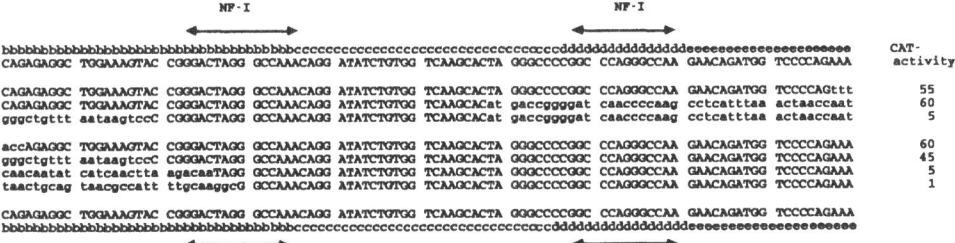

Fig.3. Nucleotide sequences of deletion plasmids in the Akv repeat structure and transient expression levels of chloramphenicol acetyl transferase. The locations of the modules B, C, D and E are indicated, as well as the two Nuclear Factor I recognition sequences.
The transient expression levels  obtained after transfection of NIH 3T3 cells are given relative to that of
pAkv6-cat.

otides from the 5' side of the B module of this plasmid gives a
strong drop in enhancer activity. The next line in  fig. 3 shows
the structure of a plasmid where only these 18 nucleotides are
deleted from the 99 bp. repeat. In this case enhancement is
maintained. However, a continued deletion of seven more nucleotides from the 5' end gives a severe reduction of the enhancement, which is then completely abolished by a further deletion
of four nucleotides. This crucial region in the B module contains a sequence  with homology to the Nuclear Factor I recognition site (indicated as NF-I in fig.3 ). It was therefore obvious to search for a nuclear protein with binding capacity towards the B module, ever so more as band shift experiments

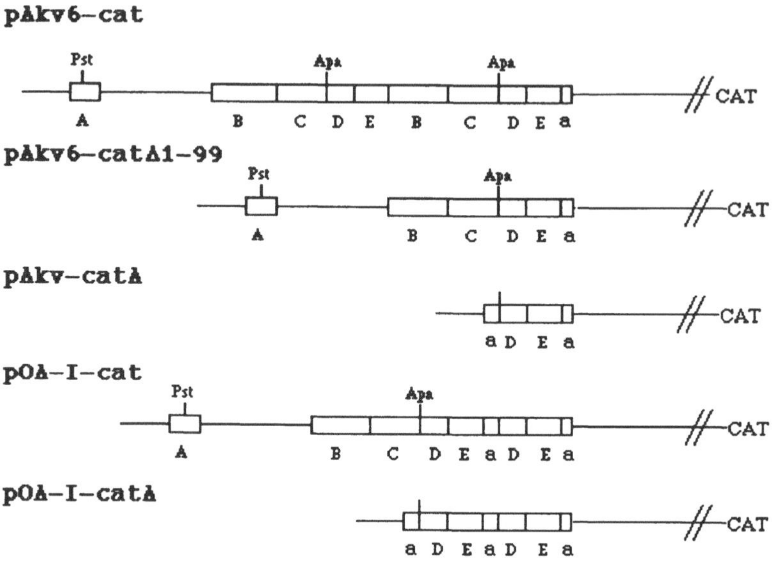

**pAkv6-cat**

**pAkv6-catΔ1-99**

**pAkv-catΔ**

**pOA-I-cat**

**pOA-I-catΔ**

Fig.4. Modular arrays and transient expression levels of a series of deletion plasmids derived from pAkv6-cat and pOA-I-cat. Transient expression levels were measured after transfections into NIH 3T3 cells. The results below are expressed relative to the value for pAkv6-cat. The values represent the averages of two cat enzyme determinations from six transfection for each of two plasmid DNA preparations:

| pAkv6-cat | pAkv6-catΔ | pAkv6-catA | pOA-I-cat | pOA-I-catA |
|-----------|------------|------------|-----------|------------|
| 100 | 40 | 0.5 | 55 | 36 |

showed that such factor(s) is present in nuclear extracts, (see below).

It should not be forgotten that although a deletion analysis as above is structurally detailed, it might be functionally crude, as it inevitably creates new sequence surroundings and possibly new distances for the cooperative proteins in a regulatory region. A deletion analysis attempting to maintain modular arrays and the surrounding sequences may therefore be functionally advantageous.

OA-I MuLV

The OA-I and the FBJ MuLV (Van Beveren et al., 1983) are both derived from spontaneous bone tumors and can induce bone tumors by injection into newborn mice (Pedersen et al., 1987). The modular arrays of their repeat regions, as shown in fig. 1, are very similar, the FBJ having three repeats and the OA-I two repeats with a DEa configuration. Furthermore they both share the very same point mutations in the D and E modules relative to Akv.

The plasmid pOA-I-cat is constructed by insertion of the

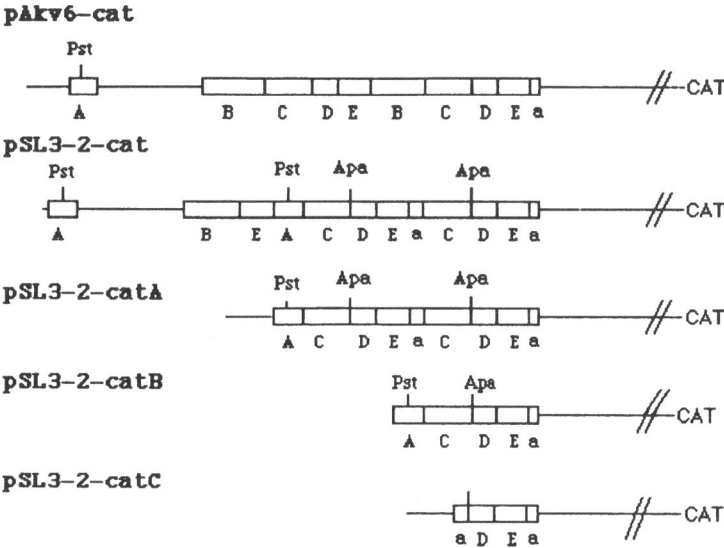

Fig.5. Modular arrays of a series of deletion plasmids derived
from pSL3-2-cat.

PstI-KpnI fragment of the OA-I LTR into pL6-cat. Restriction enzyme deletion of the PstI-ApaI fragment of pOA-I yields the
plasmid pOA-I-catA. We compared the function of OA-I U3 and the
Akv U3 in NIH 3T3 cells in transfection experiments. Fig.4 shows
that the pOA-I-cat has a somewhat lower transient expression
level than pAkv6-cat and that deletion of the aBC structure
only has a minor influence on that activity. The deletion of the
PstI-ApaI fragment from pAkv6-catΔ1-99, only leaves a single DEa
fragment in the plasmid, pAkv6-catA, (fig. 4), leading to a complete abolition of the enhancer activity of the modular region.
These results point to a significant function of the linked DEa
structure in the OA-I. The differences between the activity of
the Akv and the OA-I modular structures could also be determined
by the nucleotide differences in the modules D and E. Preliminary results from protein binding experiments to the E fragments
from Akv and OA-I seems to support such a difference.

SL3-2

The two highly leukemogenic viruses, SL3-3 and SL3-2, have
rather similar modular arrays and share the very same point mutations relative to Akv. In SL3-3, half of the B fragment is replaced by a D fragment and in SL3-2, we find one complete A
fragment between the first E and C modules. Transient expression
studies have shown that the repeat structure of SL3-3 has a relative transcriptional preference for T-lymphoid cells (Celander
and Haseltine, 1984; Short et al., 1987). It is therefore likely
that a similar preference should exist for the repeat structure
of SL3-2. To analyse such a cell preference we have produced a
number of plasmids with deletions in the SL3-2 repeat structure
(fig. 5).

Table 1. Transient expression levels of deletion plasmids
in fibroblasts and lymphoid cells.

| | pAkv6-cat | pSL3-2-cat | pSL3-2-catA | pSL3-2-catB |
|---|---|---|---|---|
| cells | | | | |
| NIH 3T3 | 360 | 100 | 50 | 21 |
| L691 | 10 | 100 | 52 | 25 |

The expression of chloramphenicol acetyl transferase was
measured 36 hours after transfections. The values for the
NIH 3T3 cells represent the averages of six transfections
for each of two preparations of plasmid DNA . The values
for the L691 cells represent the averages for two transfections.

Starting from the pSL3-2-cat plasmid, a *PstI-PstI* fragment
deletion created pSL3-2-catA and a further deletion of the *ApaI-
ApaI* fragment gave the pSL3-2-catB plasmid.

Transfection experiments introducing plasmid DNA into NIH
3T3 cells and into the lymphoid cell line L691 gave the transient expression levels shown in table 1. The relative preference of the SL3-2 enhancer structure for the lymphoid cells is
clearly demonstrated. The deletions within the repeat structure
although gradually decreasing their enhancement, still maintain
the cell type preference. Even the extensive deletion in pSL3-
2catB gives a higher activity in L691 cells than does the complete Akv structure. This points to an important element in the
SL3-2 CDE configuration, possibly linked to the point mutation
in the C module.

Thornell et al., (1988) have identified a protein, SEF1,in
nuclear extracts showing binding to the TGTGGTTA sequence in the
C module of SL3-3. This protein was found most prominent in lymphoid cells and apparently binds with higher affinity to the
SL3-3 type of sequence than to the Akv type of sequence

Table 2. Transient expression levels in different
cell lines of pAkv6-cat and pSL3-2-cat.

| cell line | NIH 3T3 | L691 | Il 247 | B16 |
|---|---|---|---|---|
| U3 region | | | | |
| Akv | 100 | 100 | 100 | 100 |
| SL3-2 | 30 | 900 | 90 | 9 |

The expression of chloramphenicol acetyl transferase
was measured 36 hours after transfection. The values
are given relative to the expression level of pAkv6-
cat transfected cells.

TGTGGT<u>C</u>A. The SEF1 recognition sequence is also found in SL3-2. SEF1 might therefore contribute to the cell type preference of SL3-2. To extend these finding to other cell types we have done transfection experiments using the B16 melanoma cell line (Tanaka et al.,1988), the IL247 lymphoma cell line (Pedersen et al., 1980) besides the NIH 3T3 and the L691 cell lines. The results in table 2 which are all normalized to the transient ex-

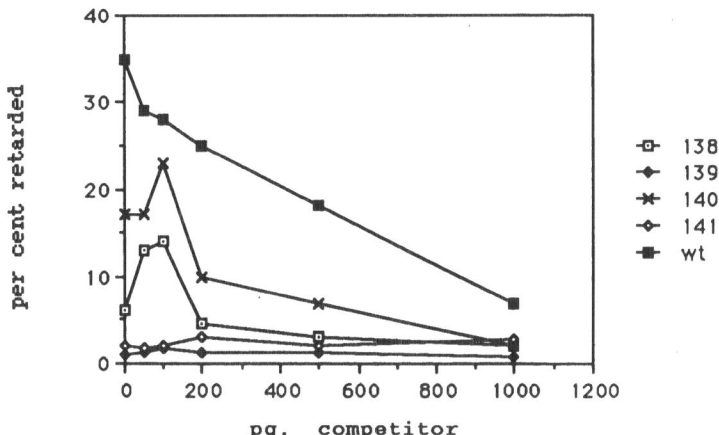

Fig.6. Band shift competition assays. Radioactively labelled 57 bp. dsDNA fragments (50 pg) were mixed with various amounts of non-radioactive Akv 99 bp. repeat competitor dsDNA. An affinity purified nuclear protein fraction was added and the mixture was subjected to polyacrylamide gel electrophoresis. The radioactive bands were identified by autoradiography and the per cent label in the displaced bands were determined.The figure shows the per cent radioactivity in protein/DNA complex as a function of the amount of competitor added. The sequences of the fragments are:

```
      Akv :           GGGACTAGGGCCAA
      oligo 138:      CC
      oligo 139:                   GG
      oligo 140:         TT
      oligo 141       CC           GG
```

pression level governed by the Akv repeat array, show very strong variations in the transient expression levels of the SL3-2 modular configuration. Up to a 100 fold variation is seen between the L691 and the B16 cell lines. Furthermore the two lymphoid cell lines show a marked difference in the transient expression level of the SL3-2 enhancer region.

As shown above, the deletion analysis of the Akv U3 region reveals a nucleotide sequence within the B module which is essential for enhancement of promotor activity, and which have homology to a Nuclear Factor-I consensus recognition site (Nagata et al., 1982). Band shift experiments confirmed the presence in nuclear extracts of a protein(s) with affinity towards the B module. We have isolated a protein fraction from C3H mouse liver nuclei by  a) heparin sepharose chromatography, b) Mono Q ion exchange chromatography and c) DNA affinity chromatography using as ligand a synthetic B module ds DNA with the plus strand composition: CAGAGAGGCTGGAAAGTACCGGGACTAGGGCCAAAC.

Our purified fractions contain four major polypeptides with mw. around 65.000 as shown by gel electrophoresis. It has binding activity towards the two NF-I recognition sites in the Akv modular structure ( fig.3)  as judged by methylation interference and DNase footprinting. We have used this fraction to study the effect of point mutations on protein binding to a 57 bp. synthetic dsDNA fragment covering the entire B module and half of the C module. The point mutations introduced in the NF-I binding region are shown in fig 5 together with results of a band-shift competition assay using the Akv 99 bp. repeat sequence as a competitor. Some of the introduced nucleotide changes completely eliminate the binding of the proteins. Preliminary results from the introduction of these mutations into the pAkv6-catΔ1-99 plasmid indicate a rough correlation between the *in vitro* binding and transient expression activity.

All the considered U3 regions have good consensus sequences to the glucocorticoid response element. The LTRs of Moloney MuSV, (Miksicek et al, 1986) and of Akv and SL3-3, (Celander and Haseltine, 1987) have been shown specifically to respond to the glucocorticoid hormone, dexamethasone. Transient expression studies after gene transfer to the lymphoid cell line, AKSL3, showed about similar strenght of the Akv and SL3-3 enhancers in the presence, as opposed to the marked difference, seen in absence of dexamethasone, (Celander and Haseltine, 1987).


EPILOGUE

Thus it seems that there are functionally important nuclear protein attachment sites within the modules B and D (NF-I), E (GR) and C (SEF1) all of which might be essential for general and cell specific enhancement of transcription. Quinn et al. (1987) have described a protein binding to the E module present in the Gibbon ALV and likely of importance for enhancer activity. Johnson et al. (1987) have isolated a 20 kD rat liver nuclear protein (EBP20) binding to the C module of Moloney MuSV, as well as to the enhancer regions of SV40 and polyoma virus. Although some of these proteins might be identical they indicate that a vast number of interactions might occur in the modular arrangements with different combinations of sequence elements.

This also suggest a weakness in the approach of testing enhancer strength in the usual transfection experiments. Under these experimental conditions, the cells expressing the reporter gene will have taken up a large number of DNA copies containing the enhancer sequence. It is possible that this might topple a normal balance of interactions between transcriptional regulatory proteins. All our deletion plasmid are constructed to permit the insertion into a transmission vector system where helper virus free stocks of retroviral particles can be produced with RNA genomes carrying a neo-gene under the engineered LTR control. In the very same way our deletion constructs can easily be inserted into a complete viral genome, and eventually tested in their true biological environment, the mouse.

REFERENCES

Bösze, Z., Thiesen, H-J. and Charnay, P., 1986, A transcriptional enhancer with specificity for erythroid cells is located in the long terminal repeat of Friend murine leukemia virus, EMBO J., 7:1615-1623.

Celander, D. and Haseltine, W.A., 1984, Tissue-specific transcription preference as a determinant of cell tropism and leukaemogenic potential of murine retroviruses, Nature (London) 312:159-163.

Celander, D. and Haseltine, W.A, 1987, Glucocorticoid regulation of murine leukemia virus transription elements is specified by determinants within the viral enhancer region, J.Virol., 61, 269-275.

Chatis, P.A., Holland,C.A.,Hartley, J.W., Rowe, W.P. and Hopkins, N., 1983, Role for the 3' end of the genome in determining disease specificity for Friend and Moloney murine leukemia viruses, Proc.Natl.Acad.Sci, USA, 80:4408-4411.

Chatis, P.A., Holland, C.A., Silver, J. Frederickson, T.N., Hopkins, N. and Hartley, J.W., 1984, A 3' end fragment encompassing the transcriptional enhancers of nondefective Friend virus confers erythroleukemogenicity on Moloney leukemia virus, J.Virol. 52:248-254.

DesGroseillers, L. and Jolicoeur, P., 1984, The tandem repeats within the long terminal repeat of murine leukemia viruses are the primary determinant of their leukemagenic potential, J.Virol., 52:945-952.

DesGroseillers, L., Rassart, E. and Jolicoeur, P., 1983, Thymotropism of murine leukemia virus is conferred by its long terminal repeat, Proc.Natl.Acad.Sci. USA, 80:4203-4207.

Etzerodt, M., Mikkelsen, T., Pedersen, F.S., Kjeldgaard, N.O. and Jørgensen, P., 1984, The nucleotide sequence of the Akv murine leukemia virus genome, Virology, 134:196-207.

Gorman, C., 1985, High efficiency gene transfer into mammalian cells, in "DNA cloning", D.M.Glover ed., IRL Press, Oxford.

Graham, F.L., van der Eb, A.J., 1973, A new technique for the assay of infectivity of human adenovirus 5 DNA, Virology, 52:456-467.

Grosschedl, R.and Baltimore D., 1985, Cell-type specificity of immunoglobulin gene expression is regulated by at least three DNA sequence elements, Cell, 41:885-897.

Hallberg, B. and Grundström, T., 1988, Tissue specific sequence motifs in the enhancer of the leukaemogenic mouse retrovirus SL3-3, Nucl.Acid Res., 16:5927-5944.

Ishimoto, A., Adachi, A., Sakai, K. and Matsuyama, M., 1985, Long terminal repeat of Friend-MCF virus contains the sequence responsible for erythroid leukemia. Virology, 141:30-42.

Johnson, P.F., Landschulz, W.A., Graves, B.J. and McKnight, S.L., 1987, Identification of a rat liver nuclear protein that binds to the enhancer core element of three animal viruses, Genes and Devel., 1:133-146.

Laimins, L.A., Gruss, P., Pozzatti, R. and Khoury, G., 1984, Characterization of enhancer elements in the long terminal repeat of Moloney murine sarcoma virus, J.Virol., 49:183-189.

Leib-Mösch, C., Schmidt, J., Etzerodt, M., Pedersen, F.S., Hehlmann, R. and Erfle, V., 1986, Oncogenic retrovirus from spontaneous murine osteomas, II. Molecular cloning and genomic characterization, Virology, 150:96-150.

Lenz, J., Crowther, R.L., Straceski, A.and Haseltine, W.A., 1982, Nucleotide sequence of the Akv env gene, J.Virol, 42:519-529.

Lenz, J., Celander, D., Crowther, R.L., Patarca, R., Perkins, D.W. and Haseltine, W.A., 1984, Determination of the leukemogenicity of a murine retrovirus by sequences within the long terminal repeat, Nature  (London) 295:467-470.

Miksicek, R, Heber, A., Schmid, W., Danesch, U., Posseckert, G., Beato, M. and Schutz, G., 1986, Glucocorticoid responsiveness of the transcriptional enhancer of Moloney murine sarcoma virus, Cell, 46:283-290.

Nagata, K., Guggenheimer, R.A., Enomoto, T., Lichy, J.H. and Hurwitz, J., 1982, Adenovirus DNA replication in vitro: Identification of a host factor that stimulates synthesis of the preterminal protein-cCMP complex, Proc.Natl.Acad.Sci.USA, 79:6438-6442.

Pedersen, F.S., Buchhagen, D.L., Chen, C.Y., Hays, E.F. and Haeltine, W.A., 1980, Characterization of virus produced by a lymphoma induced by inoculation of MCF 247 virus, J.Virol., 35:211-218.

Pedersen, F.S., Crowther, R.L., Tenney, D.Y., Reimold, A. and Haseltine, W.A., 1981, Novel leukemogenic retroviruses isolated from a tumor cell line derived from a spontaneous AKR tumour, Nature (London), 292:167-170.

Pedersen, F.S., Etzerodt, M., Lovmand, S., Dai, H.Y., Bækgaard, A.J., Sørensen, J., Jørgensen, P., Kjeldgaard, N.O., Schmidt, J, Leib-Mösch, C., Luz, A. and Erfle, V., 1987, Transcriptional control and oncogenicity of murine leukemia viruses, in "Viral Carcinogenesis", N.O.Kjeldgaard and J.Forchhammer, eds, Munksgaard, Copenhagen.

Quinn, J.P, Holbrook, N. and Levens, D., 1987, Binding of a cellular protein to Gibbon ape leukemia virus enhancer,
Mol.Cell.Biol., 7:2735-2744.

Rosen, C.A., Haseltine, W.A., Lenz, J., Ruprecht, R. and Cloyd,
M.W., 1985, Tissue selectivity of murine leukemia virus infection is determined by long terminal repeat sequences, J.Virol,
55:862-866.

Schaffner, G., Schirm, S., Müller-Baden, B., Weber, F. and
Schaffner, W., 1988, Redundancy of Information in Enhancers as a
Principle of Mammlian Transcription Control, J.Mol.Biol.,
201:81-90.

Schmidt, J., Erfle, V., Pedersen, F.S., Rohmer, H., Schetters,
H., Marquart, K-H. and Luz,A., 1984, Oncogenic retrovirus from
spontaneous murine osteomas, I. Isolation and biological characterization, J.Gen.Virol., 65:2237-2248.

Shinnick, T.M., Lerner, R.A. and Sutcliffe, J., 1981, Nucleotide
sequence of Moloney murine leukemia virus, Nature (London),
293:543-548.

Short, M.K., Okenquist, S.A. and Lenz,J., 1987, Correlation of
leukemogenic potential of murine retroviruses with transcriptional tissue preference of the virla long terminal repeats,
J.Virol., 61:1067-1072.

Speck, N. and Baltimore, D., 1987, Six distinct nuclear factors
interact with the 75-base-pair repeat of Moloney leukemia virus
enhancer, Mol.Cell.Biol., 7:1101-1110.

Tanaka, K., Gorelik, E., Watanabe, M., Hozumi, N. and Jay, G.,
1988, Rejection of B16 melanoma induced by expression of a
transfected major histocompatibility complex class I gene,
Mol.Cell.Biol., 8:1857-1861.

Thornell, A., Hallberg, B. and Grundström, T., 1988, Differential protein binding in lymphocytes to a sequence in the enhancer of the mouse retrovirus SL3-3, Mol.Cell.Biol., 8:1625-1637.

Van Beveren, C., Van Straaten, F., Curran, T., Muller, R. and
Verma, I.M., 1983, Analysis of FBJ-MuSV provirus and c-fos
(mouse) gene reveals that viral and cellular fos gene products
have different carboxy termini. Cell, 32:1241-1255.

Vogt, M., Haggblom, S., Swift, S. and Haas, M., 1985, Envelope
gene and long terminal repeat determine the different biological
properties of Rauscher, Friend and Moloney mink cell focus-
inducing viruses, J.Virol., 55:184-192.

# NEGATIVE REGULATION OF CELL GROWTH

Claudio Schneider, Wilhelm Ansorge and Lennart Philipson

European Molecular Biology Laboratory
Postfach 10.2209
D-6900 Heidelberg, Fed. Rep. of Germany

The majority of cells in tissues, organs and the blood system of the body does not divide and stays in a resting state performing various tissue-specific functions. Some cells can remain in the quiescent state for years still maintaining the capacity for reentering the cell cycle if appropriately stimulated.

In culture, primary cells become arrested in the $G_1$ phase of the cell cycle when the environmental conditions become sub-optimal for proliferation as a result of nutrient or serum deprivation (Brooks, 1975), or high cell density (Todaro et al., 1965).

In recent years the study of growth control of animal cells has mainly focussed on the external signals inducing growth (i.e. growth factors and their receptors) and the growth-related genes induced by them. These studies have revealed a close inter-relationship between proto-oncogenes on the one hand, and cell-cycle genes, growth factors and their receptors on the other (Waterfield et al., 1983; Downward et al., 1984; Baserga and Surmacz, 1987).

Few data have accumulated about the molecular mechanisms which govern and induce the establishment of the $G_O$ phase. However, studies with heterodikaryons between senescent or quiescent cells on the one hand and actively replicating normal or transformed cells on the other (for a review see Stanbridge, 1987) provided evidence that the replicative potential of cells may be "recessive" and that inhibition of DNA synthesis can be observed in these hybrids. DNA synthesis and cell proliferation may, therefore, be regulated by an interplay between positive (inducer) and negative (suppressor) molecules produced by the cell.

Growth inhibiting mRNAs, isolated from cultured cells rendered either quiescent by serum starvation or senescent by prolonged culture were only recently identified (Lumpkin et al., 1986). The inhibition of growth of epithelial and endothelial cells in culture has also been observed with the growth factor TGFß (for a review see Massagué, 1987).

Transcriptional repression of genes has furthermore recently been proposed to be the major mechanism by which the adenovirus E1A region induces transformation in primary cells (Lillie et al., 1986). Negative elements have in addition been identified in retinoblastoma (Cavenee et al., 1984), Wilms' tumors (Koufos et al., 1984) and colorectal carcinomas (Solomon et al., 1987).

These results led us to characterize $G_0$-specific genes whose expression is negatively regulated by serum and growth factors (Schneider et al., 1988) and to develop an assay for S-entry-delay mRNA isolated from resting lymphocytes and established cell lines (Pepperkok et al., 1988a and b).

## RESULTS

### Isolation of Clones

A subtraction cDNA library specific for the arrested stage was constructed in the lambda insertion vector NM-1149 using mRNA from serum-starved (0.5%) cells. The Go-enriched library (50,000 plaques) was screened with probes from arrested and induced single-stranded cDNA and those giving a differential signal were called **g**rowth **a**rrest **s**pecific (gas) genes. When total RNA isolated from serum-starved NIH3T3 cells at various times after addition of 20% serum was analyzed on

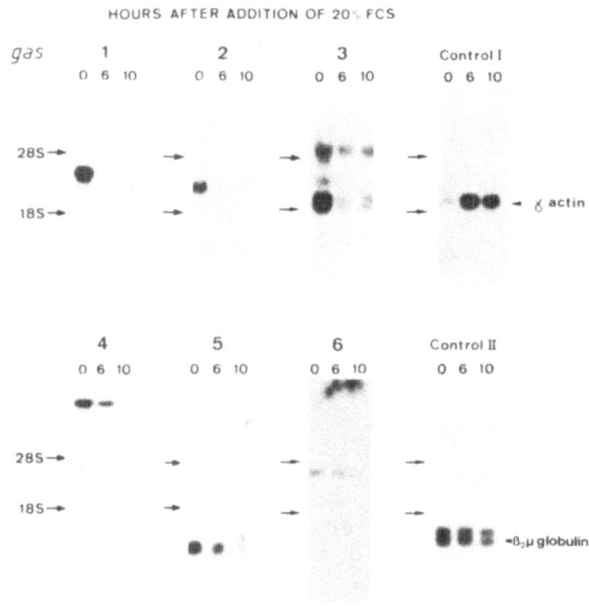

**Fig. 1.** Identification of cDNA clones accumulated at growth arrest and downregulated upon growth induction.

Northern blots using the gas 1 to 6 cDNA as probes for RNA isolated at arrest and 6 and 10 hours after addition of 10% fetal calf serum to NIH3T3 cell cultures. The Northern blots were also probed with a g-actin probe as a control for an inducible gene and b2-microglobulin to assess the total amount of RNA in each lane.

Nothern blots these cDNAs hybridized to mRNA that accumulated in the growth arrest stage, but was significantly reduced within 10 hours after stimulation of growth with 20% serum (Fig. 1). The *gas* 1 cDNA was most abundant, representing 2% of the enriched library, *gas* 5 accounted for 1%, *gas* 2-4 around 0.2 - 0.3% and *gas* 6 was found only once among the 50,000 clones. All clones detected differently sized mRNA bands ranging from 0.8 kb to 10 kb (Fig. 1) and they all disappeared at slightly different rates after serum stimulation of growth. We have subsequently concentrated on a more detailed charaterization of the *gas* 1 and *gas* 2 genes.

## Induction of *Gas* mRNAs after Density Inhibition

To study *gas* mRNA accumulation during density inhibition of cell growth NIH3T3 cells were plated in 10% FCS and maintained in the same medium for 6 days. DNA synthesis was scored by 3H-thymidine incorporation and RNA was extracted and analyzed on Northern blots from replicate cultures each day. In these experiments cells were seeded at a high density and contact inhibition of DNA synthesis was observed within 2 days

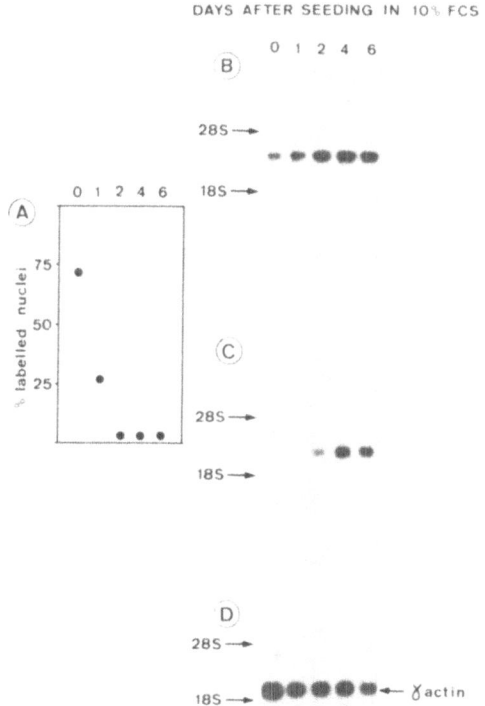

**Fig. 2.** Accumulation of *gas* gene mRNA after contact inhibition. RNA was isolated from actively growing cells kept for different times in 10% fetal calf serum.

A: DNA synthesis was scored on replicate cell cultures labelled with 3H-Thymidine for 24 hrs.; B: *gas* 1 gene expression; C: *gas* 2 gene expression; D: expression of 1 actin on the same blots as shown in B and C.

as shown in Fig. 2A. *Gas* 1 mRNA accumulated to a maximum within 2 days (Fig. 2B). The high initial level of *gas* 1 mRNA is probably due to accumulation of this RNA also during the S-phase of the growth cycle. The *gas* 2 mRNA accumulated more slowly (Fig. 2C).

## *Gas* Gene Expression in Transformed Cells

A NIH3T3 cell line transformed with v-*fos* in a retrovirus construct which lacks contact inhibition and expresses v-*fos* mRNA at high levels (E. Wagner, personal communication) was used to study *gas* 1 and *gas* 2 expression in transformed cells. The cells were plated in 10% FCS and when semiconfluent they were incubated in medium with 0.5% FCS. Fig. 3A shows that both *gas* 1 and 2 are expressed inefficiently under serum starvation in the transformed cells suggesting that they do not become arrested under these conditions. A comparison of induction in untransformed NIH3T3 cells with the same amount of RNA is included for comparison (Fig. 3B). The same pattern of *gas* 1 and *gas* 2 expression was found in an NIH3T3 cell line transformed with v-*src* which had growth properties similar to the v-*fos* transformed cells (not shown).

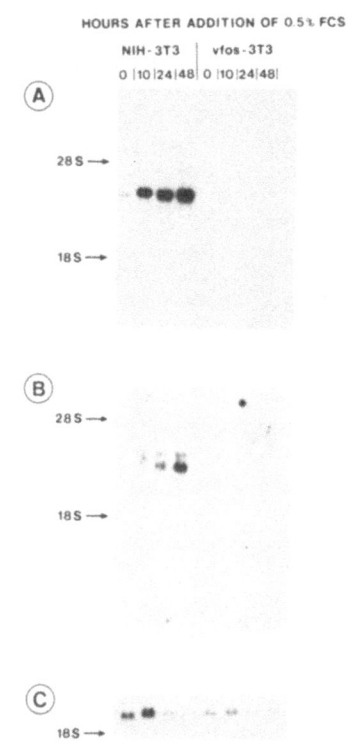

**Fig. 3.** *Gas* gene expression in transformed cells. NIH3T3 cells transformed with v-*fos* were subjected to serum starvation in the same way as the NIH3T3 cells. RNA was collected at indicated times and analyzed by Northern blots.

A: *gas* 1 expression in NIH3T3 and v-*fos* transformed NIH3T3 cells; B: *gas* 2 expression in NIH3T3 and v-fos transformed cells; C: hybridization of a GAPDH probe to the same RNA.

## Sequence of Full-Length cDNA

Full-length cDNA of several of the *gas* genes with the right polarity were cloned in a T7/T3 transcription vector and the sequences of *gas* 2 and *gas* 4 cDNAs have been completed. No homology to sequences in the Data Libraries has been identified.

## mRNA from Resting T-Cells inhibits DNA-Synthesis

An automatic microinjection was developed (Ansorge and Pepperkok, 1988; Pepperkok et al., 1988a and b) to assay for mRNA inhibition of DNA synthesis in single cells. Injection experiments were carried out with mRNA isolated from resting or activated human peripheral blood lymphocytes consisting predominantely of T-cells. Both HeLa and human fibroblasts (IMR-90) cells were used as target cells. The mRNA was injected at a concentration of 1mg/ml together with FITC-dextran (0.5%) whereupon injected cells were analysed for DNA synthesis.

Table 1 shows that significant inhibition of DNA-synthesis occurred only when mRNA isolated from resting T-cells was injected. No inhibitory activity was found with the poly A- fraction from resting cells at 2 mg/ml or after RNAse and alkaline degradation of the poly A+ fraction. When mRNA isolated from activated T-cells (stimulated to grow with TPA/Ionomycin for 72 hours) was injected, there was no inhibition. On the contrary, a 10% stimulation of DNA-synthesis was seen when mRNA from activated T-cells was injected into the human fibroblasts target cells. Neither RNA from resting nor activated T-cells were degraded.

**Table 1.** Injection of poly A+-RNA isolated from resting or activated T-cells:

| Injected poly-$A^+$-RNA (1mg/ml) from: | Percent inhibition in recipient cells (%): | |
|---|---|---|
| | HeLa | IMR-90 |
| Resting T-cells | 42 ± 7.4 | 55 ± 6.4 |
| Activated T-cells | 5 ± 1.2 | -10 ± 3.6 |

Poly A+-RNA isolated from resting or activated (for 72hrs in the presence of TPA and Ionomycin) T-cells was injected into growing HeLa cells or synchronized human fibroblasts (IMR-90). After injection the cells were incubated in the presence of 3H-thymidine for 24hrs in (i) DMEM, containing 20% FCS in the case of IMR-90 cells or (ii) DMEM, containing 10% FCS for HeLa cells. The percent inhibition (PI) was determined and the results are the mean of four separate experiments with different poly A+-RNA preparations. All preparations were checked for integrity and relative amounts of RNA as shown in Fig. 5. About 300 cells were injected in each experiment.

Fig. 4 shows, that mRNA from resting cells at a concentration of 0.2 mg/ml can still appreciably inhibit the entry into S phase. From this concentration we estimated (see legend) that around 300 mRNA copies per cell are responsible for this biological effect.

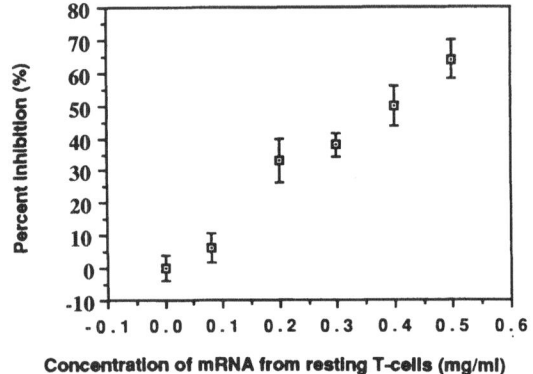

**Concentration of mRNA from resting T-cells (mg/ml)**

**Fig 4**. Various concentrations of poly A+-RNA were injected (the zero mg/ml value represents the control injection of FITC-dextran only). The data are the mean of two experiments with different mRNA preparations. 300 cells were injected in each experiment.
The minimum amount of total mRNA injected per cell to obtain a significant inhibitory effect is $10^{-11}$mg (obtained as Volume injected X Minimum mRNA concentration = $5 \times 10^{-11}$ ml x 0.2 mg/ml). This corresponds to $\approx 1 \times 10^4$ mRNA copies with 2 Kb average size. Assuming that 10 copies of the inhibitory mRNA are needed to obtain significant inhibition in one cell, the relative abundance of the inhibitory mRNA species is 0.1% which corresponds to 300 copies per resting T-cell (taking the total number of mRNA copies per cell to be $3 \times 10^5$).

Inhibition of DNA Synthesis by Resting Cell mRNA is Reversible

When HeLa cells and human fibroblasts (IMR-90) were injected with mRNA isolated from resting T-cells and labelled with 3H-thymidine for different times after injection, the inhibition is higher at 20 hours after injection than after 30 hours and at 40 hours the cells have recovered their DNA-synthesis capacity. Thus the inhibitory effect of the mRNA isolated from resting T-cells is reversible and represents a delay in DNA-synthesis. Furthermore the inhibitory effect, transferred by injecting resting mRNA, disappears when the target cells approach the G1/S boundary of the cell cylce around 10 hours after release from growth arrest of human fibroblasts. This also applies to the donor T cells and a decrease in inhibitory activity can be observed at 48 hours after activation of these cells.

## Only a Discrete Size-Class of mRNA Mediates Inhibition

To establish that a distinct class of mRNA is responsible for the inhibitory effect, the mRNA was size-fractionated on non-denaturing gels. Fig. 5A shows that the RNA in the two preparations was not degraded. They contained the equivalent amounts of ribosomal RNA revealed by ethidium bromide staining and the same amount of mRNA of a housekeeping (GAPDH) gene as revealed by Northern blots. Fig. 5B shows that the peak of inhibitory activity resides in fractions corresponding to a size of around 1.5 kb. The mRNA obtained from T-cells activated for 72 hrs with TPA and Ionomycin and sized in the same way does not show any inhibitory effect.

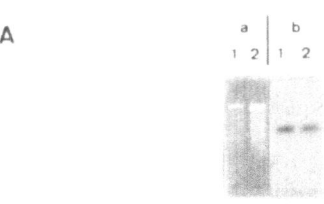

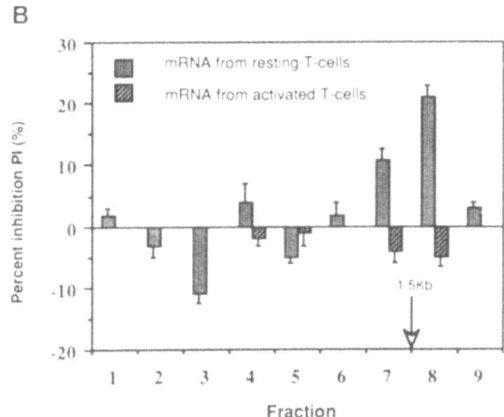

**Fig. 5.** Determination of the size of mRNA causing an inhibitory effect in recipient cells:

A.  0.5 µg of poly A+ RNA from resting T-cells (lane 1) and from 72 hrs. activated T-cells (lane 2) were run on a 3% NuSieve Agarose gel. a) represents the Ethidium Bromide stained tracks and b) represents the same two lanes as transferred to nylon-membrane (Gene Screen plus NEN) and probed with oligolabelled GAPDH (Glyceraldehyde-3-phosphate dehydrogenase) probe to check for integrity of a specific mRNA.

B.  Ten µg of poly A+-RNA were sizefractionated. The different fractions were injected into human fibroblasts. The cells were assayed for DNA synthesis 24 hrs after injection. The data shown are the mean of two experiments (300 cells injected in each). The peak of DNA synthesis inhibition in the injected cells corresponds to an RNA fraction around 1.5 kb in length.


## DISCUSSION

Since the introduction of the cell-cycle concept (Howard and Pelc, 1953), two approaches to study the growth regulation of cells have been followed. One envisages living cells as naturally quiescent needing a stimulatory encounter with one or several growth factors to be activated to cell division (Sporn and Roberts, 1985). The other considers cellular multiplication as the natural steady-state: cessation of multiplication is thus regarded as a restriction imposed on the system. In the latter case emphasis is mainly on the signals involved in arresting multiplication (Harris, 1985).

The use of an automated microinjection system (Ansorge and Pepperkok, 1988; Pepperkok et al., 1988a) with a fluorescent co-injected marker enabled us to follow every injected cell separately and analyse its DNA synthesis activity.

The results show that only the mRNA isolated from resting T-cells is able to block DNA-synthesis, both in synchronized human fibroblasts or HeLa cells, while mRNA isolated from actively growing T-cells lacks this capacity.

In a concurrent study (Schneider et al., 1988) we identified several **g**rowth **a**rrest **s**pecific (*gas*) cDNA clones from NIH3T3 cells which show a similar pattern of control, i.e. high RNA expression at arrest with a sharp decrease over 2 to 6 hours after growth stimulation. The NIH3T3 cells enter DNA synthesis after 10-12 hours compared with 72 hours for the T-cells. Among the *gas* clones, we observed either transcriptional or posttranscriptional regulation of expression (not shown). We are now identifying the inhibitory element(s) by cDNA cloning in transcription vectors and microinjection of the *in vitro* synthesized mRNA. The first step must, however, be to identify the best source for the cloning experiments. Lumpkin et al. (1986) demonstrated with microinjection that senescent human fibroblasts are at least 10-fold enriched compared to quiescent cells for growth inhibitory mRNA. We have also found this activity in quiescent NIH3T3 and human liver poly A+ RNA (Pepperkok et al., 1988b). It may, therefore, be appropriate to select RNA from a senescent or quiescent primary cell in culture to ascertain that contaminating cells do not contribute in the cloning experiments.

## ACKNOWLEDGEMENTS

We would like to thank Mrs. Nelly van der Jagt-González and Waltraud Ackermann for preparation of the manuscript.

# REFERENCES

Ansorge, W. and Pepperkok, R. (1988). Performance of an automated system for capillary microinjection into living cells. J. Biochem. Biophys. Meth. (in press).

Baserga, R. and Surmacz, E. (1987). Oncogenes, cell-cycle genes and the control of cell proliferation. Biotechnology, 5:355.

Brooks, R.F. (1975). The kinetics of serum-induced initiation of DNA synthesis in BHK21/C13 cells and the influence of exogenous adenosine. J.Cell. Physiol., 86:369.

Cavenee, W.K., Dryja, T.P., Phillips, R.A., Benedict, W.F., Godbout, R., Gallie, B.L., Murphree, A.L., Strong, L.C. and White, R.L. (1983). Expression of recessive alleles by chromosomal mechanisms in retinoblastoma. Nature, 305:779.

Downward, J., Yarden, Y., Mayes, E., Scrace, G., Totty, N., Stockwell, P., Ullrich, A., Schlessinger, J. and Waterfield, M.D. (1984). Close similarity of epidermal growth factor receptor and v-erb-B oncogene protein sequences. Nature, 307:521.

Harris, H. (1985). Suppression of malignancy in hybrid cells: The mechanisms. J. Cell Sci., 79:83.

Howard, A. and Pelc, S.R. (1953). Synthesis of deoxyribonucleic acid in normal and irradiated cells and its relation to chromosome breakage. Heredity, 6:261.

Koufos, A., Hansen, M.F., Lampkin, B.C., Workman, M.L., Copeland, N.G., Jenkins, N.A. and Cavenee, W.K. (1984). Loss of alleles at loci on human chromosome 11 during genesis of Wilms' tumor. Nature, 309:170.

Lillie, J.W., Green, M. and Green,M.R. (1986). An adenovirus E1A protein region required for transformation and transcriptional repression. Cell, 46:1043.

Lumpkin Jr., C.K., McClung, J.K., Pereira-Smith, O.M. and Smith, J.R. (1986). Existence of high abundance antiproliferative mRNA's in senescent human diploid fibroblasts. Science, 232:393.

Massagué, J. (1987). The TGF-b family of growth and differentiation factors. Cell, 49:437.

Pepperkok, R., Zanetti, M., King, R., Delia, D., Ansorge, W., Philipson, L. and Schneider, C. (1988a). An automatic microinjection system facilitates detection of growth-inhibitory mRNA. Proc. Natl. Acad. Sci., in press.

Pepperkok, R., Schneider, C., Philipson, L. and Ansorge, W. (1988b). Single cell assay with an automated capillary microinjection system. Exp. Cell Res., in press.

Schneider, C., King, R. and Philipson, L. (1988). Genes expressed at growth arrest of mammalian cells. Cell, in press.

Solomon, E., Voss, R., Hall, V., Bodmer, W.F., Jass, J.R., Jeffreys, A.J., Lucibello, F.C., Patel, I. and Rider, S.H. (1987). Chromosome 5 allele loss in human colorectal carcinomas. Nature, 328:616.

Sporn, M.B. and Roberts, A. B. (1985). Autocrine growth factors and cancer. Nature, 313:745-747.

Stanbridge, E.J. (1987). Genetic regulation of tumorigenic expression in somatic cell hybrids. Adv. Viral Oncol., 6:83.

Todaro, G.J., Lazar, G.K. and Green, H. (1965). The initiation of cell division in a contact-inhibited mammalian cell line. J. Cell. Comp. Physiol., 66:325.

Waterfield, M.D., Scrace, G.T., Whittle, N., Stroobant, P., Johnsson, A., Wasteson, Å., Westermark, B., Heldin, C.-H., Huang, J.S. and Deuel, T.F. (1983). Platelet-derived growth factor is structurally related to the putative transforming protein p28$^{v-sis}$ of simian sarcoma virus. Nature, 304:35.

THE HUMAN IMMUNOGLOBULIN K LOCUS AND ITS EVOLUTION

Hans G. Zachau

Institut für Physiologische
Chemie, Physikalische Biochemie
und Zellbiologie
der Universität München
Goethestrasse 33, 8000 München 2

## Abstract

After an introduction on the evolution of the Summer
Schools on Molecular Biology at Spetsai, the work of our
group on the human immunoglobulin Kappa (K) locus is
described. The locus consists of variable, joining and
constant ($V_K$, $J_K$, $C_K$) gene segments and comprises at
least 2 Mb (mega-base pairs) of DNA. Up to now 81 of the
perhaps about 100 $V_K$ genes and pseudogenes have been
isolated on cosmid clones. Several contigs have been
constructed covering about 1 Mb. An overall map of the
locus is based on pulsed field gel electrophoresis
experiments. The evolution of the K locus is described
in terms of $V_K$ gene duplication, amplification,
transposition and subgroup interdigitation. The role of
pseudogenes within the K locus and of the pseudogenes
which have been transposed to locations outside of the
locus is discussed.

## Introduction on the evolution of the Summer Schools on Molecular Biology at Spetsai

The Summer Schools on Molecular Biology at Spetsai have
come of age now. I have seen them evolving from their
beginning in 1966 to present times. They have continued to be
attractive to young scientists. In 1988 we had again about
three times as many applicants than we could accept.

The programs of the Schools included in the previous
years and will do so in 1988 lectures on topics of cell
biology, immunology and virology in addition to what has now
become classical molecular biology. The organizers of the
Schools drew up programs of what they considered at the time
most interesting in the field, not forgetting, of course,
their own research topics. I have taken the "German part" in
the rotation of French, British, German and, more recently,
American organizers and have been in charge, together with my
colleagues, foremost Horst Feldmann, of six Summer Schools.

Two of them had to be held "in exile", that is at Erice, Sicily. My research interest was in tRNA and protein synthesis in the beginning. It has changed via chromosomal RNA to chromatin and repetitive DNA in the '70s and via the chromatin at the immunoglobulin genes to these genes themselves in the '80s. But plenty of other topics were offered in our Schools as in the others and we believe that, in general, balanced programs have been achieved.

When I first heard about the Workshop on Evolutionary Tinkering, I thought to talk about the evolution of the Spetsai Summer Schools. The story should be fascinating since the development of molecular biology is mirrored in the programs of the Schools. Many of the main proponents of the field have lectured in Spetsai and have shaped the character of the Schools. Besides science, there have been other things which decisively influenced the series of Summer Schools ranging from politics to science policy. It is also interesting to observe that individual Schools attained a certain national flavor depending on the nationality of the organizers. Obviously, a talk and even more a manuscript on the evolution of the Spetsai Schools would have been a demanding task. I therefore welcomed the decision that we should talk and write about one of our current research interests. I shall describe in the following briefly the work of our group in Munich on the human immunoglobulin K locus and then consider some aspects of the evolution of the locus.

## Why do we work on the human immunoglobulin K locus?

After a few years of work on the mouse immunoglobulin genes of the K type, we started in 1982 to study the corresponding human genes. We had in mind to elucidate one of the immunoglobulin loci in order to contribute to the solution of the GOD question, i.e. the question of the generation of antibody diversity; with other words, we wanted to find out how many potentially functional genes are present in the germline DNA. The rest of the antibody diversity as found at the protein level must have been formed by somatic processes as recombination of gene segments and somatic mutations. Once one knows the number of germline genes one can estimate the proportion of the antibody repertoire which is not inherited but acquired by somatic processes. The human K locus offered itself as an object for such a study since early work by Bentley and Rabbitts (1981) had indicated that the number of $V_K$ genes is small. The number of the $V_K$ genes which we actually found was considerably higher than originally assumed, but our supposition that the human K locus is at least not more complex than the $\lambda$ or heavy chain locus still seems to hold. Working on a human locus one is, of course, at a disadvantage to those colleagues working on loci of inbred mouse strains. But up to now heterozygocity has not proven to be a serious handicap in our work. On the other hand, results on a human locus may attain direct medical importance when it comes to the study of immuno-deficiencies, of genes for autoantibodies etc.

## The approaches

    We started work on the human K locus by a "bottom-up
approach": isolate gene probes for the four different
subgroups (Jaenichen et al., 1984; Klobeck et al., 1984 and
1985), screen cosmid libraries with these probes, construct
so-called contigs of overlapping cosmid clones (Pech et al.,
1984 and 1985) and eventually bridge the gaps between the
contigs by chromosomal walking (Pohlenz et al., 1987, Klobeck
et al., 1987; Straubinger et al., 1988). In the course of
the work about 400 different cosmid clones have been
analyzed. This line of work was

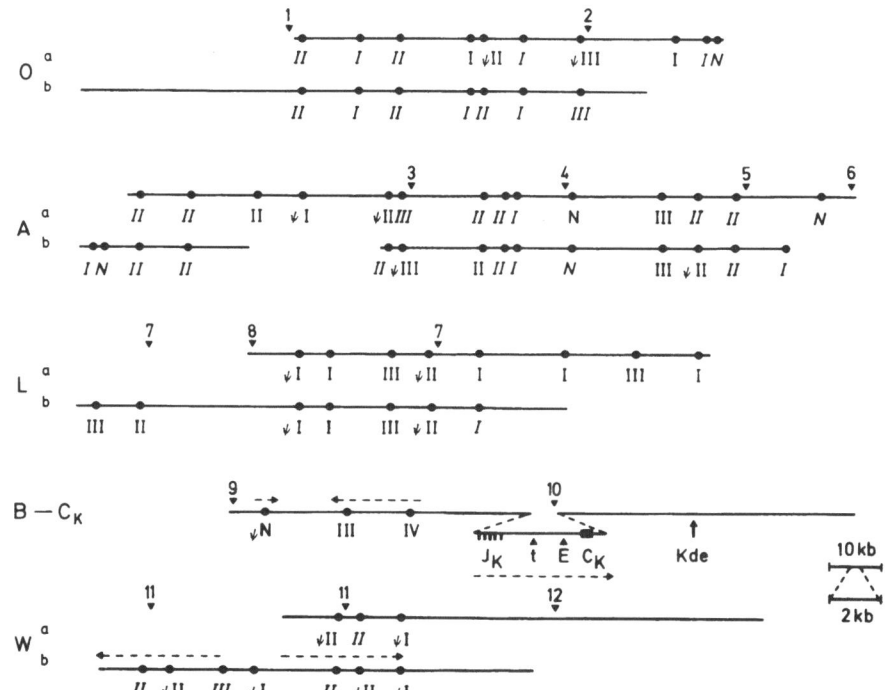

Fig. 1   Human $V_K$ gene regions and the $J_K C_K$ gene region. Genes
         are shown as dots and the subgroup assignments are
         given. The transcriptional orientation of the genes
         is from left to right; an inverted repeat in Wb and
         the opposite polarity of the B region genes are
         indicated by dashed arrows. Taken from Lorenz et al.
         (1987).

complemented more recently with the "top-down" approach of
ordering the various contigs by pulsed field gel
electrophoresis (Lorenz et al., 1987) and by as yet
unsuccessful attempts with parts of the K locus cloned in
yeast artificial chromosomes (Burke et al., 1987).

The K locus as it presents itself today comprises at least 2 Mb of DNA. The major part of the locus is duplicated, i.e. most $V_K$ genes occur in two very similar but not identical copies per haploid genome. They have opposite transcriptional polarities, a feature which is interesting with respect to the mechanism of V-J rearrangement (Klobeck et al., 1987, Lorenz et al., 1987). 81 $V_K$ genes have been isolated from the K locus up to now; 46 of them were sequenced and half of them turned out to be pseudogenes. We know from blot hybridization experiments with digests of

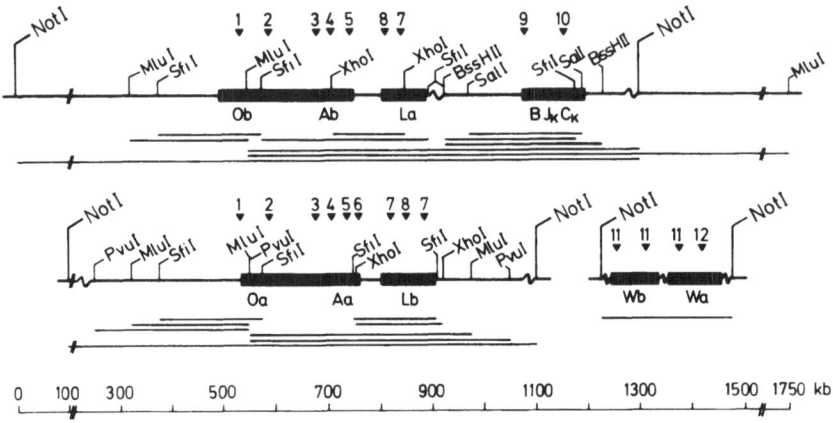

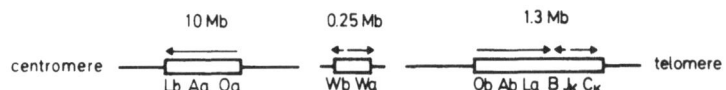

Fig. 2   Maps of three parts of the human K locus as constructed on the basis of pulsed field gel electrophoresis data. In the bottom panel a simplified scheme of the K locus is shown emphasizing the transcriptional polarities (horizontal arrows). For the localization of the W regions see text. Taken from Lorenz et al. (1987).

genomic DNA that a few more $V_K$ genes exist in germline DNA, and we estimate that the total number of $V_K$ genes may be 100 or somewhat smaller (A. Meindl, unpublished). How many of these genes are functional and how many are pseudogenes will be known only when all of them have been sequenced and when the transcripts of the genes have been extensively studied. The known contigs and their assembly in a simplified scheme are shown in Figs. 1 and 2. The early part of the work was summarized by Zachau et al. (1984) and some aspects of our more recent work are described in an article for a monograph (Zachau, 1988).

## What can we tell about the evolution of the human K locus?

The early evolution of the immunoglobulins in the context of the superfamily of recognition molecules is a much discussed topic (eg. Williams and Barcley, 1988). In addition to the immunoglobulins and the T cell receptors, several cell surface proteins belong to the superfamily. Evolutionary trees have been

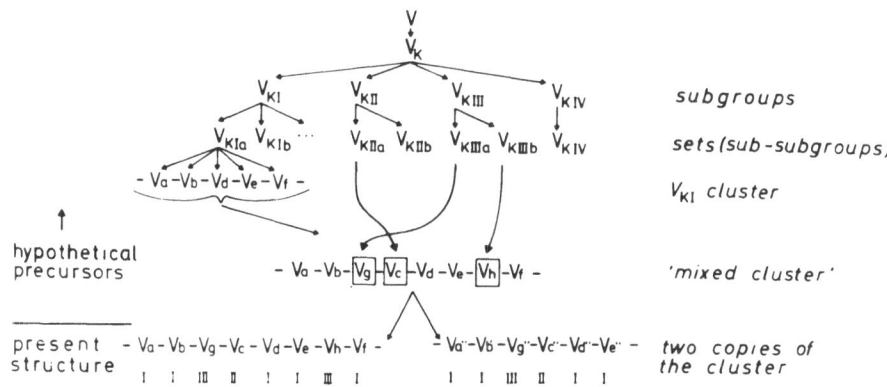

Fig. 3    Model of the evolution of part of the human K locus. A cascade of amplification and transposition steps is shown which contains only the minimal number of branch points. Taken form Pech and Zachau (1984).

constructed which not only connect the various members of the superfamily but also trace the present day variable and constant parts of the immunoglobulin back to a common ancestor or even to a primordial element of half the size of V or C. The evolution of the J and D elements and with that the evolution of combinatorial and junctional diversity is a particularly interesting problem. The evolution of the immune superfamily of genes has also been discussed from a population genetics point of view by Dover and Strachan (1986).

Our work on the human K locus contributes to the understanding of the later stages of evolution. Before considering the ur-$V_K$ gene I should mention one interspecies aspect: we isolated some $V_K$ genes from human DNA libraries with mouse $V_K$ gene probes (Jaenichen et al., 1984); sequence comparisons confirmed that some mouse $V_K$ genes are as similar to the human genes as are human $V_K$ genes of different subgroups to each other. These similarities are, of course, an outcome of the functional constraints of the antibody molecules.

The classification of the human $V_K$ genes in four subgroups rests on the extensive comparisons of protein sequences by Kabat et al. (1987 and earlier editions). The members of one subgroup are usually more than 80 % similar in sequence and $V_K$ genes of the various subgroups differ from each other by more than 20 %. There is no obvious way to relate one of the human $V_K$ subgroups to a mouse $V_K$ subgroup. A formal way to describe the evolution of the human $V_K$ genes is to postulate an ur-$V_K$ gene from which the ur-$V_K I$, $V_K II$, $V_K III$ and $V_K IV$ genes were derived. The $V_K IV$ gene stayed a one-member subgroup while the others were amplified. It was a surprise when we found that within the K locus the genes of different subgroups are not located in clusters, but are interdigitated (Pech and Zachau, 1984). We therefore had to postulate extensive transpositions in addition to amplification events (Fig.3).

The duplication of large parts of the K locus must have been a relatively late event because the duplicate copies of the $V_K$ genes are very similar to each other (0.5-3.5 % different at the DNA level). Also inactivating alterations of some genes by point mutations or the insertion of repetitive sequences are frequently found in both $V_K$ gene copies. There may also be gross differences between the duplicated copies of the K locus since the homologies between some duplicated regions break off towards the ends of the contigs.

It is certainly too naive to assume that the $V_K$ locus was duplicated and the duplication persisted only because many $V_K$ genes of the originally monomeric locus had been inactivated. One would expect that the various somatic processes can compensate for at least some deficiencies in the germline repertoire. It is also clear that some duplicated $V_K$ genes code for identical protein sequences and therefore the duplication has not increased the antibody repertoire in these cases. The duplication seems also not to be essential for a functioning immune system since we found in a group of 20 individuals an apparently healthy individual whose K locus is not duplicated. What may offer an evolutionary advantage is the opposite transcriptional polarity of the duplicated $V_K$ genes (Fig.2). If all $V_K$ and the $J_K C_K$ genes were arranged in identical orientation large parts of the $V_K$ locus would be deleted when a $V_K$ gene distant to the JC region is rearranged. The opposite polarities of the distant $V_K$ gene copies allows for an inversion mechanism of V-JC rearrangement in which the $V_K$ genes in between the rearranged one and JC are saved. The ability of the locus to rearrange related gene copies either by a deletion or by an inversion mechanism may offer a functional advantage in the series of aberrant rearrangements which frequently take place before a functional V-JC joint is formed.

During the duplication/amplification events in the K locus some genes went astray: non-processed $V_K$ genes were found also outside of the locus which resides on chromosome 2 (at 2p11). Such so-called orphons are located on chromosomes 1, 22 and others (Lötscher et al., 1986 and 1988). They are pseudogenes because they carry some defects in their sequence and because they are no longer in the neighborhood of the $J_K C_K$ region. An indirect influence on the V gene repertoire by gene conversion or recombination can, of course, not be excluded. An analysis of the sequences at the transition points between the transposed gene regions and the host chromosomes may give insights into the mechanism of their transposition and, hopefully, also into some of the mechanisms which shaped the K locus in evolution.

Recent experiments indicate that the W regions (Fig.1, Fig.2) are orphon-like; they reside on chromosome 2 but they seem to be located outside of the K locus itself (F.-J. Zimmer, unpublished).

Gene conversion-like events are generally believed to play an important role in the evolution of multigene families. In a locus coding for a family of recognition molecules, gene conversion may be evolutionary valuable because it can contribute to both, the generation of diversity and to the conservation of those features of the genes which are essential for their functioning. In the K locus numerous sequence homologies between adjacent and distant genes, gene flanks, minigenes and intergenic sequences have been found, some of which were discussed in a recent review (Zachau, 1988). In general, it cannot be decided between (1) how much of such an homology is a left-over from the original duplication or transposition event and (2) how much gene conversion has contributed to create or maintain the homology. It will be interesting to investigate the structure of the K locus of primates and relate some of its features to pre- and post-speciation events. In such a study one has to keep in mind, of course, that the possibility of gene conversion-like events in multigene families makes calculations of the exact evolutionary age of a mutation or duplication event a rather hazardous undertaking.

In summary, the detailed analysis of the human K locus is not only interesting with respect to the problems of antibody diversity but promises also insights into the evolutionary dynamics of the genome.

## ACKNOWLEDGEMENTS

The work of our laboratory was supported by Bundesministerium für Forschung und Technologie and Fonds der Chemischen Industrie.

## References

Bentley, D.L. and Rabbitts, T.H., (1981) Cell 24, 613-623.
"Human $V_K$ immunoglobulin gene number: implications for the origin of antibody diversity"

Burke, D.T., Carle, G.F., Olson, M.V. (1987) Science 236, 806-812.
"Cloning of large segments of exogenous DNA into yeast by means of artificial chromosome vectors"

Dover, G.A. and Strachan, T. (1987) in: G. Kelsoe and D.H. Schulze, eds., Univ. Texas Press, 15-33

"Molecular drive in the evolution of the immune superfamily of genes: the initiation and spread of novelty"

Jaenichen, H.R., Pech, M., Lindenmaier, W., Wildgruber, N., and Zachau, H.G. (1984) Nucl. Acids Res. 12, 5249-5263.
"Composite human $V_K$ genes and a model of their evolution"

Kabat, E.A., Wu, T.T., Reid-Miller, M., Perry, H., and Gottesman, K.S. (1987) NIH Publication, Bethesda, MD.
"Sequences of proteins of immunological interest"

Klobeck, H.G., Solomon, A., and Zachau, H.G. (1984) Nature 309, 73-76.
"Contribution of human $V_{KII}$ germ-line genes to light-chain diversity"

Klobeck, H.G., Bornkamm, G.W., Combriato, G., Mocikat R., Pohlenz, H.D., and Zachau, H.G. (1985) Nucl. Acids Res. 13, 6515-6529.
"Subgroup IV of human immunoglobulin K light chains is encoded by a single germline gene"

Klobeck, H.-G., Zimmer, F.-J., Combriato, G., and Zachau, H.G. (1987) Nucl. Acids Res. 15, 9655-9665.
"Linking of the human immunoglobulin $V_K$ and $J_KC_K$ regions by chromosomal walking"

Lorenz, W., Straubinger, B., and Zachau, H.G. (1987) Nucl. Acids Res. 15, 9667-9676.
"Physical map of the human immunoglobulin K locus and its implications for the mechanisms of $V_K$-$J_K$ rearrangement"

Pech, M., Jaenichen, H.R., Pohlenz, H.D., Neumaier, P.S., Klobeck, H.G., and Zachau, H.G. (1984) J. Mol. Biol. 176, 189-204.
"Organization and evolution of a gene cluster for human immunoglobulin variable regions of the kappa type"

Pech, M. and Zachau, H.G. (1984) Nucl. Acids Res. 12, 9229-9236.
"Immunoglobulin genes of different subgroups are inter-digitated within the $V_K$ locus"

Pech, M., Smola, H., Pohlenz, H.D., Straubinger, B., Gerl, R., and Zachau, H.G. (1985) J. Mol. Biol. 183, 291-299.
"A large section of the gene locus encoding human immunoglobulin variable regions of the kappa type is duplicated"

Pohlenz, H.-D., Straubinger, B., Thiebe, R., Pech, M., Zimmer, F-J., and Zachau, H.G. (1987) J. Mol. Biol. 193, 241-253.
"The human $V_K$ locus: characterization of extended immunoglobulin gene regions by cosmid cloning"

Straubinger, B., Huber, E., Lorenz, W., Osterholzer, E., Pargent, W., Pech, M., Pohlenz, H.D., Zimmer, F.-J., and Zachau, H.G. (1988) J. Mol. Biol. 199, 23-34.
"The human $V_K$ locus: characterization of a duplicated region encoding 28 different immunoglobulin genes"

Williams, A.F., and Barcley, N.A. (1988) Annual Reviews of Immunology 5, 381-383.
"The immunoglobulin superfamily - domains for cell surface recognition"

Zachau, H.G., Pech, M., Klobeck, H.G., Pohlenz, H.D., Straubinger, B., Falkner, F.G. (1984) Hoppe-Seyler's Z. Physiol. Chem. 365, 1363-1373.
"Wie entstehen die Antikörper?" 9. Lipmann-Vorlesung.

Zachau, H.G. (1988) in: The Immunoglobulin Genes, T. Honjo, F.W. Alt and T. Rabbitts eds. Academic Press, London, in press.
"Immunoglobulin light chain genes of the K type in man and mouse"

# MURINE ORNITHINE TRANSCARBAMYLASE: STRUCTURE AND EXPRESSION

Steven E. Scherer, Gabor Veres, William J. Craigen,
Stephen N. Jones and C. Thomas Caskey

Institute for Molecular Genetics, Howard Hughes
Medical Institute, Baylor College of Medicine
One Baylor Plaza, Houston, Texas 77030, USA

## INTRODUCTION

The hepatic enzyme ornithine transcarbamylase (OTC) (EC 2.1.3.3) catalyses the condensation of carbamyl phosphate and ornithine to form citrulline (Gisolia and Cohen, 1952; Jones et al., 1961). This is the second step of the mammalian urea cycle involved in the detoxification of ammonia and biosynthesis of urea. OTC is encoded on the X chromosome (Lindgren et al., 1984) and expressed primarily in liver and to a lesser extent in small bowel. The enzyme is synthesized in the cytoplasm as a 41 kD precursor and then transported to the mitochondria where a 32 amino acid leader peptide is cleaved and trimers associate within the matrix to form the active enzyme (Conboy et al., 1979; Mori et al., 1982). OTC deficiency is the most common urea cycle defect in humans and results in a severe and often fatal neonatal ammonia intoxication in hemizygous males (Walser, 1983). Female heterozygotes display a variable phenotype based on the severity of the mutation and random X-chromosome inactivation. Treatment of affected infants has been confined to restricted protein intake, peritoneal dialysis and administration of amino acid acylating agents such as benzoate, butyrate, or phenylacetate (Batshaw, 1982; Batshaw, 1981; Brusilow et al. 1984). Long-term outlook, particularly for affected males is poor, evidenced by the high incidence of mental retardation and cerebral palsy while hyperammoneimic episodes often lead to coma and death (Msall et al., 1984). Prenatal diagnosis has been largely confined to restriction fragment length polymorphism (RFLP) analysis (Fox et al., 1986; Nussbaum et al., 1986) as amniocytes do not express the enzyme, there are no metabolic products of the disorder present in amniotic fluid, and fetal liver biopsy poses a high risk to both mother and fetus.

It is the aim of this laboratory to develop tissue-specific gene transfer therapies for this disorder. Prerequisites to reach this objective include the isolation and characterization of the locus and analysis of the control elements necessary for its tissue-specific expression.

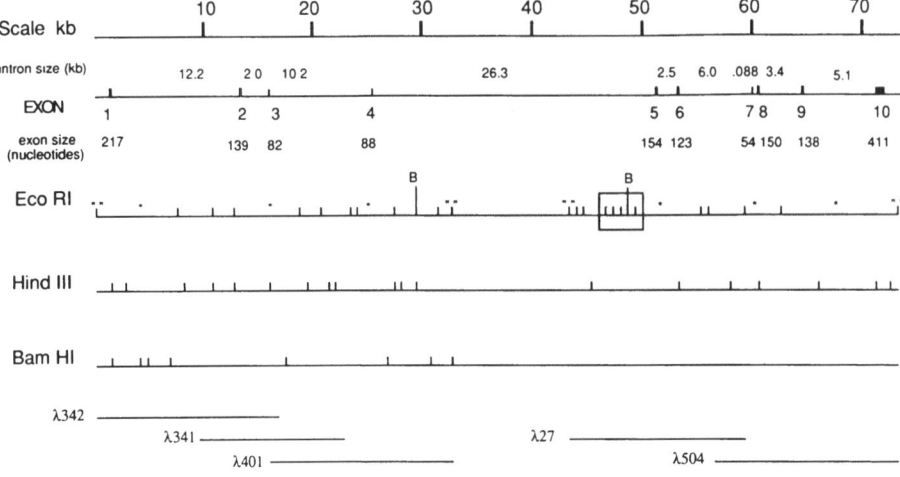

Figure 1. Summary of the OTC gene organization. The overall structure
of the gene is represented in the second line. Exons appear as vertical
bars and their size is noted below their number. Intron sizes are
denoted above the line. Composite Eco RI, Hind III and Bam HI restriction maps appear below the gene map and the lambda clones used to
generate the maps appear at the bottom of the figure. The two sites
labelled "B" in the Eco RI map represent Bgl I sites used to map the gap
distance between clones λ401 and λ27. The asterisks denote the Eco RI
fragments that were sub-cloned for localization of intron-exon
boundaries. The boxed Eco RI sites are impossible to accurately order
using the mapping techniques cited herein due to their number and size.
The Eco RI sites in quotation marks are artificial sites created during
construction of the lambda library.


GENE STRUCTURE

    As a first step in isolating genomic clones containing OTC
sequences, we cloned the mouse OTC cDNA by screening a λ gt 11 library
that was generated from C57BL/6 mouse liver polyadenylated RNA with a
partial rat OTC cDNA probe (kindly provided by R. Nussbaum, Univ. of
Pennsylvania). We then used the mouse OTC cDNA to screen a genomic
library constructed in bacteriaphage Charon 4A by partial HaeIII digestion followed by the addition of EcoRl linkers (courtesy of R. Seidman,
Harvard University). Twelve positive clones were isolated (Benton and
Davis, 1977), seven were unique and four were found to overlap without
redundancy. The rapid restriction enzyme mapping technique of Rackwitz,
et al. (1984) utilizing end-labelled oligonucleotides complementary to
the 3' vector COS-site was used to order the restriction fragments within
the phage inserts. Accurate fragment sizes were determined by complete
single restriction endonuclease digestion and ambiguities resolved by
double digestion. Overlapping genomic sequences between clones were
confirmed by hybridization with selected genomic fragment probes
(Southern, 1975; Maniatis et al., 1982). A complete map of the gene is
presented in Figure 1 (Scherer et al., 1988). The OTC gene is composed
of 10 exons and distributed over approximately 70 kilobases (kb). All
intron-exon boundaries have been sequenced and conform closely to
consensus sequences established for other eukaryotic genes (Breathrock
and Chambon, 1981). The structural organization of both the rat and
human OTC genes has also now been established and conform closely to the
mouse gene in size (75 kb and 73 kb respectively) (Takiguchi et al.,
1987; Hata et al., 1988) while the exon break points are identical.


122

In 1976, De Mars et al. described a radiation induced mouse mutant characterized in hemizygous males and homozygous females by stunted growth and a paucity of fur for the first two weeks of life. Furthermore, they found highly elevated levels of orotic acid in the urine, an 80% reduction in endogenous OTC activity and an elevated pH optima for the mutant OTC enzyme. This sparse fur *(spf)* mouse displays the same OTC deficiency phenotype as many human patients and would serve as an ideal recipient for gene transfer therapy. Previous biochemical and enzymological studies (Briand et al., 1982; Qureshi et al., 1979) had suggested that the *spf* phenotype could be due to a point mutation which was consistent with our data showing no difference in mRNA size between *spf* and wild type mice on Northern blots. After sequencing the wild-type OTC cDNA (see Figure 2) we were interested in the molecular basis of this mutation and set out to find it by combining two relatively new techniques: RNase A cleavage (Myers et al., 1985) and polymerase chain

```
            met leu ser asn leu arg ile leu leu asn asn ala ala leu arg lys gly his thr ser    20
mouse 5'    ATG CTG TCT AAT TTG AGG ATC CTC CTC AAC AAT GCA GCT CTT AGA AAG GGT CAC ACT TCT    64
       GAAG ___

val val arg his phe trp cys gly lys pro val gln ser gln val gln leu lys gly arg asp leu leu thr leu    45
GTG GTT CGA CAT TTT TGG TGT GGG AAG CCA GTC CAA AGT CAA GTA CAG CTG AAA GGC CGT GAC CTC CTC ACC TTG    139

lys asn phe thr gly glu glu ile gln tyr met leu trp leu ser ala asp leu lys phe arg ile lys gln lys    70
AAG AAC TTC ACA GGA GAG GAG ATT CAG TAC ATG CTA TGG CTC TCT GCA GAT CTG AAA TTC AGG ATC AAG CAG AAA    214

gly glu tyr leu pro leu leu gln gly lys ser leu gly met ile phe glu lys arg ser thr arg thr arg leu    95
GGA GAA TAT TTA CCT TTA TTG CAA GGG AAA TCC TTA GGA ATG ATT TTT GAG AAA AGA AGT ACT CGA ACA AGA CTG    289

ser thr glu thr gly phe ala leu leu gly gly his pro ser phe leu thr thr gln asp ile[his]leu gly val    120
TCC ACA GAA ACA GGC TTT GCT CTG CTG GGA GGA CAC CCT TCC TTT CTT ACC ACA CAA GAC ATT[CAC]TTG GGT GTG    364

asn glu ser leu thr asp thr ala arg val leu ser ser met thr asp ala val leu ala arg val tyr lys gln    145
AAT GAA AGT CTC ACA GAC ACC GCT CGT GTC TTA TCT AGC ATG ACA GAT GCA GTG TTA GCT CGA GTG TAT AAA CAA    439

ser asp leu asp thr leu ala lys glu ala ser ile pro ile val asn gly leu ser asp leu tyr his pro ile    170
TCA GAT CTG GAC ACC CTG GCT AAA GAA GCA TCC ATC CCA ATT GTC AAT GGA CTG TCA GAC TTG TAT CAT CCT ATC    514

gln ile leu ala asp tyr leu thr leu gln glu his tyr gly ser leu lys gly leu thr leu ser trp ile gly    195
CAG ATC CTG GCT GAT TAC CTT ACA CTC CAG GAA CAC TAT GGC TCT CTC AAA GGT CTT ACC CTC AGC TGG ATA GGG    589

asp gly asn asn ile leu his ser ile met met ser ala ala lys phe gly met his leu gln ala ala thr pro    220
GAT GGG AAC AAT ATC TTG CAC TCT ATC ATG ATG AGT GCT GCA AAA TTC GGG ATG CAC CTT CAA GCA GCT ACT CCA    664

lys gly tyr glu pro asp pro asn ile val lys leu ala glu gln tyr ala lys glu asn gly thr lys leu ser    245
AAG GGT TAT GAG CCA GAT CCT AAT ATA GTC AAG CTA GCA GAG CAG TAT GCC AAG GAG AAT GGT ACC AAG TTG TCA    739

met thr asn asp pro leu glu ala ala arg gly gly asn val leu ile thr asp thr trp ile ser met gly gln    270
ATG ACA AAT GAT CCA CTG GAA GCA GCA CGT GGA GGC AAT GTA TTA ATT ACA GAT ACT TGG ATA AGC ATG GGA CAA    814

glu asp glu lys lys lys arg leu gln ala phe gln gly tyr gln val thr met lys thr ala lys val ala ala    295
GAG GAT GAG AAG AAA AAG CGT CTT CAA GCT TTC CAA GGT TAC CAG GTT ACG ATG AAG ACT GCC AAA GTG GCT GCG    889

ser asp trp thr phe leu his cys leu pro arg lys pro glu glu val asp asp glu val phe tyr ser pro arg    320
TCT GAC TGG ACA TTT TTA CAC TGT TTG CCT AGA AAG CCA GAA GAA GTG GAT GAT GAA GTA TTT TAT TCT CCA CGG    964

ser leu val phe pro glu ala glu asn arg lys trp thr ile met ala val met val ser leu leu thr asp tyr    345
TCA TTA GTG TTC CCA GCA GCA GAG AAT AGA AAG TGG ACA ATC ATG GCT GTC ATG GTA TCC CTG CTG ACA GAC TAC    1039

ser pro val leu gln lys pro lys phe ***    354
TCA CCT GTG CTC CAG AAG CCA AAG TTT TGA TGCCTGTCAAAAGGAAAAAAACAGAAACAAAACAATAACAATAACAACAACAACAACA    1127

AAAACCCCTCTGTTCTTTAGCAATAGAATAAGTCAGTTTATGTGGGAAAGAGAAGAATTTAAAATTGTAAACACATCCCTAGTGCATGGTATGATTATG    1224

TAATTGCTTTGCTATTATGAGAATTTCTTAAAGCTTTTAGTTTAAGTGCCTGGCATTTTATTATCCTGCTTGACTTGGTTTAAACACTCTTCTTCAATTT    1318

ACAACCTCTGAATGACATTTGGGTATCATATTAATTATCATACACATTTCCTTCCACTAAACATTAAACACTTTGCTTACAATGTCTAAGTCATAAAAA₂₀    1437
```

Figure 2. Nucleotide and amino acid sequence of the mouse OTC cDNA. The predicted amino acid sequence of the mouse OTC is shown above the nucleotide sequence. The translation initiation codon is double underlined; the boxed amino acid and triplet code indicates the *spf* mutation (CAC - AAC) site. Underlined regions show the oligomer sequences used in the PCR reaction. The first primer at the 5' end is complemetary to the (-) strand and the second one is complementary to the (+) strand of the mouse OTC cDNA.

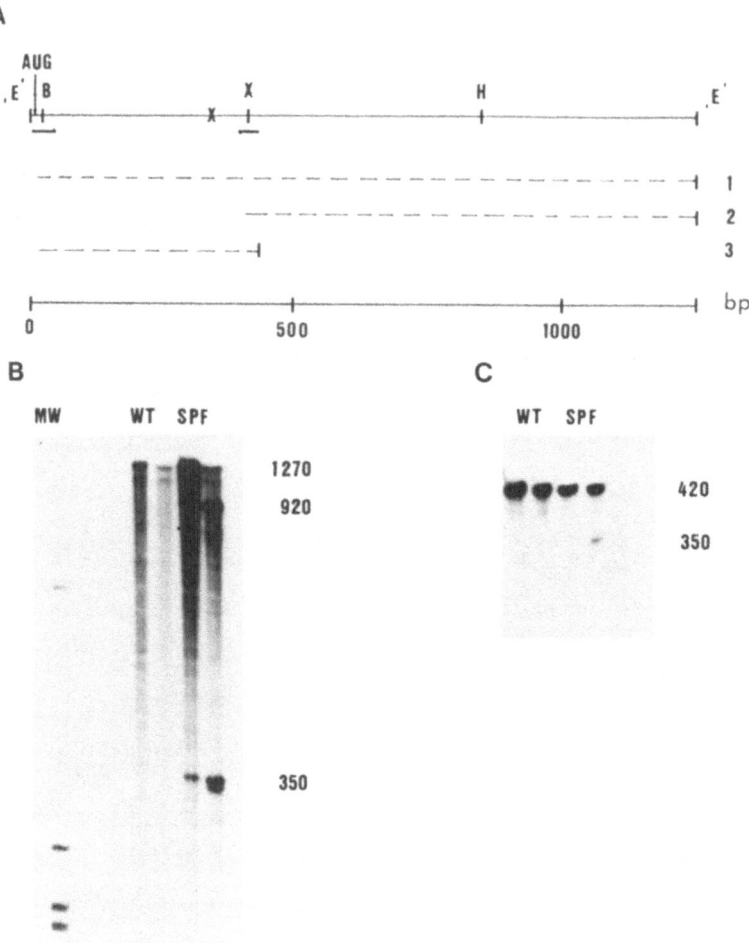

Figure 3.  RNase A cleavage analysis of wild-type and sparse fur mouse
OTC RNA. (A) Diagram of the RNA probes used in the cleavage analysis.
(B) Autoradiograph of the RNase A digestion products obtained with probe
1.   Probe 2 provided full protection for both wild-type and *spf* mRNA.
(C) RNase A cleavage pattern obtained with the 420 bp 5' probe 3'.   The
restriction sites are ,E', Eco RI linkers; B, Bam HI; X, Xho I; H, Hind
III. -X- is the site of the *spf* mutation.   The arrows show the binding
sites for the primers used in the PCR reaction.

reaction amplification (Saiki et al., 1985).   RNase A cleavage was used
to localize the mutation by radiolabelling an *in vitro* synthesized 1270
base pair (bp) wild-type antisense RNA probe and this probe was then
annealed to total RNA isolated from both wild-type and *spf* mice.   The
products of the annealing reaction were then treated with RNase A to
cleave single stranded RNA and any internal mismatches between the wild-
type probe and the *spf* OTC mRNA.   As can be seen in Figure 3, the wild-

type RNA is fully protected while partial cleavage of an internal mismatch produces two additional bands in the *spf* lanes. Truncated RNA probes were then used to locate the internal cleavage site relative to the ends of the OTC mRNA. To facilitate molecular cloning and sequencing of the region around the RNase A cleavage site, a first strand cDNA was generated from poly A$^+$ *spf* RNA with random hexanucleotide primers and reverse transcriptase. Two synthetic 25-base oligonucleotides containing unique restriction endonuclease recognition sites were then used as specific primers to amplify a 420 bp segment of the cDNA. After ten rounds of amplification, the products were cleaved with appropriate restriction endonucleases and ligated into an M13 sequencing vector. Two independent clones were sequenced and found to be identical to the wild-type sequence with the exception of a C to A transversion present 348 bp downstream of the translation initiation codon. This is within two nucleotides of the position estimated by RNase A cleavage and results in the replacement of a histidine residue with an asparagine residue at amino acid position 117. To ensure that the mutation we identified was responsible for the *spf* phenotype, we examined the OTC activities of extracts prepared from COS cells which had been transiently transfected with the expression vector p91023 (Wong, et al., 1985) containing both wild type OTC cDNA and an OTC cDNA bearing the aforementioned transversion. The results mirrored those obtained previously by others (DeMars et al., 1976) showing a shift in pH optima between OTC enzyme isolated from wild-type and *spf* mouse livers (Veres et al., 1987).


THE OTC PROMOTER

Transcription Initiation and Tissue Specificity

In our initial effort to characterize the control elements responsible for the tissue-specific expression of OTC, we subcloned a 757 bp fragment from the 5' most end of our genomic phage clones which contains 735 bp of 5' untranslated sequence to use in identifying the transcription start site and for promoter activity experiments. Initial experiments using S1 nuclease protection (Berk and Sharp, 1974) and primer extension (Agarwal et al., 1981) established two minor start sites 108 bp and 177 bp upstream of the translation start site and a major site 136 bp upstream. Preliminary data using RNase A cleavage analysis points to the 108 bp upstream site as being the major transcription initiation point. We have yet to reconcile these data although others have had trouble finding this site in both the rat and human OTC genes (Takiguchi et al. 1987; Hata et al. 1986). Sequencing of the promoter failed to reveal any canonical transcriptional control sequences except for a possible TATA box located approximately 280 bp upstream of the translation start site.

To test whether these same upstream sequences are sufficient to direct tissue-specific transcription, we constructed a plasmid containing the 757 bp promoter fragment in both orientations adjacent to the bacterial-specific reporter gene, chloramphenicol acetyl transferase (CAT). These constructs together with the positive controls pSV2-CAT and pRSV-CAT and the negative control pSVO-CAT were transfected into both a human hepatoma cell line, HepG2 and NIH 3T3 mouse fibroblasts. In NIH 3T3 cells, CAT expression (Gorman et al., 1982) was limited to the positive controls while in the Hep G2 cells the CAT activity was also observed in cells transfected with the OTC promoter-CAT construct in the forward orientation. No CAT activity was observed in either cell line transfected with the negative control (see Figure 4). Quantitation of the CAT activities showed only a 5% CAT activity for the OTC promoter relative to pSV2 and pRSV-CAT. This low CAT activity may correspond to the decreased level of OTC activity found in HEP G2 cells (5% relative to

levels found in crude liver extract). This is supported by the finding
that transfection of a second hepatoma line Hepa 1-6 with the OTC
promoter-CAT construct yields no CAT activity and correspondingly lacks
any cellular OTC activity. These data demonstrate that the 757 bp
promoter fragment is sufficient to direct tissue-specific transcription
(Veres et al., 1985).

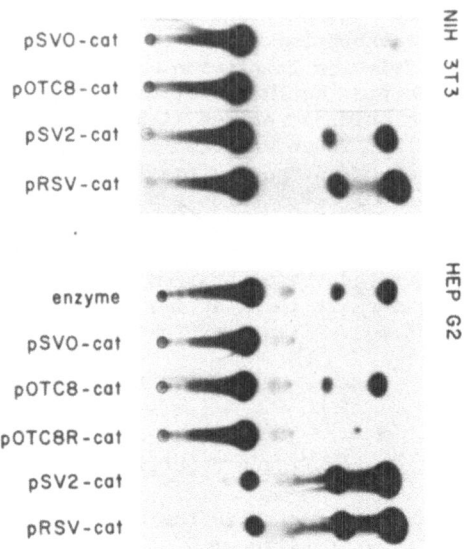

Figure 4. Functional analysis of the mouse OTC promoter region in NIH
3T3 mouse fibroblast and Hep G2 human hepatoma cells. Cells were
transfected with the indicated plasmids, and chloramphenicol acetyltransferase activity was measured 48 h later. Enzyme control lane represents
0.05 unit of purified chloramphenicol acetyltransferase enzyme activity
as a comparison.

Transgenic Mice

Studies of other liver specific genes such as albumin (Pinkert et
al., 1987) and α-fetoprotein (Godbout et al., 1988) have indicated that
hepatic specific control sequences may be found up to 12 kb upstream of
the initiation codon. For this reason, we cloned a further 3.2 kb of
upstream sequence before proceeding with tissue-specificity studies in
vivo. The new 4kb OTC promoter was linked to the CAT gene and microinjected into C57 BL/6 mouse embryos which in turn were implanted into
pseudopregnant female mice (Hogan et al., 1986). Three of twenty
offspring picked up the transgene and two were found to express CAT. One
line expresses CAT in liver and small intestine while the other shows
expression in small intestine only, possibly due to the site of integration. In no mice was the CAT gene expressed in inappropriate tissues
(Veres, unpublished results).

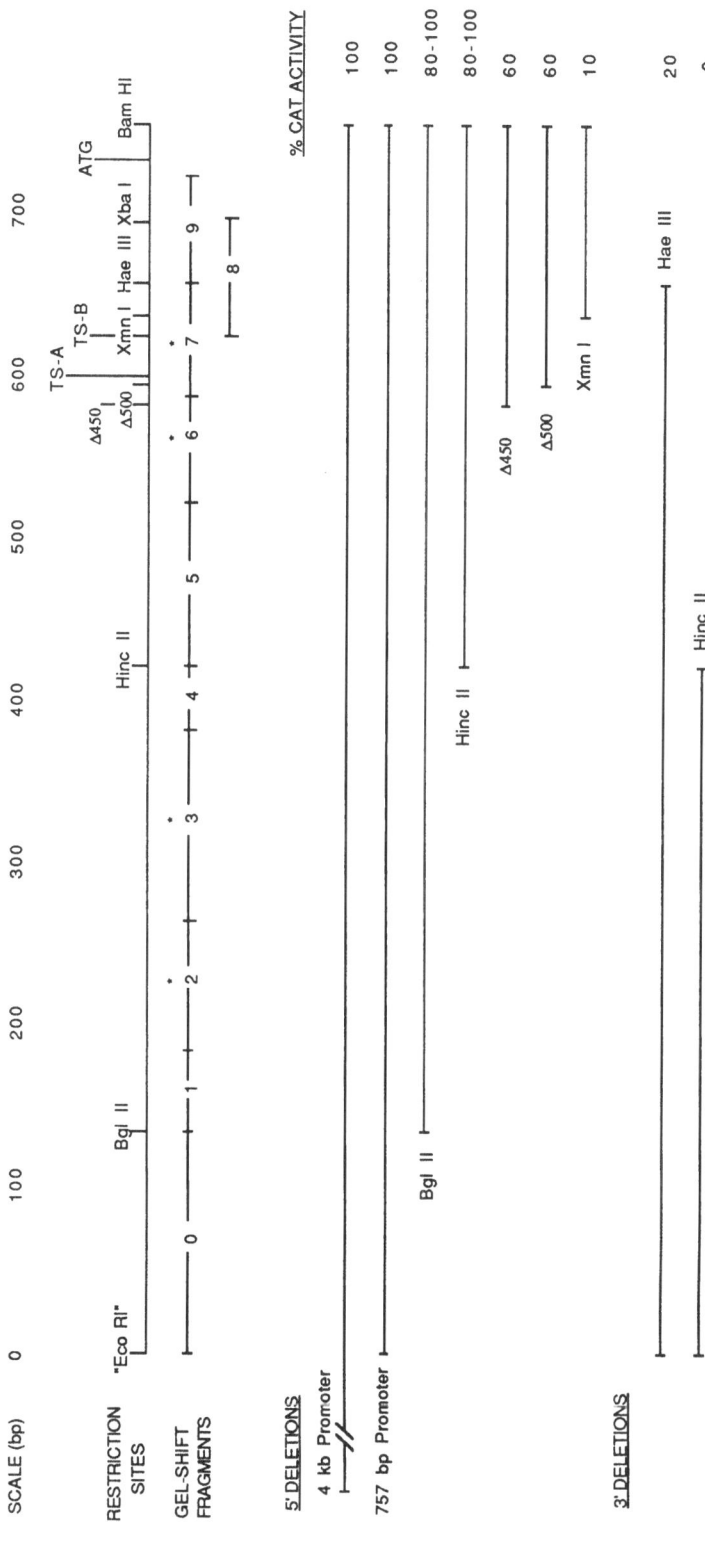

Figure 5. OTC promoter analysis and gel retardation fragment results. Restriction endonuclease sites and Bal 31 nuclease digestions appear above the second line. TS-A is the transcription start site according to S1 nuclease and primer extension analysis and TS-B is the transcription start site according to RNase A cleavage analysis. ATG is the translation initiation codon. The numbered fragments used in gel retardation analysis are depicted in the third line. Those with a * above them are shifted in the presence of liver nuclear extract.

127

Recently, we have microinjected a construct made up of the 757 bp mouse OTC promoter linked to the human OTC cDNA which in turn is linked to SV40 polyadenylation signals into embryos from matings between homozygous female and hemizygous male *spf* mice. We have strong phenotypic evidence that two of nineteen transgenic offspring are expressing the OTC transgene thus reversing their mutation. These are preliminary observations and require further genetic and enzymological data before conclusions can be drawn (Jones - unpublished results).

## Promoter deletion analysis

A series of deletion mutants has been generated in the 757 bp OTC promoter-CAT gene construct using either Bal 31 nuclease digestion or convenient restriction endonuclease sites. The constructs were transfected into the hepatoma cell line PLC-5F and after forty-eight hours, these cells were quantitatively assayed for CAT activity relative to the full length 757 bp and 4kb promoters (See Figure 5)together with positive and negative controls. Preliminary data suggests that sequences located downstream of the Hinc II site at nucleotide position -312 with respect to the initiation codon are absolutely required for OTC promoter activity and that sequences between the Hinc II and the Bgl II site at -598 may also be important. We found no difference in CAT expression levels between the 757 bp and 4 kb promoters (Veres et al., manuscript in preparation). A further set of deletion mutants have now been generated and await transfection and assay.

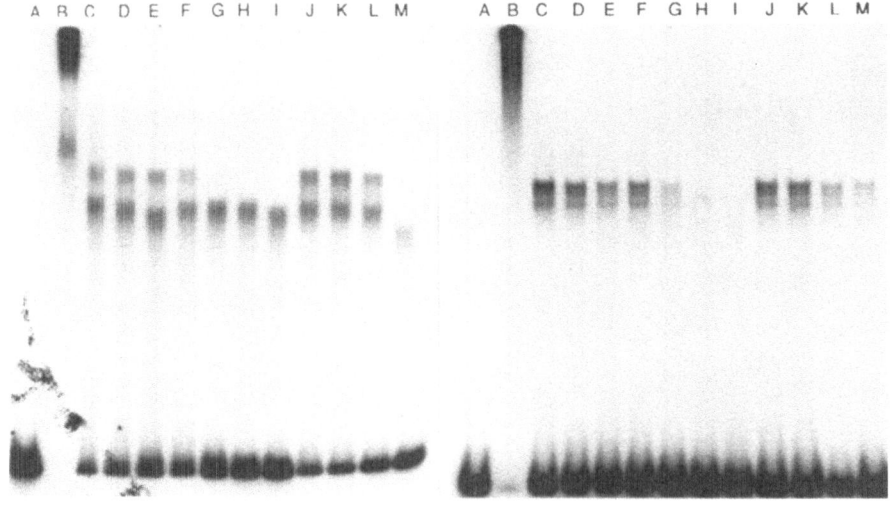

Figure 6. Gel retardation analysis of OTC promoter fragments 6 and 7 (see Figure 6). Fragment 6 appears on the left and fragment 7 on the right. In each case, 1 ng of probe fragment was incubated under the following conditions: Lane A, probe fragment alone, Lane B, no poly dI*, dC, Lane C, 2 μg poly dI*dC, Lane D, 4 μg poly dI*dC, Lane E 6 μg poly dI*dC. Lane F, specific comeptitor (SC) present at 1:1 molar ratio with respect to the probe, Lane G, SC at 10:1 molar ratio, Lane H, SC at 100:1 molar ratio, and Lane I, SC at 1000:1 molar ratio. Lane J, non-specific competitor (NSC) (Hae III digested pUC 9) at 1:1 molar ratio with respect to the probe, Lane K, NSC at 10:1 molar ratio, Lane L, NSC at 100:1 molar ratio and Lane M, NSC at 1000:1 molar ratio. The incubation buffer is according to Singh et al. (1986). The diminution of signal seen in Lane M may be due to oversaturation of the incubation mixture with DNA.

## Gel Retardation Analysis

Tissue-specific gene expression may be due to the binding of positive
trans-acting nuclear protein factors present in the appropriate cell
nucleus or the binding of negative factors present in inappropriate
tissues or may involve some combination of both. Furthermore, possible
effects of chromatin structure must be considered as nucleosome placement, methylation and bending may influence the accessibility of these
protein factors to their DNA target sequences. In a further effort to
narrow down cis-acting regulatory elements necessary for the tissue-
specific expression of OTC, the 757 bp OTC promoter was divided into ten
fragments ranging in size from 41 to 137 bp. Each fragment was subjected
to gel retardation analysis (Singh et al., 1986) using protein extract
derived from mouse liver nuclei. Results depicted in Figure 5 suggest
that two separate regions of the promoter bind nuclear factors present in
liver cells and thus may be important in OTC transcription. (An example
of positive gel retardation fragments is presented in Figure 6). One
segment begins at the putative transcription start site and extends 140
bp upstream. This region contains sequences showing a high degree of
homology to control sequences present in other liver-specific genes
including mouse transthyretin and $\alpha$-1-antitrypsin, rat albumin (Costa et
al., 1988) and human $\alpha$-1-antitrypsin (Monaci et al., 1988). The second
begins at nucleotide position -243 with respect to the cap site and
extends a further 204 bp upstream. We are currently running gel retardation assays on these same fragments using nuclear extracts derived from
tissues that do not express OTC such as kidney and brain. We are also
proceeding with DNase I footprinting analysis to delineate these cis-
acting sequences at the nucleotide level. Once this is accomplished, we
hope to isolate the trans-acting factors themselves, as well as the genes
that encode them as we work our way back into the regulatory pathway
responsible for the tissue-specific expression of OTC.

## REFERENCES

Agarwal, K.L., Brunstedt, J., and Noyes, B.E., 1981, A general method for
detection and characterization of a mRNA using an oligonucleotide
probe, *J. Biol. Chem.*, 256:1023.

Batshaw, M.L., Brusilow, S., and Waber, L., 1982, Treatment of inborn
errors of urea synthesis, *N. Eng. J. Med.*, 30:1387.

Batshaw, M.L., Painter, M.J., Sproul, G.T., Schafer, I.A., Thomas, G.H.,
and Brusilow, S., 1981, Therapy of urea cycle enzymopathies: three
case studies, *Johns Hopkins Med. J.*, 148:34.

Benton, W.D. and Davis, R., 1977, Screening λ gt recombinant clones by
hybridization to single plaques *in situ*, *Science*, 196:180.

Berk, A.J., and Sharp, P.A., 1974, Sizing and mapping of early adenovirus
mRNA's by gel electrophoresis of S1 endonuclease-digested hybrids,
*Cell*, 12:721.

Breathnach, R., and Chambon, P., 1981, Organization and expression of
eucaryotic split genes coding for proteins, *Ann. Rev. Biochem.*
50:349.

Briand, P., Francois, B., Rabier, D., and Cathlineau, L., 1982, Ornithine
transcarbamylase deficiencies in human males: kinetic and immuno-
chemical classification, *Biochem. Biophys. Acta.*, 704:100.

Brusilow, S.W., Danney, M., and Waber, L.J., 1984, Treatment of episodic
hyperammoneimia in children with inborn errors of urea synthesis, *N.
Eng. J. Med.*, 310:630.

Conboy, J.G., Kalousek, F., and Rosenberg, L.E., 1979, *In vitro* synthesis
of a putative precursor of mitochondrial ornithine transcarbamylase,
*Proc. Natl. Acad. Sci. USA*, 76:5724.

Costa, R.H., Grayson, D.R., Zanthopoulos, K.G., and Darnell, S.E., 1988, A liver-specific DNA binding protein recognizes multiple nucleotide sites in regulatory regions of transthyretin, α 1-antitrypsin, albumin, and simian virus 40 genes, Proc. Natl. Acad. Sci., USA, 85:3840.

DeMars, R., LeVan, S.L., Trend, B.L., and Russel, L.B., 1976, Abnormal ornithine carbamoyltransferase in mice having the sparse-fur mutation, Proc. Natl. Acad. Sci. USA, 73:1693.

Fox, J.E., Hack, A.M., Fenton, W.A., and Rosenberg, L.E., 1986, Identification and application of additional restriction fragment length polymorphisms at the human ornithine transcarbamylase locus, Am. J. Hum. Genet., 38:841.

Godbout, R., Ingram, R.S., and Tilghman, S.M., 1988, Fine structure mapping of the three mouse α-fetoprotein gene enhancers, Mol. Cell. Biol., 8:1169.

Gorman, C.M., Merlino, G.T., Willingham, M.D., Pastan, I., and Howard, B.H., 1982, The Rous sarcoma virus long terminal repeat is a strong promoter when introduced into a variety of eukaryotic cells by DNA-mediated transfection, Proc. Natl. Acad. Sci. USA., 79:6777.

Grisolia, S., and Cohen, P.P., 1952, The catalytic role of carbamylglutamate in citrulline biosynthesis, J. Biol. Chem. 198:561.

Hata, A., Tsuzuki, T., Shimada, K., Takiguchi, M., Mori, M., and Matsuda, I., 1988, Structure of the human ornithine transcarbamylase gene, J. Biochem. 103:302.

Hata, A., Tsuzuki, T., Shimada, K., Takiguchi, M., Mori, M., and Matsuda, I. 1986, Isolation and characterization of the human ornithine transcarbamylase gene: structure of the 5' end-region., J. Biochem. 100:717.

Hogan, B.L.M., Constantini, F., and Lacy, L., 1986, "Manipulating the Mouse Embryo: A Laboratory Manual," Cold Spring Harbor Laboratory, Cold Spring Harbor.

Jones, N.E., Anderson, A.D., Anderson, C., and Hodes, S., 1961, Citrulline synthesis in rat tissues, Arch. Biochem. Biophys., 95:499.

Lindgren, V., Martinville, B., Horwich, A.L., Rosenberg, L.E., and Francke, U., 1984, Human ornithine transcarbamylase locus mapped to band Xp 21.1 near the Duchenne muscular dystrophy locus, Science, 226:698.

Maniatis, T., Fritsch, E.F., and Sambrook, J., 1982, "Molecular Cloning: A Laboratory Manual," Cold Spring Harbor Laboratory, Cold Spring Harbor.

Monaci, P., Nicosia, A., and Cortese, R., 1988, Two different liver specific factors stimulate in vitro transcription from the human α 1 - antitrypsin promoter, EMBO J. 7:2075.

Mori, M., Miura, S., Morita, T., Takiguchi, M., and Tatibana, M., 1982, Ornithine transcarbamylase in liver mitochondria, Mol. Cell Bioch. 49:97.

Msall, M., Batshaw, M.L., Suss, R., Brusilow, S.W., Mellitos, E.D., 1984, Neurologic outcome in children with inborn errors of urea synthesis: outcome of urea cycle enzymopathies., N. Eng. J. Med., 310:1500.

Myers, R.M., Larin, Z., and Maniatis, T., 1985, Detection of single base substitutions by ribonuclease cleavage at mismatches in RNA:DNA duplexes, Science, 230:1242.

Nussbaum, R.L., Boggs, B.A., Beaudet, A.L., Doyle, S., Potter, J.L., and O'Brien, W.E., 1986, New mutation and prenatal diagnosis in ornithine transcarbamylase deficiency, Am. J. Human. Genet. 38:149.

Pinkert, C.A., Ornitz, D.M., Brinster, R.L. and Palmiter, R.D., 1987, An albumin enhancer located 10 kb upstream functions along with its promoter to direct efficient, liver specific expression in transgenic mice, Genes Dev., 1:268.

Quereshi, I.A., Letarte, S., and Ouellet, R., 1979, Ornithine transcarbamylase deficiency in mutant mice 1. Studies on the characterization of enzyme defect and suitability as animal model of human disease., <u>Pediat. Res</u>., 13:807.

Rackwitz, H.R., Zehetner, G., Frischauf, A.M., and Lehrach, H., 1984, Rapid restriction mapping of DNA cloned in lambda phage vectors, <u>Gene</u>, 30:195.

Ryall, J., Nguyen, M., Bendayan, M., and Shore, G.C., 1985, Expression of nuclear genes encoding the urea cycle enzymes, carbamoyl-phosphate synthetase I and ornithine carbamoyl transferase, in rat liver and intestinal mucosa., <u>Eur. J. Biochem</u>., 152:287.

Saiki, R.K., Bugawan, T.L., Horn, G.T., Mullis, K.B., and Erlich, 1986; Analysis of enzymatically amplified $\beta$-globin and HLA-DQ$\alpha$ DNA with allele-specific oligonucleotide probes, <u>Nature</u>, 324:163.

Scherer, S.E., Veres, G. and Caskey, C.T., 1988, The genetic structure of mouse ornithine transcarbamylase, <u>Nucl. Acids. Res</u>., 16:1593.

Singh, H., Sen, R., Baltimore, D., and Sharp, P.A., 1986, A nuclear factor that binds to a conserved sequence motif in transcriptional control elements of immunoglobulin genes, <u>Nature</u>, 319:154.

Southern, E.M., 1975, Detection of specific sequences among DNA fragments separated by gel electrophoresis, <u>J. Mol. Biol</u>. 98:503.

Takiguchi, M., Murakami, T., Miura, S., and Mori, M., 1987, Structure of the rat ornithine transcarbamylase gene, a large, X chromosome-linked gene with an atypical promoter, <u>Proc. Natl. Acad. Sci. USA</u>, 84:6136.

Veres, G., Gibbs, R.A., Scherer, S.E., and Caskey, C.T., 1987, The molecular basis of the sparse fur mutation, <u>Science</u>, 237:415.

Veres, G., Craigen, W.J., and Caskey, C.T., 1985. The 5' flanking region of the ornithine transcarbamylase gene contains DNA sequences regulating tissue-specific expression, <u>J. Biol. Chem</u>., 261:7588.

Walser, M., 1983, Urea cycle disorders and other hereditary hyperammonemic syndromes, <u>in</u>: "The Metabolic Basis of Inherited Disease," J.B. Stanbury, J.B. Wyngaarden, D.S. Fredrickson, J.L. Goldstein, M.S. Brown, eds., McGraw-Hill, New York.

Wong, G.G., Witek, J.S., Temple, P.A., Wilkens, K.M., Leary, A.C., Luxenberg, D.P., Jones, S.S., Brown, E.L., Kay, R.M., Orr E.C., Shoemaker, C., Golde, D.W., Kaufman, R.J., Hewick, R.M. Wang, E.A., and Clark, S.C., 1985, Human GM-CSF: molecular cloning of the complementary DNA and purification of the natural and recombinant proteins, <u>Science</u>, 228:810.

COMPOSITIONAL PATTERNS IN VERTEBRATE GENOMES:

CONSERVATION AND CHANGE IN EVOLUTION

Giorgio  Bernardi (1), Dominique Mouchiroud (2),
Christian Gautier (2) and Giacomo Bernardi (1)

(1) Laboratoire de Génétique Moléculaire
Institut Jacques Monod, 2 Place Jussieu
75005 Paris (France)

(2) Laboratoire de Biométrie, U.A. 243
Université Claude Bernard Lyon I
69622 Lyon (France)

## INTRODUCTION

Three approaches have recently provided new insights into the organization and evolution of the nuclear genomes of vertebrates. They are all based on the compositional properties of genome segments ranging in size from about 1 Kb, for coding sequences, to 100 Kb, for DNA fragments. The rationale for these approaches is the fact that vertebrate genomes are mosaics of isochores, which are evolutionarily relevant structures (see the following paragraph).

In the first approach, large DNA fragments (in the 30-100 Kb size range) were fractionated according to their GC levels.This allowed the study of their compositional distribution (by plotting the relative amounts of DNA fractions against their GC levels), and the localization of specific sequences to these fractions. Such investigations (see Bernardi et al., 1985, for a review) revealed that : (1) vertebrate genomes are made up of very long DNA segments (estimated to be larger than 200-300 Kb), the isochores, that are compositionally fairly homogeneous (at least above sizes of 3 Kb), and belong to a small number of classes characterized by different GC levels; (2) GC-rich isochores represent about one third of the genome of warm-blooded vertebrates, whereas they are absent from, or poorly represented in, most cold-blooded vertebrates; lesser, but highly significant differences in the compositional distribution of isochores were found between mammals and birds, and also, to a smaller extent yet, between most mammals and murids; (3) in warm-blooded vertebrates, GC-rich isochores are located in Giemsa light bands (or in Reverse bands) of metaphase chromosomes and replicate early in the cell cycle, whereas GC-poor isochores are located in Giemsa dark bands and replicate late in the cell cycle;  in  cold-blooded vertebrates, metaphase chromosomes show poor Giemsa banding or no banding at all (see Medrano et al., 1988); (4) GC levels of coding sequences and their different codon positions are linearly correlated with those of the corresponding introns and of the intergenic non-coding sequences in which they are embedded; (5) the CpG doublets (the potential methylation sites) are very low in GC-poor, but almost "statistically abundant" (i.e., not under-represented compared to other doublets) in GC-rich coding sequences (as is also the case for the genomes of mammalian and avian viruses; Bernardi, 1985; Bernardi and Bernardi, 1986); (6) genes are mainly concentrated in the GC-richest isochores of the genomes of warm-blooded vertebrates.

In the second approach, data from sequence banks were used (Bernardi et al., 1985; Bernardi and Bernardi, 1985, 1986; Mouchiroud et al., 1987, 1988; Perrin and Bernardi, 1987) to analyze the GC levels of individual coding sequences and the average GC levels of first, second and third codon positions within such sequences (see Fig. 1 for an example of compositional distribution of third codon positions). These investigations have shown that (1) in cold-blooded vertebrates, the compositional distribution curves of coding sequences and of DNA fragments are roughly symmetrical; (2) in warm-blooded vertebrates, GC-rich coding sequences represent the majority of coding sequences, whereas GC-rich DNA fragments correspond to a minor part of the genome; (3) the compositional distributions of coding sequences are different in mammals and birds (chicken), and also, but less so, in most mammals on the one hand, and murids on the other.

In a third approach, we have compared the GC levels of pairs of homologous coding sequences from vertebrates and of their different codon positions (Perrin and Bernardi, 1987; Mouchiroud and Gautier, 1988; Mouchiroud et al., 1988). These comparisons have provided information on the evolutionary conservation of regional compositional patterns (as we shall call here the compositional distributions of DNA fragments, of coding sequences and of their different codon positions), and shown how such patterns can shift in evolution.

Here, we shall investigate in more detail the conservation and the shifts of compositional patterns of vertebrate genomes using more recent sets of data than those previously analyzed. We shall then discuss the mechanisms and the causes of these phenomena. A more detailed presentation of this work will appear elsewhere (Bernardi et al., 1988).

## Two types of compositional patterns in mammals

We have recently identified two types of compositional distributions in the available mammalian coding sequences. These types will be called here the "general" distribution and the "murid" distribution. Indeed, when coding sequences are examined (as a whole, or in their different codon positions), those of several mammals (mainly man, artiodactyls and rabbit) show similar broader distributions, and higher average GC values compared to those of murids (Mouchiroud et al., 1987, 1988; data for third codon positions are shown in Fig. 1).

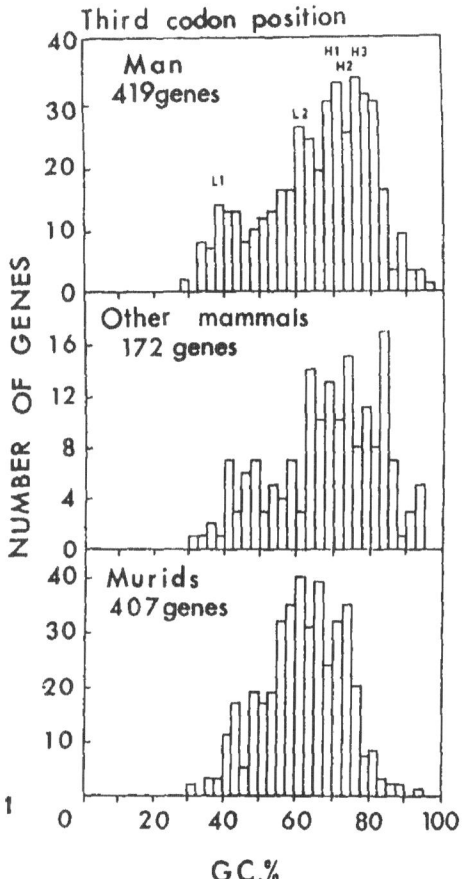

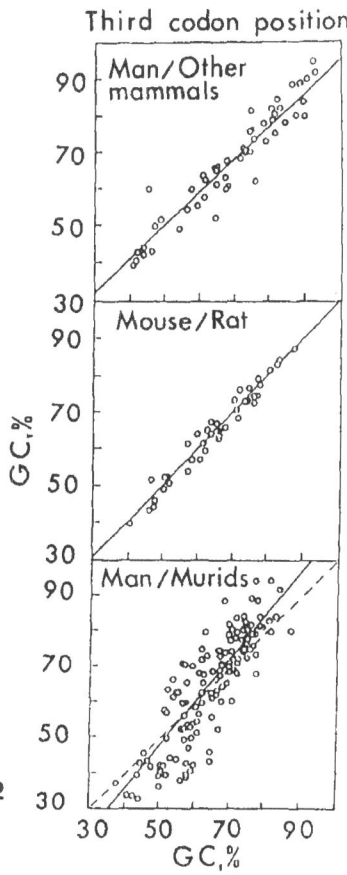

Fig. 1. (left) Compositional distribution of third codon positions for genes (i) from man; (ii) from mammals other than man, murids and hamster; and (iii) from murids (rat and mouse). A 2.5% GC window was used. Tentative identifications of different compositional classes of coding sequences corresponding to the compartments of the human genome (L1,L2, H1, H2 and H3) are indicated following Mouchiroud et al.(1987, 1988).

Fig. 2. (right) Relationships between GC contents of third codon position for pairs of homologous genes (i) from man (ordinate) and other mammals (except murids and hamster; abscissa); (ii) from mouse (ordinate) and rat (abscissa); and (iii) from man (ordinate) and murids (abscissa).

These similarities and differences are confirmed by comparisons of GC levels of homologous coding sequences, and of their different codon positions. Indeed, these GC levels show linear relationships, passing through the origin and showing unity slopes, in the case of both the "general" and the "murid" distribution (see Fig. 2 for plots of third codon positions). When homologous sequences from the "general" and from the "murid" distributions are compared with each other, linear relationships with high correlation coefficients are still found, but points show a larger scatter and the slope is significantly different from unity in the case of third codon positions (Mouchiroud et al., 1988; see Fig. 2). This difference is due to deviations mainly affecting the two opposite ends of distributions. In murids, GC-poor and GC-rich coding sequences (and their different codon positions) are less GC-poor and less GC-rich, respectively, than in other mammals. It should be stressed that "minor" shifts correspond to orderly changes in the genome, since the order of coding sequences by increasing GC content is very largely conserved in the two distributions (Mouchiroud et al., 1988).

Parallel investigations have shown that the compositional distribution of large DNA fragments (in the 30-100 Kb size range) of mouse is narrower (Salinas et al., 1986; Zerial et al., 1986) and, in contrast with that of coding sequences, centered on a slightly higher GC level, compared to man. When the distribution of DNA fragments from rat and mouse, or from man and mouse, are plotted against each other, the results are very similar to those obtained in Fig. 2. In the first case, the distributions are identical; in the second, the distribution of human DNA fragments starts at lower GC values and ends at higher ones, compared to mouse.

## The causes of the conservation of mammalian compositional patterns : compositional conservation of the base substitution process and negative selection at the isochore level

The extraordinary conservation of compositional patterns just described can only be due to two factors, the base substitution process itself and/or selection. It is conceivable that, when averaged over large time intervals, base substitutions exhibit a certain degree of compositional conservation. A complete absence of compositional biases in mutations is, however, most unlikely, in view of the demonstrated influence of nearest-base composition, particularly in coding sequences where composition is averaged over a few hundred bases only. Under these circumstances, the additional intervention of the second factor, selection, appears to be inescapable, as already suggested (Bernardi and Bernardi, 1986).

Selection of individual point mutations, however, can hardly be the explanation for the findings under consideration, because this would require assigning a significant advantage or disadvantage to any event affecting one base pair out of about 3.10 . Even if this is conceivable for a very small number of sites, where base substitutions change critical amino acids or critical

136

nucleotides in tRNAs, rRNAs and in signal sequences, it cannot be the general rule in vertebrate genomes, where more than 90% of DNA is noncoding, because this would cause an unbearable mutational load. In contrast, a negative, "stabilizing", selection process, acting at a regional level, and eliminating deviations from a narrow range of values, presumably corresponding to functionally optimal regional compositional patterns, appears to be the only plausible explanation for our results. This regional level can be identified with isochores, since isochores are, in fact, the genome segments exhibiting compositional conservation.

It should be stressed that isochores correspond to individual or contiguous chromosomal domains, the "chromatin loops", since these are each estimated to comprise 30-300 Kb of supercoiled DNA in mammals. In turn, chromatin loops are supposed to correspond to replicons and to contiguous transcription units.

## Compositional shifts in the evolution of vertebrate genomes

Two "major" shifts in compositional patterns occurred in the evolution of vertebrate genomes. Indeed, massive changes in extended genome segments led (i) to the formation, in mammals and birds, of GC-rich isochores that are absent or scarcely represented in the vast majority of cold-blooded vertebrates (Bernardi et al., 1985); and (ii) to changes in the compositional distributions of coding sequences (Bernardi et al., 1985; Mouchiroud et al., 1987; see Fig. 3). These shifts were completely independent from each other, since the paleontological record indicates that mammals derived from therapsids over 200 million years ago, birds from dinosaurs about 150 million years ago. In agreement with this conclusion, compositional patterns of birds and mammals are different from each other (see Fig.3 and below) and from those of reptiles. Even if the precise durations of the shifts are not known, they certainly were much shorter than the time of existence of mammals and birds.

The main mechanism by which the "major" compositional shifts were achieved was a directional fixation of point mutations (Perrin and Bernardi, 1987). Indeed, when homologous coding sequences from 21 genes of cold-blooded vertebrates and from 41 genes of warm-blooded vertebrates were compared, most of them showed GC increases mainly in third codon positions, but also in first and second positions (implying, therefore, amino acid changes). A smaller number of homologous genes showed no change in GC levels and corresponded to genes located in GC-poor isochores of warm-blooded vertebrates. (Three exceptional cases, in which GC levels were lower in warm-blooded vertebrates, corresponded, in all likelihood, to genes located in the very scarce GC-rich isochores that arose in some cold-blooded vertebrates).

The situation just described can also be seen in comparisons of homologous coding sequences from man and Xenopus. These show that most third codon positions of man are higher in GC, a second smaller set of positions is equal, and just one position is lower (Fig. 4). Expectedly, no significant linear correlations were found in this case between GC levels of codon positions from homologous genes (contrary to what was found for mammalian genes; Fig. 2).

Several additional phenomena accompanied major shifts: (i) gene translocations led to the very high gene concentrations found in the GC-richest compartments of genomes from warm-blooded vertebrates; an example of this phenomenon is that of and globin genes, which are clustered in _Xenopus_, but underwent translocations to different chromosomes in both mammals and birds; this process was accompanied, or followed, by GC increases in the gene (in mammals), or in both and genes (in birds); (ii) large changes in chromosome structure took place, as indicated by the appearance of Giemsa and Reverse bands; moreover, the Giemsa light (or Reverse) bands apparently acquired a more complex structure relative to Giemsa dark bands, since their DNA is a mosaic of the different GC-rich isochores and also seems to have a less dense packing (G. Bernardi, unpublished observations); (iii) non-methylated CpG-rich islands, associated with genes (Bird, 1987), that are absent, or very poorly represented, in fishes and amphibians, became abundant and concentrated in GC-rich isochores of warm-blooded vertebrates (B. Aissani, G. Bernardi and G. Bernardi, unpublished observations).

A "minor" compositional shift, involving more moderate changes (towards both GC increases and GC decreases) in less extended genome sections, separated murids (as well as cricetids and spalacids) from most other mammals (Salinas et al., 1986; Zerial et al., 1986; Mouchiroud et al., 1987, 1988; Mouchiroud and Gautier, 1988). Amplification and insertion of GC-rich interspersed repeats, like Alu sequences (Bernardi et al., 1985; Zerial et al., 1986) apparently also played a role in this shift. In contrast with "major" shifts, in "minor" shifts correlations are still found between GC levels of third codon positions from homologous genes (compare Fig. 2 and 3), and the order of coding sequences by increasing GC is largely preserved.


## The causes of compositional shifts in genome evolution : negative and positive selection at the isochore level

An explanation for the compositional shifts (Bernardi and Bernardi, 1986) is that they are mainly due to both negative and positive, "directional", selection acting at the isochore level, (positive selection at individual nucleotides is likely to be so rare that it will be neglected in the present discussion).

In the case of compositional shifts, negative selection should be visualized as eliminating only compositional deviations directed towards lower GC contents instead of any compositional deviation(as in the conservative evolutionary process discussed before). On the other hand, positive selection is the only explanation that can also account for the diverse events accompanying the major shifts in compositional patterns (gene translocations into GC-rich isochores, chromosome restructuring, formation or increase of unmethylated CpG islands). Moreover, positive selection is an explanation that rests on demonstrated functional advantages.

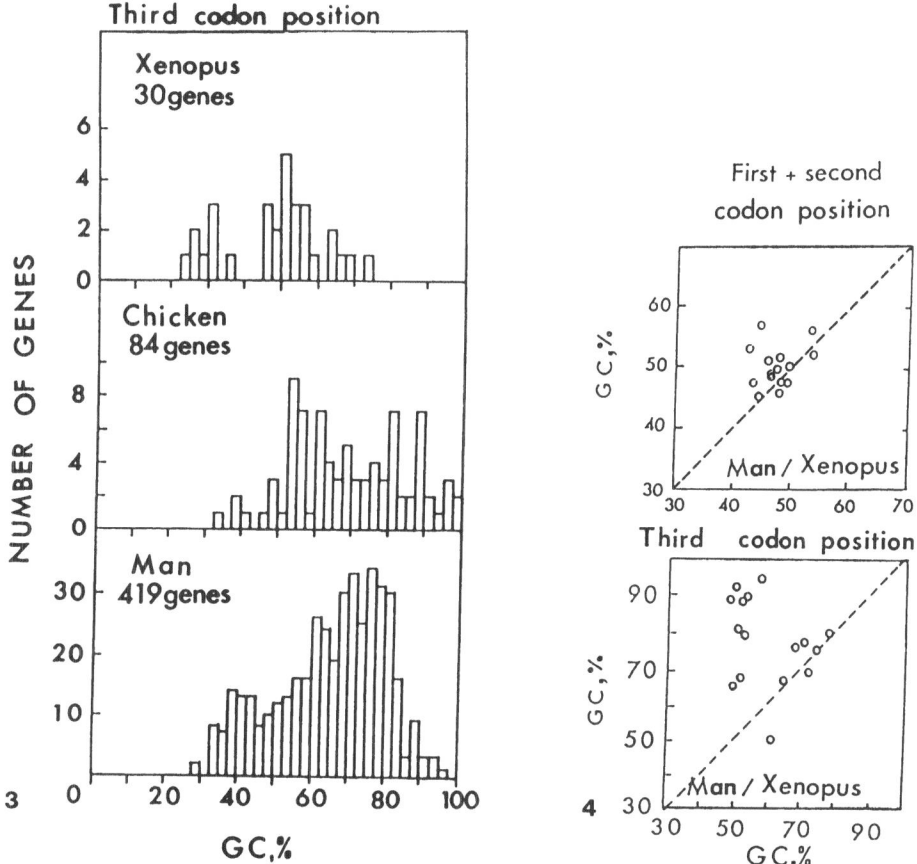

Fig. 3. (left) Compositional distribution of third codon positions for genes from <u>Xenopus</u>, chicken and man. Other indications as in Figure 1.

Fig. 4. (right) GC contents of first + second and third codon positions of pairs of homologous genes are plotted against each other for man and <u>Xenopus</u> (16 genes). Data for man correspond to the ordinate, data for Xenopus to the abscissa.

Indeed, as already pointed out (Bernardi and Bernardi, 1986), the increase in body temperature that accompanied these shifts, is associated with GC increases that have advantageous consequences : (i) GC increases in first and second codon positions lead to amino acid changes that confer (Argos et al., 1979) thermal stability to proteins; indeed, GC increases in coding sequences of vertebrates have been shown to be accompanied by increases in stabilizing amino acids (like alanine and arginine) and by decreases in destabilizing amino acids (like serine and lysine) in the encoded proteins (Bernardi and Bernardi, 1986); interestingly, reports on choices of thermally stabilizing amino acids and of GC-rich codons in thermophilic organisms are rapidly accumulating in the literature; (ii) GC increases in introns, in third codon positions and in DNA segments corresponding to untranslated regions contribute, in addition, to the thermal stability (Wada and Suyama, 1986) of primary transcripts and mRNAs; (iii) GC increases in intergenic, noncoding sequences can also conceivably help in stabilizing DNA structures, possibly through changes in DNA-protein interactions.

It should be stressed that, if our discussion was centered so far on temperature as the main selection factor responsible for the major compositional shifts of vertebrate genomes, it is only because precise selective advantages can be identified in this case. Our general suggestion is, however, that regional negative and positive selection is due to the functional advantages associated with the changes in compositional patterns. These advantages may be of a very different nature, and may be elusive because of the interplay of many factors. Thus, we do not have yet, for example, an explanation for the minor shifts leading to the murid pattern.

## The selectionist-neutralist controversy

At this point, it may be interesting to reconsider the selectionist-neutralist controversy which has continued for the past twenty years (Kimura, 1968, 1983), in the light of the present results.

First of all, it should be stressed that previous investigations (including those on which the neutral theory was built) practically only dealt with the evolution of mammalian proteins and genes; they only concerned, therefore, the accumulation of mutations in the conservative mode of evolution (we neglect here the differences between the general compositional pattern of mammals and the murid pattern). In other words, these investigations missed (i) the compositional conservation of homologous coding sequences (and isochores), that characterize the conservative mode of evolution; and (ii) the compositional shifts of homologous coding sequences (and of isochores), that took place at the transition between cold-blooded and warm-blooded vertebrates.

As a consequence, the selectionist-neutralist controversy was only considered at the level of individual base substitutions. On this basis, the conclusions were drawn (Kimura, 1983) that (i) "the great majority of evolutionary changes at the molecular level are caused not by Darwinian selection acting on advantageous mutations, but by random fixation of selectively neutral or nearly neutral mutants"; and (ii) "only a minute fraction of DNA changes in evolution are adaptive in nature". It is obvious that these conclusions cannot be considered any longer as having a general validity in genome evolution, since they could not take into account some aspects of the problem.

Indeed, positive selection is not only the very rare process operating on individual base substitutions, but also a regional process largely underlying compositional shifts, that appear to have an adaptive value. It should be noted that the transitional mode of genome evolution is rare and concerns only a very small part of the genome in mammals. It is, however, increasingly more frequent and more extensive when moving from warm-blooded to cold-blooded vertebrates, to invertebrates, plants, unicellular eukaryotes and prokaryotes, as indicated by the increasing spread of genome compositions of these organisms. This phenomenon may be related to the increasingly variable environmental conditions to which these genomes are submitted.

Moreover, even in the conservative mode of evolution, there are compositional constraints that affect the fixation of mutations. These particular "selective molecular constraints" not only contradict the "randomness in the pattern of substitutions" predicted by the neutral theory (Kimura, 1983), but also are so pervasive that the definition of a neutral mutation rate (namely, of a substitution rate reaching the maximum value set by the mutation rate) has proven elusive so far (see Zuckerkandl, 1986). A number of individual substitutions occurring in the conservative mode of evolution (which comprises most changes in the evolution of vertebrates) may, however, conceivably approach neutral, nearly neutral or slightly deleterious mutations, as described by the neutral theory.

Under these circumstances, while the overall process of genome evolution can be fully understood only within a selectionist framework, the neutralist view and the selectionist view (as presented here) appear to be more complementary than contradictory.

REFERENCES

Argos, P., Rossmann, M.G., Grau, U.M., Zuber, A., Franck, G. and Tratschin, J.D., 1979, Thermal stability and protein structure, Biochemistry, 18:5698.

Bernardi, G., 1985, The organization of the vertebrate genome and the problem of the CpG shortage, in: "Biochemistry and Biology of DNA methylation", G.L. Cantoni and A. Razin, eds., p. 3, Alan Liss, New York.

Bernardi, G. and Bernardi, G., 1985, Codon usage and genome composition, J. Mol. Evol., 22:363.

Bernardi, G., Mouchiroud, D., Gautier, C. and Bernardi, G., 1988, Compositional patterns in vertebrate genomes : conservation and change in evolution, J. Mol. Evol., in press.

Bernardi, G., Olofsson, B., Filipski, J., Zerial, M., Salinas, J., Cuny G., Meunier-Rotival, M. and Rodier, F., 1985, The mosaic genome of warm-blooded vertebrates, Science, 228:953.

Bernardi, G. and Bernardi, G., 1986, Compositional constraints and genome evolution, J. Mol. Evol., 24:1.

Bird, A.P., 1987, CpG islands as gene markers in the vertebrate nucleus, Trends in Genetics, 3:342.

Kimura, M., 1968, Evolutionary rate at the molecular level, Nature, 217:624.

Kimura, M., 1983, The neutral theory of molecular evolution, Cambridge University Press, Cambridge, England.

Medrano, L., Bernardi, G., Couturier, J., Dutrillaux, B. and Bernardi, G., 1988, Chromosome banding and genome compartmentalization in fishes, Chromosoma, 96:178.

Mouchiroud, D., Fichant, G. and Bernardi, G. 1987, Compositional compartmentalization and gene composition in the genomes of vertebrates, J. Mol. Evol., 26:198.

Mouchiroud, D. and Gautier, C., 1988, High codon-usage changes in mammalian genes, Mol. Biol. Evol., 5:192.

Mouchiroud, D., Gautier, C. and Bernardi, G., 1988, The compositionaldistribution of coding sequences and DNA molecules in man and murids, J. Mol. Evol., in press.

Perrin, P. and Bernardi, G. 1987, Directional fixation of mutations in vertebrate evolution, J. Mol. Evol., 26:301.

Salinas, J., Zerial, M., Filipski, J. and Bernardi, G., 1986, Gene distribution and nucleotide sequence organization of the mouse genome, Eur. J. Biochem., 160:469.

Wada, A. and Suyama, A. 1986, Local stability of DNA and RNA secondary structure and its relation to biological function, Prog. Biophys. Mol. Biol., 47:113.

Zerial, M., Salinas, J., Filipski, J. and Bernardi, G., 1986, Gene distribution and nucleotide sequence organization in the human genome, Eur. J. Biochem., 160:479.

Zuckerkandl, E. 1986, Polite DNA: functional density and functional compatibility in genomes, J. Mol. Evol., 24:12.

# ORGANIZATION AND EVOLUTION OF GENOMES

# AS SEEN FROM A MEGABASE PERSPECTIVE

Charles R. Cantor, Larysa Pevny and Cassandra L. Smith

Departments of Genetics and Development, Microbiology and Psychiatry
College of Physicians and Surgeons
Columbia University
New York, NY 10032 USA

## INTRODUCTION

It seems appropriate that the evolutionary pattern of gene expression is the focus of this special workshop to commemorate many years of molecular biology advanced study institutes on Spetses. These institutes have witnessed the evolution of molecular biology from the earliest days, where only basic processes in simple organisms were studied, to the present, where no biological issue or organism seems too complex to withstand analysis by molecular methods. For those of us privileged to participate in a number of such institutes, the process has been one of continual scientific growth and stimulation. The methods and approaches described below were not even conceived of a decade ago. Yet, already they are sufficiently standard to require almost mandatory inclusion as one introduces newcomers to the state of the art of molecular biology.

It is currently impossible to do prospective experiments to test most ideas about evolutionary mechanisms and events. Thus, inevitably, one is forced to do retrospective comparisons between present day species, or available fossil remains, to try to reconstruct events long past. Morphological features are frequently excellently preserved in the fossil record, but in comparing them one must often be fairly subjective, and the factors which determined these features are very distant from the molecular events of mutation and repair that must form the fundamental input to evolutionary pathways. In contrast, comparisons at the nucleic acid or protein sequence level are fairly fundamental but, with the data sets currently in hand, they sample only a minute fraction of the genomes of the organisms included. Thus, such comparisons may carry the risk of being too myopic.

In the past few years, it has become possible to examine, at low resolution, the physical structure of entire genomes. This is done by combining available genetic map data with the analysis of large DNA fragments to construct a restriction map of entire chromosomal DNA molecules (Smith et al., 1987a; Fan et al., 1988). Such maps provide an overview of the entire genome, and they can potentially reveal patterns of organization that might be missed in more detailed studies of small genome regions. Here we will review the methods used to construct these maps. The kinds of restriction patterns seen to date in selected mapping studies will then be described. Finally the potential of such studies, in the future, to provide evolutionary insights will be discussed.

## LARGE DNA METHODS

Large DNA techniques have dramatically changed strategies for dealing with DNA, because they allow the manipulation of molecules up to about 10 Mb in size (Smith et al., 1987c). This is more than a hundred times larger than previously possible. These techniques include methods for preparing intact chromosomal DNA, for specifically cleaving this DNA into large fragments, for separating fragments by size (Schwartz et al., 1983; Schwartz and Cantor, 1984; Carle and Olson, 1984; Vollrath et al., 1987), for determining the original order of fragments in the intact chromosome (Cantor et al., 1987; Smith et al., 1987b), and for cloning relatively large DNA pieces (Burke et al., 1987).

Intact DNA molecules even from very small chromosomes range from a few hundred kb to many Mb in size. These molecules cannot be handled in liquid because of their susceptibility to shear breakage. However they can be rendered remarkably stable by embedding them in agarose gels. This is done by casting live cells into agarose blocks, and then using appropriate combinations of enzymes and chemicals to remove all cellular constituents except the DNA. The procedures work because DNA molecules larger than 50 kb show negligible diffusion through the 100 nm pores in a typical agarose gel while proteins and small molecules diffuse readily.

Cleavage of chromosomal DNA into large specific fragments is accomplished by allowing restriction enzymes to diffuse into the gel and act on the immobilized DNA. It is almost always possible to find one or more enzymes that will cut a particular genome into fragments in the 100 kb to 1 Mb size large optimal for current DNA mapping methods. The resulting fragments (or even the original intact chromosomal DNAs, if they are not too large) can be size fractionated by pulsed field gel electophoresis (PFG). In this method, separation is achieved by the differential ability of various size DNAs to change direction in an agarose gel. In this electrophoretic technique the electric field direction is periodically alternated. The time scale of these alternations acts like a zoom lens and sets the size scale for DNA to be fractionated. By changing the time scale from seconds to hours one tunes the separation range from thousands to millions of base pairs.

The usual method for making macrorestriction maps starts from a set of genetically mapped and cloned DNA fragments (Smith et al., 1987a). The order of DNA markers on a genetic and physical map must be the same, although relative distances between adjacent markers on the two maps can fluctuate by orders of magnitude. Available DNA markers are used to identify corresponding large DNA fragments. These then serve as bench marks, and strategies like partial digestion, comparisons of polymorphisms in different cell lines, and specialized DNA probes like linking clones, which bridge two adjacent large fragments, are used to move out from the initial bench marks. It is far more difficult to finish such maps than to start them, but in spite of this complete maps are already available for a number of simple organisms, including the bacterium _E. coli_ and the yeast _S. pombe_. In addition, substantial regions of the human genome have been mapped near genes of particular medical interest (Lawrance et al., 1986; Drumm et al., 1988; Poustka et al., 1988).

One major limitation of macrorestriction mapping is that it doesn't directly provide a source of purified DNA for continued study. This would be best accomplished in principle by cloning intact large DNA fragments. The advent of yeast artificial chromosome vectors (YACs) is an important advance (Burke et al., 1987). Today such vectors easily handle inserts a few hundred kb in size; hopefully in the near future their range can be extended to cover all of the fragments represented in a typical macrorestriction map.


FREQUENCY OF SITES

Any restriction map is a cartoon of the DNA sequence itself, since it represents the location of particular sites on the DNA cleaved by enzymes. The ability of such a map to provide a useful genome overview will depend on the number and distribution of these sites and whether this has any functional or evolutionary significance. Let us first consider the number of sites. Only two enzymes are currently known with 8-base recognition sites, Not I and Sfi I. For a genome with 50% A+T, such enzymes should yield DNA fragments averaging 64 kb in size; this is 16 times larger than the average expected for 6-base specific enzymes. If these predictions were borne out in practice, large DNA techniques would be of very limited utility since there would be no easy way to generate large DNA fragments from intact chromosomes. (In principle various methylase-nuclease combinations ought to be useful in overcoming these limitations (McClelland et al., 1985). But in practice such methods have still not achieved the reliability needed for routine laboratory use.

Fortunately Not I, Sfi I and many other restriction enzymes appear to have sites much rarer than predicted by simple statistics (Smith et al., 1987b; Smith et al., 1987d). Deviations in A+T content away from 50% are also quite common, and this can be used to help select potentially useful enzymes. Certain other combinations of base sequences are also quite rare in selected organisms. This includes termination codons in *E.coli* and GATC in Halobacteria. These factors all lead to the convenient observation that a detailed search will almost always be rewarded with enzymes that cut a region of interest into fragments averaging significantly more than 100 kb in size.

In higher vertebrates, extensive methylation of the sequence CpG occurs. This has severe consequences for the distribution of sites cut by enzymes that contain one or more copies of this sequence as part of their recognition site. As far as we know 5-meCpG is the only significant site of DNA methylation in mammals. Deamination of 5-meC is a relatively frequent mutagenic event, since the resulting base is T, which cannot be corrected by the dU mismatch repair system (Barker et al., 1984). As a result, most CpG has disappeared from mammalian genomes. What remains is 80% methylated, and most 6 or 8 base specific restriction enzymes currently known cannot cleave at the methylated sequence. Thus, these enzymes sample the genome only at unmethylated CpG which, it turns out, is mostly clustered in CpG rich segments (Bird, 1986; Smith et al., 1987b). These have been called HTF islands because the methylation sensitive enzyme Hpa II produces tiny fragments from these regions. The HTF islands tend to be located near genes, often within 1 kb upstream from a gene expressed in all tissues, i.e. a housekeeping gene. Thus macrorestriction maps of mammalian genomes will resemble, at first glance, maps of the locations of housekeeping genes.

The consequences of the rarity of CpG, and its extensive methylation, are severe for restriction enzyme digestions. Table 1 shows the potential frequency of certain enzyme cutting sites (ignoring methylation) and the predicted frequency when clustered methylation free islands are considered. It is evident that the latter is quite close to observed fragment sizes while the former is quite different.

# DISTRIBUTION OF SITES

The usefulness of macrorestriction enzyme cutting frequencies, or finished macrorestriction maps for evolutionary comparisons will clearly depend on the factors that alter these frequencies. For example, infection by a mobile element carrying a rare site would rapidly produce a very distorted picture of the genome as a whole. In practice, restriction enzyme fragmentation patterns appear to be quite useful in analyzing the relationship among archaebacterial species and monitoring events taking place in populations of parasitic protozoa. However sometimes organisms thought to be closely related, like *Salmonella typhimurium* and *E. coli* can look quite different in their overall genome restriction site pattern (Smith, 1988).

Table 1. Predicted and observed macrorestriction fragment sizes

| Enzyme | kb screened[a] | Site Clusters Seen[a] | Site Clusters Expected[b] | Expected Fragment Size (kb) | Observed Fragment Size(kb)[c] |
|--------|-----------|-----------|-----------|---------|---------|
| Not I | 5360 | 16 | 4 | 1340 | 1000 |
| Mlu I | 5360 | 24 | 6 | 893 | 1000 |
| Sfi I | 3330 | 34 | 34[d] | 100[d] | 250 |
| BssH II | 2000 | 32 | 8 | 250 | 200 |
| Sac II | 2000 | 25 | 6.25 | 320 | --- |

[a] Data from Smith et al., 1987d

[b] Assuming 25% unmethylated

[c] Average estimated by Smith et al., 1987b

[d] Ignoring any Sfi I methylation inhibition

When the distribution of sites recognized by enzymes that cut mammalian genomes infrequently is observed, a number of very peculiar features stand out. First, the size of any particular fragment detected by a single copy DNA probe, is often very different in different cell lines and tissues. This is a consequence of the fact that CpG methylation of particular sites varies substantially from cell to cell in a manner that often correlates quite closely with the expression of genes. Second, the overall pattern of fragments generated by several of these enzymes is biphasic (Figure 1). This is not simply a consequence of the clustering of unmethylated CpG's in HTF islands. That tends to make the sizes of fragments cut from a given region by different enzymes similar, since there is a good chance sites will fall within the same pair of HTF islands. Rather it must reflect a more long range inhomogeneous feature of mammalian genomes since any uniform organizational pattern ought to produce a quasi-Gaussian distribution of HTF islands.

It is virtually certain that the biphasic distribution of fragment sizes seen with Not I, Mlu I, and other enzymes reflects the mosaic nature of mammalian genomes first noticed by the pattern of bands seen when stained metaphase chromosomes are examined in the light microscope. Other indications of this pattern occur in the ability to fractionate extensive regions of the genome into discrete density classes, isochores, and association of particular types of genes and patterns of replication timing with these direct physical indications of genome inhomogeneity (Bernardi et al., 1985; Holmquist, 1987). While it is not yet possible to be precise at this stage, it now appears that light staining Giemsa regions correspond to GC-rich isochores, domains rich in early replicating housekeeping genes and dense concentrations of HTF islands which, at least on chromosome 21, produce predominantly the small class of macrorestriction fragments (Smith et al., 1988). Conversely, dark Giemsa bands are A-T rich isochores, with relatively few housekeeping genes. They are late replicating and on chromosome 21 they yield predominantly the large class of macrorestriction fragments. How this mosaic genome evolved, and what its functional consequences are, is a fascinating topic for current speculation and further detailed analysis.

As large DNA fragments are mapped to specific regions of mammalian genomes it becomes possible to use them to analyze the distribution pattern of functionally discrete DNA units. For example one can ask where particular classes of repeated DNA are located, where centromeres and telomeres are located, where replication origins are located, and so on. Answers to these questions are just beginning to emerge, and they may contain quite a number of surprises. For example chromosome bands near telomeres may be rich in HTF islands, particularly so in human cells. SINES, like the human Alu sequence, may be excluded from certain genome regions including centromeres (Moyzis, 1988). One recent report suggests that the locations of SINES and LINES like the human Kpn I sequence, are anticorrelated (Korenberg and Rykowski, 1988). The data in Figure 2 do not support this notion for human chromosome 21.

Earlier, the heterogeneity in CpG methylation pattern was described,

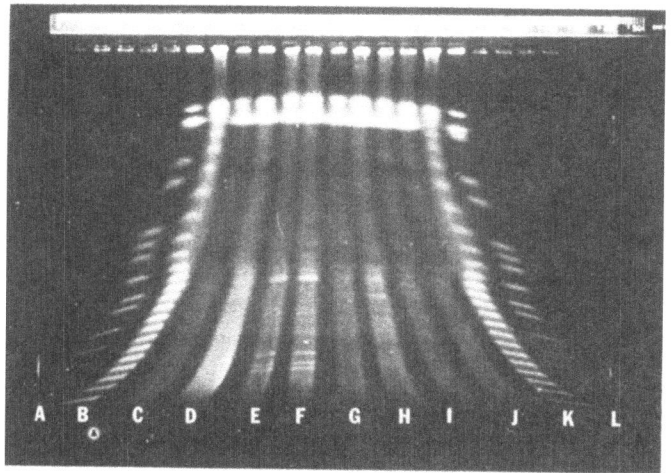

Figure 1. Restriction nuclease digests of different mammalian cell lines have been fractionated by PFG in 1% agarose using a 100 sec. pulse time for 40 hrs at 10 Vcm$^{-1}$ and then stained with ethidium bromide. Note the overall biphasic size distribution and the presence of repeated large DNA fragments in some cell lines. The lanes contain: A, L, S. cerevisiae chromosomal DNA; B, K, lambda DNA concatemers; C, D, Not I digest of a mouse cell line; E, F, Mlu I digest of a mouse cell line; G, M, Mlu I digest of a mouse cell line containing human chromosome 21; I, J, Not I digest of a mouse cell line containing human chromosome 21.

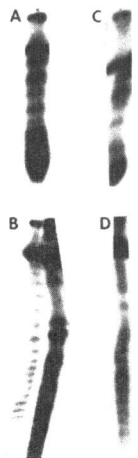

Figure 2. Distribution of SINES and LINES on human chromosome 21. A hybrid cell line containing only human chromosome 21 was digested with Not I and the resulting fragments separated by PFG, blotted, and then hybridized, successively with: A, B, Alu (SINE) and C, D, Kpn I (LINE) probes. The separation at top resolves the largest DNA fragments, in the 1 to 7 Mb size range; while the one at the bottom resolves fragments less than 1 Mb. The left lane in B shows lambda DNA concatemers.

and the high mutation rate of the CpG sequence was mentioned. The two factors act in concert to yield the very large extent of restriction fragment length polymorphism seen with large DNA fragments. The epigenetic nature of much of this polymorphism precludes its ready use in genetic linkage studies. However it may contain clues for the functional organization of the genome, and thus could be useful, eventually, in evolutionary studies. We have observed that the polymorphism is patchy both in the tip of human chromosome 4, in the neighborhood of the gene for Huntington's Disease, and along the long arm of human chromosome 21. If this observation is general it would suggest that domains of methylation are regulated across chromosome segments smaller than bands (at least those bands seen in typical low resolution metaphase chromosome spreads) but larger than individual genes.

What is missing in almost all current studies is a correlation between patterns seen along the linear DNA of each chromosome and how that DNA is actually arranged in the nucleus during gene expression, DNA synthesis, and various steps in cell division. We know that quite remarkable changes in the overall size and number of chromosomes can be tolerated in closely related species. However the pattern of gene organization on mammalian genomes is quite well preserved between mouse and man, for example, as soon as one looks at finer resolution. How this relates, at all, to the need to arrange the DNA in an orderly fashion to orchestrate gene function is largely a matter of speculation. However it is clear that, as more is learned about the linear order of DNA fragments along the chromosomes, we will be approaching the study of higher order structure function relationships.

A fascinating hint of what may lie in store is the presence of ethidium bromide patterns seen even in total genomic digests with Not I and Mlu I that are fractionated by PFG. Occasional bright discrete bands are seen which must represent a highly reiterated large DNA fragment. These bands are quite specific to certain species and even certain cell types as shown by the example in Figure 1. Whether this represents a simple sequence-dependent change in DNA methylation involving millions of base pairs of DNA or whether it reflects some higher order property of DNA organization in the nucleus remains for future studies to reveal.

In addition to bright bands, in the highest resolution DNA seperations, most mammalian samples show many discrete lighter bands. The origin of these bands is not understood. One possibility is that certain spacings between HTF islands are preferred, perhaps as a consequence of the need to organize them at the time of gene expression.

ACKNOWLEDGMENT

This work was supported by grants from the NIH (GM 14825), the NCI (CA 39782), the DOE (DE-F602-87ER-GD582), and the Hereditary Disease Foundation.

REFERENCES

Barker, D., Schaetter, M., and White, R., 1984, Restriction sites containing CpG show a higher frequency of polymorphism in human DNA. Cell, 36: 131.

Bernardi, G., Olofsson, B., Filipski, J., Zerial, M., Salinas, J., Cuny, G., Meunier-Rotival, M., and Rodier, F., 1985, The mosaic genome of warmblooded vertebrates. Science, 228: 953.

Bird, A.P., 1986, CpG-rich islands and the function of DNA methylation. Nature, 321: 209.

Burke, D.T., Carle, G.F., and Olson, M.V., 1987, Cloning of large segments of exogenous DNA into yeast by means of artificial chromosome vectors. Science, 236: 806.

Cantor, C.R., Smith, C.L., and Argarana, C., 1987, Strategies for finishing physical maps of macro-DNA regions, in: "Integration and Control of Metabolic Processes: Pure and Applied Aspects," O.L. Kon and M.C.-M. Chung, P.L.H. Hwang, S.-F. Leong, K.H. Loke, P. Thiyagarajah, P.T.-H. Wong, eds., Cambridge University Press, Cambridge, England. p. 427.

Carle, G.F., and Olson, M.V., 1984, Separation of chromosomal DNA molecules from yeast by orthogonal-field-alternation gel electrophoresis. Nucl. Acids Res., 12: 5647.

Drumm, M.L., Smith, C.L., Dean, M., Cole, J.L., Iannuzzi, M.C. and Collins, F.S., 1988, Physical mapping of the cystic fibrosis region by pulsed-field gel electrophoresis. Genomics, 2: 346.

Fan, J.B., Chikashige, Y., Smith, C.L., Niwa, O., Yanagida, M., and Cantor, C.R., 1988, Construction of a Not I restriction map of the fission yeast Schizosaccharomyces pombe genome. m.s. in preparation.

Holmquist, G.P., 1987, Role of replication time in the control of tissue specific gene expression. Am.J. Hum. Genet., 40: 151.

Korenberg, J.R. and Rykowski, M.C., 1988, Human genome organization: Alu, lines, and the molecular structure of metaphase chromosome bands. Cell, 53: 391.

Lawrance, S.K., Srivastava, R., Chorney, M.J., Rigas, B., Vasavada, H., Gillespie, G.A., Smith, C., Cantor, C., Collins, F.S., Weissman, S.M., 1986, Molecular Approaches to the characterization of megabase regions of DNA:  Applications to the human major histocompatibility complex.  Cold Spring Harbor Symp. Quant. Biol., 51: 123.

McClelland, M., Nelson, M., and Cantor, C.R., 1985, Purification of Mbo II methylase (GAAGma) from Moraxella bovis:  site specific cleavage of DNA at nine and ten base pair sequences. Nucl. Acids Res., 13: 7171.

Moyzis, R., 1988, Personal communication.

Poustka, A., Lehrach, H., Williamson, R. and Bates, G., 1988, A long range restriction map encompassing the cystic fibrosis locus and its closely linked genetic markers.  Genomics, 2: 337.

Schwartz, D., Saffran, W., Welsh, J., Haas, R., Goldenberg, M., and Cantor, C. 1983. New techniques for purifying large DNAs and studying their properties and packaging.  Cold Spring Harbor Symp. Quant. Biol., 47: 189.

Schwartz, D.C. and Cantor, C.R., 1984, Separation of yeast chromosome-sized DNAs by pulsed field gradient gel electrophoresis.  Cell, 37:67.

Smith, C.L., 1988, Unpublished results.

Smith, C.L., Econome, J.G., Schutt, A., Klco, S., and Cantor, C.R.,1987a,  A physical map of the Escherichia coli K12 genome.  Science, 236: 1448.

Smith, C.L., Lawrance, S.K., Gillespie, G.A., Cantor, C.R., Weissman, S.M., and Collins, F.S., 1987b, Strategies for mapping an cloning macro-regions of mammalian genomes, in: "Methods in Enzymology," M. Gottesman, ed., Academic Press, New York. 151: 461.

Smith, C.L., Matsumoto, T., Niwa, O., Klco, S., Fan, J-B., Yanagida, M., and Cantor, C.R., 1987c. An electrophoretic karyotype for *Schizosaccharomyces pombe* by pulsed field gel electrophoresis. *Nucl. Acids Res.*, 15: 4481.

Smith, C.L., Yu, M.T. and Cantor, C.R., 1988. Unpublished results.

Smith, D.I., Golembieski, W., Gilbert, J.D., Kizyma, L. and Miller, O.J., 1987d, Overabundance of rare-cutting restriction endonuclease sites in the human genome. *Nucl. Acids Res.*, 15: 1173.

Vollrath, D., and Davis, R.W., 1987, Resolution of DNA molecules greater than 5 megabases by contour-clamped homogenous electric fields. *Nucl. Acids Res.*, 15: 7865.

# LINE-1 SEQUENCES: HUMAN TRANSPOSABLE ELEMENTS

Maxine F. Singer

Laboratory of Biochemistry, National Cancer Institute
Bethesda, MD. 20892 and Carnegie Institution of
Washington, Washington, D.C., 20005

The LINE-1 family of interspersed repeated DNA sequences that is ubiquitous in placental mammals and marsupials is now known to include active transposable elements. This was first suspected some years ago because 1) LINE-1 segments are generally surrounded by short direct repeats that were either known or presumed to be target site duplications and 2) alleles that differ by the presence or absence of a LINE-1 segment were known (reviewed in Bellis et al., 1987). Proof of transposability in the human genome was provided by two recent reports. First, two unrelated boys suffering from Hemophilia A contain LINE-1 insertions in a Factor VIII gene exon, insertions that do not occur in the X-chromosomes of the patients' mothers (Kazazian et al.,1988). Second, one myc allele in a human breast adenocarcinoma contains a LINE-1 insertion, an insertion that is absent in the myc alleles in the patient's normal tissue (Morse et al., 1988).

It is now also apparent that mammalian LINE-1s are part of a newly recognized class of transposable elements that is wide-spread in plants and animals (Table 1). No fewer than 5 families of this class occur in Drosophila melanogaster and one of these, the I-factor, is associated with a hybrid dysgenesis syndrome distinct from that caused by P elements. As described below, evidence suggesting that these elements transpose by means of a self-encoded reverse transcriptase is accumulating. Thus, they appear to be active moveable elements, not passively produced retrogenes. We have proposed that this class of elements be termed class II retrotransposons to distinguish them from the more familiar retrotransposons such as Ty of yeast, copia of Drosophila, and IAP of mouse (class I retrotransposons) (Fanning Singer, 1987a).

The common features of class II retrotransposons are compared with those of the retroviral-like class I retrotransposons in Figure 1 (reviewed in Fanning and Singer, 1988; Finnegan, 1988; Kingsman et al., 1988). The similarities between the two classes include long open reading frames (ORFs), one of which contains regions predicting peptides that are highly conserved among known reverse transcriptases. Also recognizable in the class II element ORFs are stretches that predict polypeptides similar to known retroviral and class I proteases and nucleic acid binding proteins. None of the predicted class II proteins or enzymatic activities have yet been identified.

TABLE 1

Class II Retrotransposons

| Element | Species | Approximate copy number | Reference |
|---------|---------|------------------------|-----------|
| LINE-1 | Mammals, marsupials | $10^4$–$10^5$ | Fanning and Singer, 1988 |
| I | D. melanogaster | 10–15 | Finnegan, 1988 |
| F | D. melanogaster | 50 | DiNocera, 1988 |
| G | D. melanogaster | 10–20 | DiNocera, 1988 |
| Doc | D. melanogaster | 40 | Schneuwly et al., 1987 |
| Jockey | D. melanogaster | 50 | Mizrokhi et al., 1988 |
| Ingi | T. brucei | 200 | Kimmel et al., 1987 Murphy et al., 1987 |
| R2 | B. mori | | Xiong and Eickbush, 1988 |
| Cin-4 | Z. mays | 50–100 | Schwarz-Sommer et et al., 1987 |

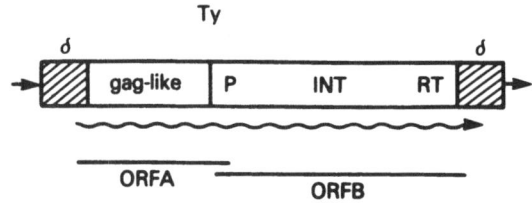

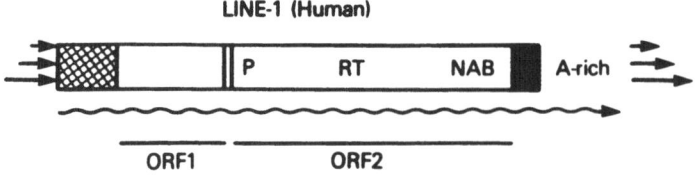

Fig. 1. Comparison of typical class I (Ty of yeast, Kingsman et al, 1988) and class II (human LINE-1, Skowronski et al., 1988) retrotransposons. The wavy lines are transcripts. Small arrows are target site duplications which are a fixed length for class I and variable lengths for class II. P is protease homology (Y. Sakaki, personal communication); RT is reverse transcriptase homology; INT is intergrase coding region; NAB is a region with homology to nucleic acid binding proteins (Fanning and Singer, 1987b).

Several class II elements contain at least two recognizable reading frames that are, like those in class I and in retroviruses, either overlapping (Loeb et al., 1987) or separated by in-frame stop codons (Scott et al., 1987; Skowronski et al., 1988). The substantial structural differences between the two classes include 1) the absence of LTRs from class II, 2) the presence in class II of an A-rich region at the 3' end of one strand (generally termed the 3' end of an element), 3) the variable size target site duplications surrounding class II elements compared to the fixed size associated with each family of class I element, 4) the enormous copy number of class II elements in mammalian genomes, (Burton et al., 1986), and 5) the high frequency of class II retrotransposons that are truncated (generally at the 5' end) or otherwise rearranged (e.g., inversions and internal deletions).

Insertions of class I retrotransposons into new genomic loci, like the insertion of retroviral proviruses, is likely to occur through 1) formation of nearly full length, polyadenylated transcripts, 2) reverse transcription dependent on element-encoded reverse transcriptase, 3) insertion of the cDNA into a staggered break in genomic DNA that is formed by action of an element-encoded endonuclease, and 4) repair of the gaps at the junction between the cDNA and genomic insertion site producing the target site duplication (reviewed by Kingsman et al., 1988). Current models for class II retrotransposon insertion are analogous, primarily because of the similarity between the elements and processed pseudogenes and the predicted reverse transcriptase activity. Clearly, however, the details of the class II insertion mechanism must differ substantially in detail from the class I mechanism because of the structural differences in the elements.

One expected difference is the likely absence, from class II retrotransposons, of a self-encoded endonuclease that makes a fixed size staggered break. Class II elements may be inserted in staggered breaks in genomic DNA that are caused by non-specific endonucleases encoded either within the element or in other genes. This would explain the variable size of the target site duplication surrounding different elements within one genome.

Other important differences in the transposition mechanism are expected to arise from the absence of LTRs from class II elements. How, for example, can full length copies of class II elements be generated by the model in the absence of LTRs to provide redundant information? One possible explanation that is beginning to acquire experimental support is the presence, within the element, of an RNA polymerase II promoter that directs transcription initiation upstream of itself, at the 5' end of the element. Thus, full length, polyadenylated, cytoplasmic transcripts that begin at the 5' end of LINE-1 and the 5' end of Jockey occur in a human teratocarcinoma cell line (Skowronski et al., 1988) and Drosophila cells (Mizrokhi et al., 1988), respectively. Formation of the Jockey transcripts is inhibited by low concentrations of alpha-amanitin. Possibly similar transcripts have been detected in mouse lymphoid cells (Dudley, 1987). Also, the approximately 800 bp region that precedes a 1200 bp ORF (ORF1) at the 5' end of a human LINE-1 cDNA (see Figure 1), permits the expression of a fused chloramphenicol transacetylase gene upon transfection into a variety of primate cell lines (S. Mongkolsuk, G. Swergold, and M.F. Singer, unpublished experiments) and similar results have been obtained with Jockey (Mizrokhi et al., 1988). Moreover, Drosophila I-factors can transpose intact in the absence of any identifiable flanking promoter (Finnegan, 1988).

Although the synthesis of an active reverse transcriptase by a class II retrotransposon element remains to be demonstrated, the data just summarized suggest that the proposed transposition mechanism is plausible. Therefore it is interesting to consider some features of mammalian LINE-1 families and their evolution in the context of the proposed transposition model. The discussion will refer to the properties of primate LINE-1s. Other mammalian LINE-1s appear to be similar although not precisely the same in all details.

There are about $4\times10^3$ full length (approximately 6 kbp) and more than $5\times10^4$ truncated LINE-1 elements in the human genome (Hwu et al., 1986). Truncated forms can be as short as 70 bp and can include internally rearranged elements and elements with significant deletions. Two hypotheses have been considered to explain the prevalence of truncated and rearranged elements: 1) interruption of reverse transcription and 2) alteration of existing elements by recombinational events over time. Several observations are consistent with the occurrence of aborted reverse transcription. One is the prevalence of 5' truncations, which would follow from an interruption of first-strand synthesis on a template of the observed polyadenylated, cytoplasmic RNAs. There is no apparent reason to expect the polarized truncation from secondary recombinational events. In addition, target site duplications surround many (though not all) truncated LINE-1s which would be unlikely if such truncated elements derived from secondary deletions. Most convincing, however, is the fact that the two LINE-1 insertions causing Hemophila A are both truncated and one also contains an internal inversion (Kazazian et al., 1988). Assuming that the two patients are not mosaics (a likely but not rigorously proven fact), then these insertions may well represent the initial transposition products rather than rearrangements that occurred over multiple cell generations. Thus, at least some of the truncated elements probably arise during reverse transcription.

According to the proposed model, replicative transposition of class II retrotransposons can, in principle, generate additional 'active' elements because the internal promoter will be conserved. However, it appears that in the human genome and probably in other mammalian genomes, most LINE-1 sequences are not competent to act as independent moveable elements. Truncated elements that lack the internal promoter are presumably not readily transcribed. In addition it is likely that only a subset of full length LINE-1s can be transcribed. Thus, among the 19 cDNAs cloned from the teratocarcinoma cytoplasmic, polyadenylated RNA, none contain the extra 132 bp that occur in about 50% of unit length LINE-1s between residues 782 and 783 in the human consensus sequence (Hattori et al., 1985; Scott et al., 1987). This suggests that this subset of LINE-1s (subset 132) is not transcribed, at least in these cells. It may be that some mechanism other than reverse transcription will be needed to explain the abundance of subset 132; gene conversion or DNA-mediated transpositions are possibilities. Another indication that not all full length LINE-1s are transcribed comes from sequence analysis of the cDNAs from the teratocarcinoma RNA (Skowronski et al., 1988). The cDNAs have a distinctive consensus sequence in their 3' 204 bp long trailer (subset T, for transcribed). Full length LINE-1s carrying the consensus sequence of random genomic copies (subset U) rather than the cDNA consensus (subset T) in the 3' trailer appear to be transcribed into full length, polyadenylated, cytoplasmic RNA less frequently, if at all. The full length LINE-1 downstream of the human beta-globin gene has both the 132 bp insert and the genomic 3' trailer sequence (subset U/132) (Hattori et al., 1985). It is not known whether the 132 bp insert and the genomic 3' trailer consensus always or even often coexist in individual LINE-1s.

If active elements are transposed through an RNA intermediate, then newly inserted LINE-1s would be expected to belong to subset T. Indeed, both new insertions associated with the Factor VIII mutations belong to subset T (Kazazian et al., 1988).

The presence in human DNA of at least two classes of full length elements that differ in the 5' leader region (i.e., plus or minus the 132 bp segment) is typical of other mammals. Among primates, two classes are reported in the prosimian Galago garnetti: one has, starting about 730 bp from the 5' end, a series of 73 bp tandem repeats that are missing in the second class; the additional sequences in the 5' leader regions also diverge (Lloyd and Potter, 1988). In mice, at least three different types of 5' leader have been identified and two of·these have tandem repeats (Loeb et al., 1987; Fanning, 1983; Jubier-Maurin et al., 1987). Both the sequences of the repeats and their placement in the 5' leader differ. It is not known which, if any, of these subsets are transcribed in the respective genomes. These results emphasize the striking divergence of the LINE-1 5' leader regions. Within a species, this divergence is much more marked than the divergence found in LINE-1 ORFs, which is generally limited to about 10 percent. The same tendency is seen between species. Only very short homologous regions can be found when comparing the 5' leader of the slow loris (Nycticebu coucang, Hattori et al., 1986) or galago with that of humans. Similarly, there is little homology between the 5' leaders of mouse and rat LINE-1s. This is in marked contrast to the extensive homology in ORF2 and even the more limited conservation of regions of ORF1.

The ability to be transcribed does not assure that a LINE-1 element is competent as an independent moveable element. Among the elements that can give rise to full length polyadenylated, cytoplasmic transcripts, a large proportion appear to be untranslatable. Indeed, most of the cDNAs from human teratocarcinoma cells have a closed ORF1 or closed ORF2, or both (Skowronski et al., 1988). These ORFs carry mutations causing translation termination or frame shifts. Once transcribed, transposition of such elements could be mediated through a reverse transcriptase acting in trans. When more is known about the mechanism of reverse transcription, examples of transcripts that lack signals required for cDNA synthesis may also be found. Overall, it appears likely that the transposition mechanism (e.g., reverse transcription) as well as common mutagenic processes together maintain a low level of competent LINE-1s.

The presence of class II retrotransposons in plants, invertebrates, and vertebrates suggests a powerful tendency toward conservation as well as a very ancient origin. In each of the organisms analyzed, one or a few active elements appear to be conserved and be competent to generate a limited number of additional active elements as well as, at least in mammals, high numbers of inactive ones. Alternatively, horizontal transmission of the elements could explain their distribution. This seems less likely in view of the absence of any indication of 'infective agents' derived from these elements.

If the idea of ancient origin and conservation is to be taken seriously, we are compelled to look for plausible explanations. The usual list of the roles of transposable elements does not provide much help. For example, transposable elements, including class II retrotransposons, are known to be agents of insertional mutagenesis (Finnegan, 1988; Kazazian et al., 1988) and to alter gene expression of neighboring genes (Fanning et al., 1985; Yancopoulos et al., 1986; Laimins et al., 1986). But none of these roles are associated with positive selective pressure for conservation. The "selfish gene" concept too may be unsatisfactory in view of the ubiquity of class II retrotransposons. The central evolutionary question thus remains open. It is possible that one or more function(s) provided by these elements was from early times, or later became, essential in some species.

Another interesting point relates to the evolutionary relation between the various elements that depend on reverse transcription: class I and class II retrotransposons, retroviruses, hepatitis B virus, cauliflower mosaic virus. Comparing the reverse transcriptase homology region, the class II retrotransposons are significantly more similar to one another than to the other elements (Hattori et al., 1986; Burke et al., 1987; Finnegan, 1988). This, together with their distribution, suggests a rather ancient divergence of the class II elements from the others. This prompts the question: what and where was the common ancestor of the reverse transcriptases? It may be that the class II retrotransposons, ancient as they are, and lacking as they do the sophistication of LTRs or envelope genes, are a clue.

Well before class II retrotransposons were recognized, Temin suggested that retroviruses might have originated from cellular transposable elements (Temin, 1980). His model starts with two copies of a small moveable element like a bacterial insertion sequence. It is interesting now to consider an alternative possibility that class II retrotransposons were the progenitors of class I elements, retroviruses, and the DNA viruses that replicate through reverse transcription. LTRs could have evolved through schemes similar to that proposed by Temin, but from a 5' leader of a class II element. Concomitantly, the internal promoter of the class II elements also could have been modified or replaced so that transcription begins downstream rather than upstream. The mechanism of reverse transcription would also have evolved to arrive at the complex series of reactions utilized by retroviruses and, presumably, retrotransposons. This proposal would make unnecessary Temin's suggestion that reverse transcriptase evolved from a DNA polymerase. Rather, it would suggest that reverse transcriptase is the more ancient enzyme. This possibility is of interest in connection with recent proposals suggesting that RNA preceded DNA as an informational molecule during early evolution (Darnell and Doolittle, 1986).

REFERENCES

Bellis, M., Jubier-Maurin, V., Dod, B., Vanlerberghe, F., Laurent, A-M., Senglat, C., Bonhomme, F., and Roizes, G., 1987, Distributions of two recently inserted long interspersed elements of the L1 repetitive family at the Alb and Beta-h3 loci in wild mice populations, Mol. Biol. Evol., 4: 351-363.
Burke, W.D., Calalang, C.C., and Eickbush, T.H., 1987,The site-specific ribosomal insertion element type II of Bombyx mori (R2Bm) contains the coding sequence for a reverse transcriptase-like enzyme, Mol. Cell. Biol., 7: 2221-2230.
Burton, F.H., Loeb, D.D., Voliva, C.F., Martin, S.L., Edgell, M.H., and Hutchison III, C.A., 1986, Conservation throughout mammalia and extensive protein coding capacity of the highly repeated DNA long interspersed sequence 1, J. Mol. Biol., 187: 291-304.
Darnell, J.E., and Doolittle, W.F., 1986, Speculations on the early course of evolution, Proc. Natl Acad. Sci., 83: 1271-1275.
DiNocera, P., 1988, Close relationship between non-viral retroposons in Drosophila melanogaster, Nucleic Acids Res., 16: 4041-4052.
Dudley, J.P., 1987, Discrete high molecular weight RNA transcribed from the long interspersed repetitive element L1Md. Nucleic Acids Res., 15: 2581-2592.

Fanning, T.G., 1983, Size and structure of the highly repeated BAM H1 element in mice, Nucleic Acids Res., 11: 5073-5091.

Fanning, T.G., Morris, D.W., Cordiff, R.D., and Brodshaw Jr, H.D., 1985, Characterization of an endogenous retrovirus-repetitive DNA chimera in the mouse genome, J. Virol., 53: 998-1000.

Fanning, T.G., and Singer, M.F., 1987a, LINE-1: a mammalian transposable element, Biochim. Biophys. Acta 910: 203-212.

Fanning, T.G., and Singer, M.F., 1987b, The LINE-1 DNA sequences in four mammalian orders predict proteins that conserve homology to retroviral proteins, Nucleic Acids Res., 15: 2251-2260.

Finnegan, D.J., 1988, I-Factors in Drosophila melanogaster and similar elements in other eukaryotes, in: Transposition, Symposium 43, The Society of General Microbiology. Eds: A.J. Kingsman, K.F. Chater and S.M. Kingsman. pp. 271-285.

Hattori, M., Hidaka, S., and Sakaki, Y., 1985, Sequence analysis of a KpnI family member near the 3'-end of the human beta-globin gene, Nucleic Acids Res., 13: 7813-7827.

Hattori, M., Kuhara, S., Takenaka, O., and Sakaki, Y., 1986, The L1 family of long interspersed repetitive DNA sequences in primates may be derived from a sequence encoding a reverse transcriptase related protein, Nature, 321: 625-628.

Hwu, H.R., Roberts, J.W., Davidson, E.H., and Britten, R.J., 1986, Insertion and/or deletion of many repeated DNA sequences in human and higher ape evolution, Proc. Natl. Acad, Sci. USA. 83: 3875-3879.

Jubier-Maurin, V., Wincker, P., Cuny, G., and Roizes, G., 1987, The relationships between the 5'-end repeats and the largest members of the L1 interspersed repeated family in the mouse genome, Nucleic Acids Res. 15: 7395-7410.

Kazazian Jr., H.H., Wong, C., Youssoufian, H., Scott, A.F., Phillips, D.G., and Antonarakis, S.E., 1988, Haemophilia A resulting from de nova insertion of L1 sequences represents a novel mechanism for mutation in man, Nature 332: 164-166.

Kimmel, B., Ole-Moiyoi, O.K., and Young, J.R., 1987, Ingi, a 5.2 kbp dispersed sequence element from Trypanosoma brucei that carries half of a smaller modile element at either end and has homology with mammalian lines, Mol. Cell. Biol. 7: 1465-1475.

Kingsman, A.J., Adams, S.E., Fulton, S.M., Malim. M.H., Rathjen, P.D. Wilson, W., and Kingsman, S.M., The yeast retrotransposon Ty and related elements, in: Transposition, Symposium 43, The Society for General Microbiology. Eds: A.J. Kingsman, K.F. Chater, and S.M. Kingsman. 1988, pp. 223-246.

Laimins, L., Holmgren-Konig, M., and Khoury, G., 1986, Transcriptional "Silencer" element in rat repetitive sequences associated with the rat insulin 1 gene locus, Proc. Natl. Acad. Sci. USA 83: 3151-3155.

Lloyd, J.A., and Potter, S.S., 1988, Distinct subfamilies of primate L1Gg retroposons, with some elements carrying tandem repeats in the 5' region, Nucleic Acids Res. 16: 6147-6156.

Loeb, D.D., Padgett, R.W., Hardies, S.C., Shehee, W.R., Comer, M.B. Edgell, M.H., and Hutchison III, C.A., 1986, The sequence of a large L1Md element reveals a tandemly repeated 5'-end and several features found in retrotransposons, Mol. Cell. Biol. 6: 168-182.

Mizrokhi, L.H., Priimagi, A.F., Georgieva, S.G., and Ilyin, Y.V., 1988, Drosophila mobile element Jockey is similar to Lines and is transcribed by RNA polymerase II from the internal promotor, Cell in press.

Morse, B., Rothberg, P.G., South, V.J., Spandorder, J.M., and Astrin, S.M., 1988, Insertional mutagenesis of the myc locus by a LINE-1 sequence in a human breast carcinoma, Nature 333: 87-90.

Murphy, N.B., Pays, A., Tebabi, P., Coquelet, H., Guyaux, M. Steinert, M., and Pays, E., 1987, Trypanosoma brucei repeated element with unusual structural and transcriptional properties, J. Mol. Biol. 195: 855-871.

Schneuwly, S., Kuroiwa, A., and Gehring, W.J., 1987, Molecular analysis of the dominant homeotic Antennapedia phenotype, EMBO. J. 6: 201-206.

Schwarz-Sommer, Z., Leclercg, L., Gobel, E., and Haedler, H., 1987, Cin4, an insert altering the structure of the Al gene in Zea mays, exhibits properties of nonviral retrotransposons, EMBO J. 6: 3873-3880.

Scott, A.F., Schmeckpeper, B.J., Abdelrazik, M., Comey, C.T., O'Hara, B., Rossiter, J.P., Colley, T., Heath, P., Smith, K.D., and Margolet, L. 1987, Origin of the human L1 elements: proposed progenitor genes deduced from a consensus DNA sequence, Genomics 1: 113-125.

Skowronski, J., Fanning, T.G., and Singer, M.F. 1988, Unit-length LINE-1 transcripts in human teratocarcinoma cells, Mol. Cell. Biol. 8: 1385-1397.

Temin, H. 1980, Origin of retroviruses from cellular moveable genetic elements, Cell 21: 599-600.

Xiong, Y., and Eickbush, T.H. 1988, The site-specific ribosomal DNA insertion element R1Bm belongs to a class of non-long-terminal repeat retrotransposons, Mol. Cell. Biol. 8: 114-123.

Yancopoulos, G.D., DePinho, R.A., Zimmerman, K.A., Lutzker, S.G., Rosenberg, N., and Alt, F.W. 1986, Secondary genomic rearrangement events in pre-B cells: VhDJh replacement by a LINE-1 sequence and directed class switching, EMBO J. 5: 3259-3266.

THE EVOLUTIONARY ORIGIN OF GLYCOSOMES:

HOW GLYCOLYSIS MOVED FROM CYTOSOL TO ORGANELLE IN EVOLUTION

P. Borst and B. W. Swinkels

The Netherlands Cancer Institute
Plesmanlaan 121, 1066 CX Amsterdam
The Netherlands

## INTRODUCTION

One of the most remarkable forms of biochemical tinkering in evolution, is the movement of biochemical pathways from one cellular compartment to another. A striking example is provided by glycolysis, one of the most invariant and conserved biochemical pathways in nature. Textbooks state that this pathway is found in the cytosol in all eukaryotes, but in 1977 trypanosomes were found to be the exception to this rule (1,2). As shown in Fig. 1, glycolysis down to 3-P-glycerate is located in a peroxisome-like organelle, the glycosome, and only the last two steps, the conversion of 3-P-glycerate into pyruvate, occur in the cytosol. There is no doubt that this organization of glycolysis is eminently sensible: bloodstream trypanosomes rely entirely on glycolysis for ATP production. The segregation of the pathway in an organelle which takes up 4% of the trypanosome volume will reduce diffusion times of substrates and should allow a faster rate of glycolysis than would be possible if substrates and enzymes were freely distributed over the entire cytosol. Indeed, trypanosomes have by far the highest rate of glycolysis in nature.

How did the trypanosome manage to get all nine enzymes, depicted in the glycosome in Fig. 1, into an organelle? Or rather, how was this accomplished in an early ancestor of the Kinetoplastida, since all representatives of this order appear to have glycosomes (4)? The minimal number of proteins required to operate a successful glycosomal branch of glycolysis may even exceed nine, because we doubt whether the transport of phosphorylated sugars through the glycosomal membrane, required to make the scheme in Fig.1 work, can be accomplished at sufficient rates without transport proteins. Minimally, one would require a protein facilitating exchange of dihydroxyacetone-P and glycerol-P and another one exchanging inorganic phosphate for 3-P-glycerate. The transfer in evolution of such a complex pathway from cytosol to glycosomes is a remarkable achievement. How does nature handle such a task?

In this paper we summarize our present ideas on this question. To provide some substance to a speculative story, we start out by briefly summarizing how glycosomal enzymes are thought to enter glycosomes now.

### Topogenesis of glycosomal enzymes

Glycosomes are bona fide members of the microbody family and they contain many of the enzymes uniquely found in peroxisomes in other organisms (5). One would therefore expect that topogenesis of glycosomal enzymes would follow the route established for peroxisomal enzymes in liver and fungi. This is indeed the case: glycosomal enzymes are made in their mature form on cytosolic polyribosomes and imported post-translationally into the organelle (6). Glycosomes contain no endogenous DNA or protein-synthesizing machinery. All proteins are therefore imported.

To identify topogenic sequences in glycosomal proteins, we have isolated genes for glycosomal enzymes and compared the deduced amino acid sequences with those of their cytosolic counterparts. Most informative has been the 3-P-glycerate kinase (PGK) sequence (7), because this is one of the few glycolytic enzymes present both in glycosome and cytosol in Kinetoplastida. Fig. 2 shows that T. brucei contains three genes that hybridize with PGK gene probes. Gene B codes for the cytosolic PGK, gene C for the glycosomal PGK and gene A for a conserved PGK-like protein of unknown function and cellular location.

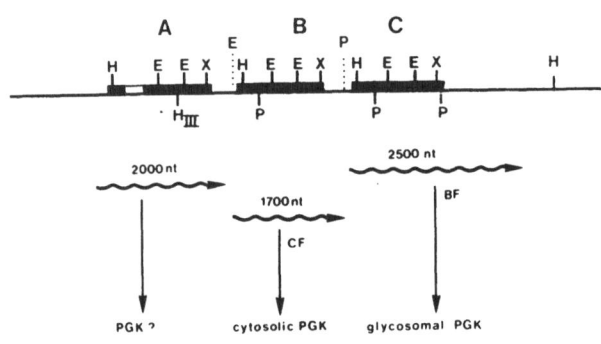

Fig. 1
Simplified diagrammatic representation of the compartmentation of glycolysis in T. brucei (from 3). Abbreviations: G-6-P, glucose-6-P; FDP, fructose-1,6-bisP; DHAP, dihydroxyacetone-P; GAP, glyceraldehyde-3-P; SHAM, salicylhydroxamic acid; G-3-P, glycerol-3-P; 1,3-DPGA, 1,3-diP-glycerate; PGA, P-glycerate; Pi, inorganic phosphate.

The sequences of the cytosolic(c) and glycosomal(g) PGKs are remarkably homologous, with 95% identity at the nucleotide level and 93% at the amino acid level. The two major differences between c and gPGKs that strike the eye are:
- the glycosomal enzyme has a C-terminal extension of 20 amino acids lacking in cPGKs (and all other PGKs in nature).
- in a total of 29 amino acid substitutions, gPGK has gained 13 positive charges over cPGK.

Most of the other glycosomal enzymes also have a high pI relative to their cytosolic counterparts (10), but lack a C-terminal extension (11). This led us to focus on the basic residues (11), although not without hesitation (see 5). Analysis of the PGK genes of another Kinetoplastid Crithidia, immediately showed, however, that we were walking up the garden path. Crithidia contains a similar array of PGK genes as T. brucei, but in this case the cPGK and gPGK are virtually identical (12). The differences are limited to:

- a C-terminal extension in gPGK of 38 amino acids
- a total of two amino-acid substitutions, near the N-terminus, both
  conservative.

As there is no charge difference between cPGK and gPGK in <u>Crithidia</u>,
topogenesis must be determined by the C-terminal extension. The most
straightforward interpretation is that the addressing signal is contained in
the extension itself, but we cannot formally exclude that the C-terminal
extension affects the folding of nascent PGK and thereby exposes an internal
topogenic signal. It should be noted, however, that the 3-D structure of PGKs
is highly conserved in evolution; that kinetoplastid PGKs contain the
conserved sequences found in the other PGKs and can be easily made to fit the
coordinates of the horse PGK (11), and that it is unlikely that a C-terminal
extension, synthesized last, could substantially change the folding of the
nascent protein. For the remaining part of this paper we therefore assume
that the C-terminal extension does contain the addressing signal for
glycosomes.

Fig.3 compares the C-terminal extensions of the <u>T.brucei</u> and <u>Crithidia</u>
PGKs with a C-terminal signal for peroxisome uptake recently identified in
firefly luciferase (13, 14). It is not clear what the putative receptor in
the glycosomal/peroxisomal membrane recognizes as common features in these
divergent sequences. Even in intensely studied signals, like the
mitochondrial addressing signals (5), this remains an unsolved question,
however.

Two other points follow from the sequence comparisons of PGK genes
summarized in Table 1. First, it is obvious that gene conversion is
continuously erasing the differences that arise between the PGK genes in the
course of evolution. Whereas the amino acid sequence difference between
<u>Crithidia</u> and <u>T. brucei</u> PGK genes is around 25%, the differences between

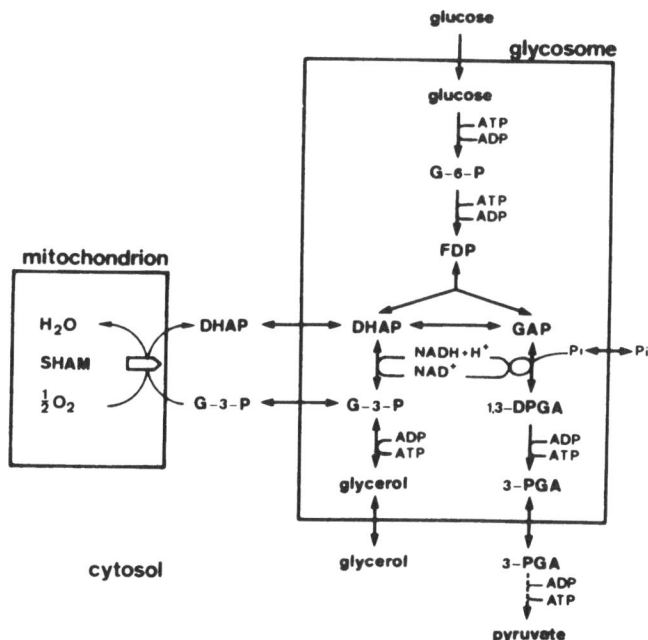

Fig. 2

A physical map of the three PGK genes of <u>T. brucei</u> and their mature
transcsripts (based on 7-9). The solid blocks indicate protein-coding
sequences. The hatched box in gene A represents a 300-bp insertion that does
not interrupt the reading frame. The wavy lines indicate the size and
position of the mature mRNAs on the map. The letters denote restriction
enzyme cleavage sites (see 7).

genes B and C of each organism is only < 1% and 7% respectively.

Secondly, our data show that the evolutionary transfer of a single enzyme from cytosol to glycosome is a relatively simple operation. Although this ties in with the results of other studies on topogenesis, it remains remarkable that enzymes with such a long history of exclusive cytosolic location, can be moved through a membrane and routed into an organelle merely by tagging on an address label.

## Failed attempts to define topogenic signals in glycosomal proteins by reverse genetics

We have used reverse genetics in an attempt to define the glycosomal addressing sequence more precisely. These attempts have failed thusfar for want of a suitable system in which the mutagenized genes can be tested. Homologous transformation does not work (yet) with trypanosomes and we therefore had to resort to heterologous in vivo systems or uptake of proteins by isolated glycosomes. As heterologous in vivo system we have used the yeast Saccharomyces cerevisiae. The gPGK gene, inserted into a yeast shuttle vector under a mitochondrial promoter, led to copious production of gPGK in yeast, but the enzyme was not imported into peroxisomes, even when peroxisome proliferation was induced by growth of the yeast on oleate. (Unpublished experiments of B. W. S., B. Distel and M. Veenhuis). Apparently the glycosomal topogenic signal is not recognized by yeast peroxisomes under these conditions.

Considerable effort was put into the development of a system to study import of proteins into glycosomes in vitro, using PGK generated by coupled transcription-translation (unpublished experiments of B. W. S.). Although gPGK associates with a glycosome-containing particle fraction from trypanosomes, whereas cPGK does not, control experiments summarized in Table 2 suggest that this association is only dependent on the positive charge in the gPGK and not on the C-terminal extension. We therefore think that the in vitro system does not measure true import and cannot be used to study topogenic signals. This conclusion contrasts with that of Dovey et al. (15), who did analogous experiments, but without the critical constructs that allow a separate evaluation of the contribution of positive charge and C-terminal extension to the association of gPGK to glycosomes. It may be somewhat optimistic to expect isolated microbodies to import proteins, as long as all isolated microbodies are fully permeable to sugar-phosphates, whereas microbodies in vivo appear to be impermeable, not only to sugar-phosphates (in trypanosomes, ref. 16), but even to protons (in yeast, ref. 17). Indeed, import of peroxisomal proteins in vivo is blocked by proton ionophores in yeast (18). More disconcerting is the fact that the glycosomal PGK does not go into yeast peroxisomes in vivo. This is unexpected since the topogenic signal in a peroxisomal luciferase of insects works in mammals and plants (13, 14), indicating that peroxisomal addressing signals are universal, like those for other organelles. As Kinetoplastida diverged from the eukaryotic lineage at an early moment, they may have developed a new postal code for microbody delivery (cf. ref. 19). We cannot exclude, however, that the picture of glycosomal topogenesis, painted in the previous section, presents an oversimplification.

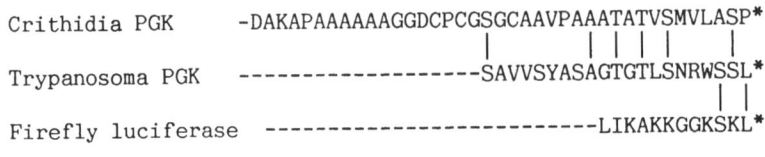

Fig. 3

Putative targeting signals for peroxisomes (glycosomes). The sequences shown are C-terminal sequences of the gPGKs of Crithidia (12), T. brucei (7) and firefly luciferase (13).

Table 1. Amino acid homologies of PGKs from T. brucei and
         Crithidia fasciculata (12)

|  | T. brucei | | C. fasciculata | |
|---|---|---|---|---|
|  | B | C | B | C |
| T. brucei B [*] | - | | | |
| T. brucei C | 93 | - | | |
| C. fasciculata B | 75 | 75 | - | |
| C. fasciculata C | 75 | 75 | 99 | - |

[*]
Gene B codes for the cytosolic PGK, gene C for the glycosomal
enzyme.


## Evolutionary origin of microbodies

How glycolysis got into microbodies in evolution is intimately tied up
with the wider question of the evolutionary origin of microbodies. This
became an issue only recently, when it was realized that microbodies only
arise from pre-existing microbodies. Although the evidence for this
interpretation is still circumstantial and incomplete (see 5), it is
sufficiently persuasive for us to base the remaining discussion on.

The organelles of eukaryotic cells are thought to have originated in
evolution either from domesticated endosymbionts or from specialized parts of
endogenous membranous structures. The endosymbiont origin of mitochondria and
chloroplasts is now generally accepted on the basis of the high sequence
homologies between some organelle components and their counterparts in
bacteria (see 20). Recent work has also delineated a plausible pathway for
the transfer of genes from organelle to nucleus, always the most awkward part
of the endosymbiont scenario. Exogenous DNA is easily integrated into nuclear
DNA and the occasional lysing mitochondrion should provide a constant source
of mitochondrial genes. In fact, bits and pieces of mtDNA have been found in
nuclei of several different eukaryotic phyla (21-23). It is also surprisingly
easy to provide a cytosolic protein with a weak (24) mitochondrial access
ticket: a substantial fraction of randomly chosen bacterial protein
sequences, tagged on to the N-terminus, will do (25, 26). Hence it is not
unreasonable that $10^9$ years of evolution would suffice to transfer more than
90% of all mitochondrial genes one by one to the nucleus and it is only a
small step to envisage microbodies as derived from endosymbionts that have
transferred all their genes to the nucleus (5, 20, 27, 28), making a genetic
system in the organelle superfluous.

There are arguments why microbodies might have been more successful in
this gene transfer than mitochondria. The few genes retained in all
mitochondrial DNAs code for highly conserved hydrophobic sub-units of the
oxidative phosphorylation system in the inner mitochondrial membrane. It has
been shown that such proteins may require large cleavable topogenic sequences
to get them in place (29, 30) and these sequences may not be as easily
obtained in evolution by simple gene rearrangements as the average
mitochondrial addressing peptide. It may also be difficult for such
hydrophobic proteins to avoid insertion into the ER, when they are
synthesized in the cytosol (31). Moreover, gene transfer from mitochondria to
nucleus was probably complicated in a later phase of evolution by an altered
genetic code in mitochondria. In contrast, microbodies do not contain complex
membrane-embedded multi-enzyme systems (see 27) and gene transfer for an
endosymbiont evolving into a microbody might therefore have been simpler. We
do not want to suggest that the membrane could not have been rather complex
initially. We rather envisage that a division of labour took place in the
ancestral eukaryote and that tough jobs were allocated to mitochondria and
chloroplasts. Microbodies ended up with the unskilled metabolic labour that
did not require a complex membrane. The remaining membrane proteins could
easily be inserted from the outside. No essential genes for membrane proteins

Table 2    Binding of nascent PGKs to isolated glycosomes
in vitro (based of unpublished results of B.W.S.)

| Gene constructs used | | Binding to glycosomes of | |
|---|---|---|---|
| | | T. brucei | Crithidia |
| T. br. gene B | ——————— | - | - |
| T. br. gene C | + + + ▮ | + | + |
| T. br. gene Bc | ▮ | - | - |
| T. br. gene Cb | + + + | + | + |
| Crith. gene B | ▮———— | - | - |
| Crith. gene C | ▮ ▬ | - | - |

The experiment measures binding to glycosomes of nascent PGKs generated
by transcription and translation in vitro. The gene constructs were placed
downstream of a SP6 promoter, transcribed with SP6 RNA polymerase and
translated in a reticulocyte system containing [$^{35}$S]-methionine. T.brucei
(T.br.) gene B codes for the cytosolic neutral PGK (see Fig. 1), gene C
for the glycosomal, positively charged (++) PGK with C-terminal extension
(filled block) The Bc construct yields a cytosolic PGK, but with a C-
terminal extension. The Cb construct yield the glycosomal PGK, but without
its C-terminal extension. Crithidia (Crith.) genes B (cytosolic PGK) and C
(glycosomal PGK) both yield neutral PGKs. Each construct yielded a labeled
protein of the expected size in denaturing gels. Crude glycosomes
consisted of a particle fraction collected for 10 minutes at 14.000 x g
and adsorption was measured at 25° C in isotonic buffer, containing 100 mM
KCl and ATP.

Adsorption in this system was not dependent on incubation time (but 5
minutes were required to separate particles from incubation mixture),
incubation temperature (same result at 0° C) or added ATP and seems only
to require positively charged complete PGK. PGK fragments were not
significantly bound.

would have to be retained in the organelle genome and hence the microbody
would be free to lose its genetic system.

An endosymbiotic origin of microbodies explains why microbody proteins
are made on free polyribosomes and imported post-translationally. It explains
how complex metabolic pathways have come to reside within the microbody: they
were there from the start. It can also account for the diversity of microbody
contents either by differential loss of pathways or by a multiplicity of
endosymbiotic ancestors. Finally, the absence of peroxisomes, mitochondria
and chloroplasts in a group of primitive protozoa, the Archezoa, is also most
easily explained by the later acquisition of all three organelles as
endosymbionts (20, 32).

There are few properties of microbodies that do not readily fit the endosymbiont hypothesis. One example is the tubular network that connects peroxisomes of liver and possibly other animal tissues (see 27). Prokaryotes never consist of tubular networks and other cell organelles of probable endosymbiotic origin, like mitochondria and chloroplasts, maintain their individuality. Peroxisomal networks are not a universal feature of all eukaryotes, however, as fungi appear to lack these (33). Moreover in organisms like baker's yeast, intensive fusion and fission of mitochondria takes place and a substantial fraction of the mitochondria is even interconnected (34). Peroxisome networks may therefore represent a late evolutionary development.

The lack in microbody proteins of cleavable N-terminal addressing signals (5), characteristic of most (but not all, see 5 and 35) mitochondrial and chloroplast proteins, does not argue against an endosymbiotic origin either. If the microbody ancestor would have transferred its genes to the nucleus at an early moment in evolution, a long period might have been available to incorporate all signal sequences within the mature protein, as this is also the case for nuclear proteins (see 5). Hence, the endosymbiont hypothesis provides a reasonable hypothesis for microbody origin.

The alternative is that microbodies represent an invention of the eukaryotic cell, like lysosomes. They could have arisen from other membranes by increasing specialization like Pallas Athene sprang from the head of Zeus, or they could have been created de novo. As an autogenous (36) or cluster clone (37) origin has long been argued for mitochondria and chloroplasts, one can hardly claim that such an origin would be impossible for microbodies.

## How did the glycolytic enzymes get into microbodies in the course of evolution?

Three scenarios can be considered:
1. One by one transfer
This scenario postulates that the acquisition of a glycosomal glycolytic system was a relatively late event in evolution and that the enzymes changed location one by one. As Cavalier-Smith puts it: "Glycosomes could have evolved similarly by the attachement of peroxisomal transit sequences to glycolytic enzymes" (20). We see formidable obstacles to such a massive immigration, as briefly specified before (5):
There is considerable circumstantial evidence that the glycosomal membrane has a low permeability to glycolytic intermediates and co-factors (2, 5), like most cellular membranes. We see no reason to assume that this has ever been different. This implies that the cytosolic glycolytic system would have to remain functional, as the glycosomal system was slowly assembled. This would therefore require gene duplication followed by acquisition of a microbody addressing signal by one of the copies. Gene duplication may have been driven by selection for increased levels of glycolysis. Even today there are two identical gene copies of GAPDH and aldolase per haploid T. brucei genome (38). It is difficult to envisage, however, what the advantage could be of a single glycosomal isoenzyme. Only after acquisition of a cooperating set of enzymes, would the glycosomal system become functional. What is the minimal set that might confer some selective advantage to its host? Fig. 1 shows that present-day glycolysis in glycosomes is beautifully balanced. The ATP utilized in producing fructose-1,6-bisP is regenerated at the PGK level; the $NAD^+$ reduced by GAPDH is reoxidized by glycerol-3-P dehydrogenase. Each of the 9 enzymes seems essential, not to mention translocators for exchange of DHAP and glycerol-3-P and for 3-P-glycerate and Pi. Even if one postulates additional translocators, allowing ADP/ATP exchange or movement of additional sugar-phosphates, it is still difficult to envisage how transfer of 2 or 3 enzymes could create a sufficiently advantageous division of labour to provide a wedge for further transfers. It does not seem realistic to assume that successful transfer of one enzyme would

occur at rates higher than 10$^{-8}$/division. There is also no reason to assume that the acquisition of a microbody uptake signal in one enzyme, would affect the acquisition in another: the transfer of 4 enzymes would therefore occur at 10$^{-32}$/division (disregarding loss of newly acquired glycosomal isoenzyme genes). Even the accidental creation of a new pathogen by recombinant-DNA experiments was calculated to be more likely in the early days of the recombinant DNA controversy.

It may be argued that transfer of parts of metabolic pathways can not be so difficult in practice as claimed here in theory, because present-day glycosomes (and microbodies in general) do contain odd bits and pieces of metabolism. Examples are two late steps in pyrimidine biosynthesis (uniquely found in glycosomes) and the first two steps in ether lipid biosynthesis (2). In these cases only two enzymes are affected, however. Moreover, it is not known whether these enzymes entered microbodies later in evolution, or were there from the start as well.

In conclusion, one by one transfer of glycolytic enzymes into pre-existing microbodies in the course of evolution seems unlikely.

2. Transfer en bloc

A glycolytic multi-enzyme complex might have been transferred en bloc, either by acquiring a signal that would allow uptake of the entire complex, or by acquiring a membrane that would allow the enwrapped complex to fuse with pre-existing microbodies. There are no precedents for de novo membrane generation. Bellion and Goodman (18) have suggested, however, that peroxisomal proteins accumulate as a complex at the peroxisomal membrane before or during import. If this would allow some proteins in the complex to enter peroxisomes without specific access ticket, this would provide a scenario for the transfer in evolution of multi-enzyme systems from cytosol to peroxisome by acquisition of a group ticket. The model of Bellion and Goodman is based on inhibition experiments in yeast with a proton ionophore that lowers cellular ATP. The fact that nascent peroxisomal proteins accumulate and associate (aggregate?) under these conditions on the outside of the peroxisomal membrane is not remarkable if peroxisomal entry requires a proton gradient and/or ATP. Nascent peroxisomal protein would bind to the putative uptake receptor and get stuck there in the presence of the ionophore. This does not imply that complex formation is a prerequisite for uptake, nor that members of the complex would not require an individual access ticket to get in. All available evidence suggests that protein unfolding is required to go through membranes (see 26) and it is therefore unlikely that piggy-backing at any scale could occur in this process. Finally, the recent work by Keller and Subramani and their associates shows that a mouse cytosolic protein like dihydrofolate reductase (DHFR) can be made to enter peroxisomes of mammals, plants and yeast by tagging on a short topogenic sequence (13, 14). It is unlikely that DHFR would be able to form complexes with nascent peroxisomal proteins. Complex formation is therefore no prerequisite for peroxisome uptake.

On the basis of these arguments we consider it unlikely that complex metabolic pathways could have entered microbodies as multi-enzyme complex, notwithstanding the results of Bellion and Goodman.

3. Retention of the glycolytic system from the ancestral endosymbiont that gave rise to microbodies.

This model that we favour has been discussed in the previous section.

Origin of glycosomes: the evidence

We have analysed the sequence of the available glycosomal enzymes in search for tracks of their evolutionary origin. Homologous sequences from a sufficient set of pro- and eukaryotes are only available for the glycosomal GAPDH (glyceraldehyde-P dehydrogenase) and TIM (triose-P isomerase). The complete sequence of the cytosolic GAPDH of T. brucei is not yet available; there is no cytosolic TIM. The glycosomal GAPDH shows about equal (52-57%)

170

Table 3. Percentage amino acid identity of TIMs
(from ref. 44)

|   | 1 | 2 | 3 | 4 | 5 | 6 | 7 | 8 |
|---|---|---|---|---|---|---|---|---|
| 1. T. brucei | - | | | | | | | |
| 2. Human | 52 | - | | | | | | |
| 3. Rabbit | 51 | 98 | - | | | | | |
| 4. Chicken | 52 | 89 | 88 | - | | | | |
| 5. Coelacanth | 48 | 82 | 82 | 80 | - | | | |
| 6. Yeast | 48 | 53 | 52 | 53 | 49 | - | | |
| 7. E. coli | 44 | 46 | 46 | 46 | 44 | 45 | - | |
| 8. B. stearothermophilus | 38 | 37 | 37 | 39 | 36 | 34 | 38 | - |

sequence identity with both pro- and eukaryotic GAPDHs (39). Even the S-loop region (residues 178-200), in which eukaryotic GAPDHs differ maximally from the GAPDHs of bacteria and chloroplasts (40, 41), gives intermediate results: the T. brucei glycosomal GAPDH shares a Trp residue at position 180 with the prokaryotic enzymes, but lacks most of the other residues, characteristic of prokaryotes. It is of interest, however, that the N-terminal 85 amino acids of the cytosolic GAPDH of T. brucei are 80% homologous to the E. coli enzyme, suggesting that this enzyme shares substantial homology (64-69%) with E. coli GAPDH like all other eukaryotic (cytosolic) GAPDHs, including the yeast enzyme (10, 39). The homology of the sequenced parts of cytosolic and glycosomal GAPDH of T. brucei is around 50%. Why eukaryotic nuclear GAPDHs should resemble that of E. coli is unclear (40), as eukaryotes are thought not to have arisen from eubacteria (20, 42). Nevertheless the fact that the cytosolic GAPDH of T. brucei resembles that of other eukaryotes, whereas the glycosomal GAPDH does not, is compatible with the idea that both enzymes come from different lineages and hence with an endosymbiont hypothesis of glycosome origin (43).

Table 3 presents the data for glycosomal TIM. In this case the sequence identity with the eukaryotic homologues is a little higher than with the prokaryotic ones and this is especially striking in the C-terminal 40 amino acids (5, 44).

Before squeezing a conclusion from these results, one should remember the inherent pitfalls in this type of analysis: First, the ancestral microbodies might have originated from the same bacterial sub-group as the ancestral eukaryotic cell. In that case, the differences between endosymbiont and host might have been small from the start and undetectable after $10^9$ years. There is no evidence on this point. Since peroxisomes are bounded by a single membrane, Cavalier-Smith (20, 32) has suggested that they have arisen from a single-membrane bounded bacterium, i. e. Gram-positive eubacteria or archaebacteria. It is equally plausible, however, that an outer membrane was originally present and lost. A second complication is the gene conversion, so strikingly illustrated by our results for PGK (8, 12). With separate genes for iso-enzymes in cytosol and microbody combined in one nucleus, these may be homogenized by gene conversion, leaving only the minimal differences required for differential topogenesis. Finally, if a nucleus contains more than one gene for TIM and one of these acquires a microbody signal, whereas the other is lost, the gene originally present in the symbiont and transferred to the nucleus would not have to be preferentially retained. On the contrary, it would seem simpler for the cell to duplicate the nuclear gene for TIM, attach a microbody uptake sequence to one of the genes and then lose the corresponding organelle gene, without ever bothering with gene transfer.

Our conclusion is that an endosymbiont origin for microbodies remains plausible. It provides a relatively simple explanation for the diversity of metabolic pathways found in microbodies of different origins, including the presence of a glycolytic system in glycosomes. Unfortunately, it may be difficult to obtain evidence that would refute this hypothesis. Because of the confounders outlined, any microbody pathway that might have been present already in the original host of the endosymbiont (like glycolysis) may not be informative. More useful may be the multi-enzyme complex of fatty acid oxidation, found in all microbodies and often also in mitochondria, but not elsewhere in the eukaryotic cell (45). No prokaryotic sequences are yet available, however. The most informative comparisons may come from analysis of metabolic pathways only found in (some) microbodies, like the glyoxylate cycle. A further search for microbodies in primitive eukaryotes, like Microsporidia (46), could also provide new clues.

In conclusion, there are no solid experimental data yet to answer the question how glycolysis managed to get into microbodies in the course of evolution. A plausible answer, however, is that glycolysis never moved in, but was part of the endosymbiont from which glycosomes developed $10^9$ years ago.


## Acknowledgements

Supported in part by a grant from the Netherlands Foundation for Chemical Research (SON) with financial aid from the Netherlands Organization for Research (NWO)

## References

1. F. R. Opperdoes and P. Borst, Localization of nine glycolytic enzymes in a microbody-like organelle in Trypanosoma brucei: The glycosome, FEBS Letters 80:360 (1977).
2. F. R. Opperdoes, Compartmentation of Carbohydrate metabolism, Ann. Rev. Microbiol. 41:127 (1987).
3. F. R. Opperdoes, The glycosome, Ann. N.Y. Acad. Sci. 386:543 (1982).
4. F. R. Opperdoes, E. Nohynkova, E. Van Schaftingen, A.-M. Lambeir, M. Veenhuis and J. Van Roy, Demonstration of glycosomes (microbodies) in the Bodonid flagellate Trypanoplasma borelli (Protozoa, Kinetoplastida), Mol. and Biochem. Parasit. 30:155 (1988).
5. P. Borst, How proteins get into microbodies (peroxisomes, glyoxysomes, glycosomes), Biochim. Biophys. Acta GSE 866:179 (1986).
6. D. T. Hart, P. Baudhuin, F. R. Opperdoes and Ch. De Duve, Biogenesis of the glycosome in Trypanosoma brucei: the synthesis, translocation and turnover of glycosomal polypeptides, EMBO J.6:1403 (1987).
7. K. A. Osinga, B. W. Swinkels, W. C. Gibson, P. Borst, G. H. Veneman, J. H. Van Boom, P. A. M. Michels and F. R. Opperdoes, Topogenesis of microbody enzymes: A sequence comparison of the genes for the glycosomal (microbody) and cytosolic phosphoglycerate kinases of Trypanosoma brucei, EMBO J. 4:3811 (1985).
8. S. M. Le Blancq, B. W. Swinkels, W. C. Gibson and P. Borst, Evidence for gene conversion between the phosphoglycerate kinase genes of Trypanosoma brucei, J. Mol. Biol. 200:439 (1988).
9. W. C. Gibson, B. W. Swinkels and P. Borst, Post-transcriptional control of the differential expression of phosphoglycerate kinase genes in Trypanosoma brucei, J. Mol. Biol. 201:315 (1988).
10. O. Misset, O. J. M. Bos and F. R. Opperdoes, Physical properties of glycolytic enzymes from Trypanosoma brucei: Remarkable differences with the mammalian counterparts, Eur. J. Biochem. 157:441 (1986).
11. R. K. Wierenga, B. Swinkels, P. A. M. Michels, K. Osinga, O. Misset, J. Van Beumen, W. Gibson, J. P. M. Postma, P. Borst, F. R. Opperdoes and W. G. J. Hol, Common elements on the surface of glycolytic enzymes from Trypanosoma brucei may serve as topogenic signals for import into glycosomes, EMBO J. 6:215 (1987).
12. B. W. Swinkels, R. Evers and P. Borst, The topogenic signal of the glycosomal (microbody) phosphoglycerate kinase of Crithidia

<u>fasciculata</u> resides in a carboxy-terminal extension, <u>EMBO J</u>. 7:1159 (1988).

13. S. J. Gould, G.-A. Keller and S. Subramani, Identification of a peroxisomal targeting signal at the carboxy terminus of Firefly Luciferase, <u>J. of Cell Biol</u>. 105:2923 (1987).

14. S. J. Gould, G.-A. Keller, and S. Subramani, Identification of peroxisomal targeting signals located at the carboxy terminus of four peroxisomal proteins, <u>J. of Cell Biol</u>. 107 (1988) in press.

15. H. F. Dovey, M. Parsons and C. C. Wang, Biogenesis of glycosomes of <u>T. brucei</u>: An in vitro model of 3-phosphoglycerate kinase import, <u>Proc. Natl. Acad. Sci. U.S.A</u>. 85:2598 (1988).

16. N. Visser, F. R. Opperdoes and P. Borst, Subcellular compartmentation of glycolytic intermediates in <u>Trypanosoma brucei</u>, <u>Eur. J. Biochem</u>. 118:521 (1981).

17. K. Nicolay, M. Veenhuis, A. C. Douma and W. Harder, A $^{31}P$ NMR study of the internal pH of yeast peroxisomes, <u>Arch. Microbol</u>. 147:37 (1987).

18. E. Bellion and J.M. Goodman, Proton ionospheres prevent assembly of a peroxisomal protein, <u>Cell</u> 48:165 (1987).

19. P. Bird, M-J. Gething, and J. Sambrook, Translocation in yeast and mammalian cells: not all signal sequences are functionally equivalent, <u>J. of Cell Biol</u>. 105:2905 (1987)

20. T. Cavalier-Smith, The simultaneous symbiotic origin of mitochondria, chloroplasts, and microbodies, <u>Ann. N. Y. Acad. Sci</u>. 503:55 (1987).

21. H. T. Jacobs, J. W. Posakony, J. W. Grula, J. W. Roberts, J.H. Xing, R. J. Britten, E. H. Davidson, Mitochondrial DNA sequences in the nuclear genome of Strongylocentrotus purpuratus, <u>J. Mol. Biol.</u> 156:609 (1983).

22. G. Gelissen, J.Y. Bradfield, B.N. White, G.R. Wyatt, Mitochondrial DNA sequences in the nuclear genome of a locust, <u>Nature</u> 301:631 (1983).

23. R. M. Wright, D. J. Cummings, Integration of mitochondrial gene sequences within the nuclear genome during senescence in a fungus, <u>Nature</u> 302:86 (1983).

24. N. Pfanner, R. Phaller and W. Neupert, How finicky is mitochondrial protein import? <u>Tr. Biochem. Sci.</u> 13:165 (1988).

25. A. Baker and G. Schatz, Sequences from a prokaryotic genome or the mouse dihydrofolate reductase gene can restore the import of a truncated precursor protein into yeast mitochondria, <u>Proc. Natl. Acad. Sci. U.S.A.</u> 84:3117 (1987).

26. D. Roise and G. Schatz, Mitochondrial presequences, <u>J. Biol. Chem.</u> 263:4509 (1988).

27. P. B. Lazarow and Y. Fujiki, Biogenesis of Peroxisomes, <u>Ann. Rev. Cell Biol.</u> 1:489 (1985).

28. C. De Duve "Peroxisomes and glyoxysomes," H. Kindl and P. B. Lazarow, eds. <u>Ann. N.Y. Acad. Sci.</u> 386:1 (1982).

29. D. P. Gearing and P. Nagley, Yeast mitochondrial ATPase subunit 8, normally a mitochondrial gene product, expressed in vitro and imported back into the organelle, <u>EMBO J.</u> 5:3651 (1986).

30. P. Nagley, L. B. Farrell, D. P. Gearing, D. Nero, S. Meltzer and R. J. Devenish, Assembly of functional proton-translocating ATPase complex in yeast mitochondria with cytoplasmically synthesized subunit 8, a polypeptide normally encoded within the organelle, <u>Proc. Natl. Acad. Sci. USA</u>, 85:2091 (1988).

31. G. Von Heijne, Why mitochondria need a genome, <u>FEBS Letters</u> 198:1 (1986).

32. T. Cavalier-Smith, Eukaryotes with no mitochondria, <u>Nature</u> 326:332 (1987).

33. M. Veenhuis, J. P. Van Dijken and W. Harder, The significance of peroxisomes in the metabolism of one-carbon compounds in yeasts, <u>Adv. Microb. Physiol.</u> 24:1 (1983).

34. Y. Yotsuyanagi, Etudes sur le chondriome de la levure I. Variation de l'ultrastructure du chondriome au cours du cycle de la croissance aérobie, <u>J. Ultrastructure Res.</u> 7:121 (1962)

35. L. A. Grivell, Protein import into mitochondria, Int. Rev. Cytol. 111:107 (1988).

36. H. R. Mahler, The exon: intron structure of some mitochondrial genes and its relation to mitochondrial evolution, in: Int. Rev. of Cytol. G.H. Bourne and J. F. Danielli, eds. 82:1 (1983), Academic Press,

37. L. Bogorad, Evolution of organelles and eukaryotic genomes, Science 188:891 (1975).

38. F. R. Opperdoes, Glycosomes may provide clues to the import of peroxisomal proteins, Tr. Biochem. Sci. 13:255 (1988).

39. P. A. M. Michels, A. Polliszczak, K. A. Osinga, O. Misset, J. Van Beeumen, R. K. Wieringa, P. Borst and F. Opperdoes, Two tandemly-linked identical genes code for the glycosomal glyceraldehyde-phosphate dehydrogenase in Trypanosoma brucei, EMBO J. 5:1049 (1986).

40. W. Martin and R. Cerff, Prokaryotic features of a nucleus-encoded enzyme, cDNA sequences for chloroplast and cytosolic glyceraldehyde-3-phosphate dehydrogenase from mustard (Sinapis alba), Eur. J. Biochem. 159:323 (1986).

41. M. C. Shih, G. Lazar, and H. M. Goodman, Evidence in favor of the symbiotic origin of chloroplasts: primary structure and evolution of tobacco glyceraldehyde-3-phosphate dehydrogenases, Cell 47:73 (1986).

42. J. A. Lake, Origin of the eukaryotic nucleus determined by rae-invariant analysis of rRNA sequences, Nature 331:184 (1988).

43. M. W. Gray, and W. Ford Doolittle, Has the endosymbiont hypothesis been proven? Microbiol. Rev. 46:1 (1982).

44. B. W. Swinkels, W. C. Gibson, K. A. Osinga, R. Kramer, G. H. Veeneman, J. H. Van Boom and P. Borst, Characterization of the gene for the microbody (glycosomal) triosephosphate isomerase of Trypanosoma brucei, EMBO J. 5:1291 (1986).

45. W. H. Kunau, S. Bühne, M. De La Garza, C. Kionka, M. Mateblowski, U. Schultz-Borchard and R. Thieringer, Comparative enzymology of β-oxidation, Biochem. Soc. Transact. 16:418 (1988).

46. C. R. Vossbrinck, J.V. Maddox, S. Friedman, B.A. Debrunner-Vossbrinck and C. R. Woese, Ribosomal RNA sequence suggest microsporidia are extremely ancient eukaryotes, Nature 326:411 (1988).

# CAN PLANT RNA VIRUSES EXCHANGE GENETIC MATERIAL?

Jean-Christophe Boyer, Marie-Dominique Morch and Anne-Lise Haenni

Institut Jacques Monod, 2 Place Jussieu - Tour 43, 75251 Paris Cedex 05, France

## 1. INTRODUCTION

The problem of the exchange of genetic material among viruses is intimately related to that of the evolution of these viruses, as are also the acquisition or the loss of genetic material. Similarly, the concept of virus evolution is associated with that of the origin of viruses.

The origin of viruses and their evolution have become subjects of intense research and discussion, as evidenced by the recent explosion of review articles on these topics (Holland et al., 1982; Reanney, 1982; Goldbach, 1986; Gibbs, 1987; King et al., 1987; Taylor and Hershey, 1987; Goldbach and Wellink, 1988; Hodgman and Zimmern, 1988; Strauss and Strauss, 1988). We have therefore preferred to restrict the subject of the present paper to one phenomenon participating in viral evolution, namely recombination; moreover, because of our current interest in plant viruses, this survey deals essentially with recombination among plant RNA viruses of (+) polarity, with only passing reference to reports of recombination among animal RNA viruses that do not replicate via DNA provirus copies.

Historically, the concept of recombination in RNA was long obliterated by the observations 1) that the enzymes involved in DNA recombination have no effect on RNA, 2) that early reports of recombination in influenza virus were in fact due to rearrangements of genome segments (Hirst, 1962), and 3) that with the exception of picornaviruses (Hirst, 1962; Pringle, 1965; McCahon and Slade, 1981; Kirkegaard and Baltimore, 1986; reviewed in King et al., 1987) and the naturally-occurring MDV RNAs originating from bacteriophage Qβ (reviewed in Van Duin, 1988) no RNA recombination between viral RNAs could be detected, even between the well-studied RNA phages (Horiuchi, 1975). Furthermore, among plant viruses, neither recombination in mixedly infected plants, nor defective interfering (DI) RNAs had been demonstrated until recently.

This paper first surveys the bases of viral evolution. It then reviews the examples of plant RNA viruses for which recombination is either supported by direct experimental evidence or strongly suggested by sequence analyses. Finally, it discusses the possible mechanisms involved in recombination. This is a theme of ever-growing interest, as homologies between plant and animal viruses become increasingly apparent with the accumulation of viral genome sequence data.

## 2. BRIEF SURVEY OF THE BASES OF VIRAL EVOLUTION

Although it was long considered that RNA virus recombination must be an extremely rare event (if it existed at all), it was clear that RNA viruses undergo mutations at a very high rate. This has been thought to reflect the error level of the RNA polymerase, since no mechanism of proof-reading is detected in RNA virus replication.

### 2.1. Point Mutations

The effect of single base changes stems from formal genetic and serial passaging studies on RNA phages, later followed by work on eukaryotic RNA viruses. It is important to bear in mind the fact that exact mutation rates are difficult to establish 1) since they may be counter-balanced by reversions favored by selection pressure, 2) since rates must be evaluated for a single base change that allows the mutant to grow to a reasonable final titer in the absence of competition with the wild-type parent virus, and 3) since site-dependent variations for mutations exist. In spite of these reservations, it is clear that mutation rates for RNA are several orders of magnitude greater ($10^{-4}$) than for DNA ($10^{-6}$ in prokaryotes). This high mutation rate may explain the small size of the genome of RNA viruses (typically $3 \times 10^3$ to $2 \times 10^4$ nucleotides), since the larger the genome, the smaller the number of mutations tolerated. This may also explain genome segmentation observed among many viruses: it allows for small genome parts, and thus enables non-viable mutated genome parts to be discarded without elimination of the total genome (Reanney, 1982).

It can therefore be expected that all progeny RNA molecules of $10^4$ bases statistically contain one point mutation. However, a large percentage of these may be phenotypically silent mutations if for instance, they affect the third position of codons. The situation is not as simple, because the base in the third position of the codons of viruses is not random for those amino acids corresponding to 2 or 4 codons (Grantham et al., 1981), suggesting that other factors intervene to maintain specific codons; it becomes even more complex in the case of viral genomes using overlapping reading frames as a strategy of expression.

In human influenza virus, the gradual accumulation of point mutations leading to single amino acid sequence changes ("antigenic drift") appears to form the basis for the alteration of antigenic sites, such that they are no longer recognized by the immune system (reviewed in Air and Laver, 1986).

There is nonetheless strong selective pressure to maintain a relatively uniform population of viral RNA molecules and to discard non viable mutants, possibly by direct competition during virus growth. An interesting aspect of selection pressure is DI-modulated virus infection. DI particles are virus deletion mutants that interfere specifically with the replication and packaging of the parental (helper) virus (reviewed in Holland et al., 1980; Perrault, 1981). It has been suggested that intracellular coevolution could occur between the genetic material of DI RNAs and that of their parental virus. From this point of view vesicular stomatitis virus (VSV) and its associated DI particles have certainly been the most thoroughly studied. Cave et al. (1985) noticed that if mice are infected with a mixture of standard infectious VSV and its DI particles, continuous dynamic interaction occurs between the two populations for several days and with a cyclic pattern. It appears that such coevolution proceeds indefinitly through multiple cycles of selection of virus mutants (referred to as Sdi) resistant to a certain DI particle, followed by mutants resistant to a newly predominant DI particle etc. In recent studies, DePolo et al. (1987) examined the molecular consequences of this noteworthy phenomenon by sequencing parts of certain Sdi genomes. It was demonstrated that the polymerase-associated (NS) genes are not usually mutated. On the other hand, the nucleocapsid (N) genes slowly accumulate stable mutations that could sometimes lead to large variations in the Sdi phenotype. Therefore, it seems likely that such recurring RNA population interactions can increase the rate of modifications of viral genomes and thus play an important role in virus evolution.

## 2.2. Recombination

This is the second important mechanism participating in virus evolution. Recombination basically concerns the formation of new combinations of genes or of linked genes resulting in new heritable characters or new combinations of such characters. Geneticists and molecular biologists often extend this definition to include the following phenomena.

1. Reassortment of genomic RNAs
2. Recombination of functional domains
3. Acquisition of exogenous genetic material.

These three basic types of "recombination" are discussed here; as will become apparent from the examples provided, they are not mutually exclusive.

### 2.2.1. Reassortment of Genomic RNAs

This corresponds to genetic rearrangement and is often referred to as "antigenic shift". It only occurs among viruses with a segmented genome, where "new" viruses can be obtained *in vivo* as well as in the test tube that possess new combinations of characteristics due to the exchange of genome parts (reviewed in Bruening, 1977; Air and Laver, 1986; Van Vloten-Doting and Jaspars, 1977; Mayo, 1987; Hiruki, 1987); these new isolates are called "pseudorecombinants". As demonstrated by back-cross experiments, the genome parts of the pseudorecombinants replicate true to type. In influenza virus, reshuffling among genome parts of two strains seems to be responsible for the appearance of "new" viruses infecting birds and man (reviewed in Webster *et al.*, 1982). The new isolates are not recombinants in the original sense of the term, since no breaking and rejoining of nucleic acid strand are considered to have occurred; for this reason, they will not be discussed further.

### 2.2.2. Recombination of Functional Domains

One of the main results of sequence comparisons among RNA viruses is the definition of protein domains conserved to various extents throughout several virus families and apparently characteristic of a given function. This section briefly describes the most significant domains and considers their combinations in various examples of viral genome since all these observations are in support of the "modular theory" of evolution which will then be presented.

*2.2.2a. Functional Domains do Exist.* A conserved domain is usually defined by a core consensus amino acid (aa) sequence thought to be required for its function. The aa sequence around this consensus shows additional homologies and specific residues that are conserved only in a limited number of viruses. These second order consenses have contributed to the grouping of virus families from animal and plant hosts into several "supergroups" or "superfamilies" (reviewed in Goldbach, 1986; Goldbach and Wellink, 1988; Hodgman and Zimmern, 1988; Strauss and Strauss, 1988). To date, the most spectacular information has been deduced from comparative studies performed with (+) RNA viruses and only these will be discussed here.

Two supergroups of (+) RNA viruses among those not requiring a DNA step are now well defined on the bases of certain homologous domains and genome organization.

The "picorna-like" viruses (reviewed in Goldbach and Wellink, 1988) share with authentic picornaviruses (polio-, encephalomyocarditis-, rhinoviruses) the following features. A small viral protein (VPg) is bound at the 5' terminus of their genomic RNAs which are polyadenylated at the 3' end. Each genomic RNA is translated as a polyprotein subsequently cleaved to yield structural and non structural viral proteins. Among plant viruses, the como- (e.g. cowpea mosaic virus: CPMV), nepo- and possibly potyviruses are members of the picorna-like supergroup.

The "Sindbis-like" supergroup (reviewed in Goldbach and Wellink, 1988) associates the animal alphaviruses (Sindbis virus, Semliki Forest virus: SFV), possibly the coronaviruses (Gorbalenya et al., 1988b) and a large variety of plant RNA viruses such as alfalfa mosaic virus (AlMV), the bromoviruses (e.g. brome mosaic virus: BMV), the cucumoviruses (e.g. cucumber mosaic virus: CMV), the furoviruses (e.g. beet necrotic yellow vein virus: BNYVV), the hordeiviruses (e.g. barley stripe mosaic virus: BSMV), the ilarviruses (e.g. tobacco streak virus, TSV), the potexviruses, the tobamoviruses (e.g. tobacco mosaic virus: TMV), the tobraviruses (e.g. tobacco rattle virus: TRV), the tymoviruses (e.g. turnip yellow mosaic virus: TYMV; Morch et al., 1988) and possibly the carmoviruses (e.g. carnation mottle virus: CarMV) and the tombusviruses (e.g. cucumber necrosis virus: CNV). The genomes all have a cap structure at their 5' end. They all produce subgenomic RNAs in combination with various additional translation strategies to express their proteins (Morch et al., 1987).

The first domain considered here has been found in all (+) RNA viruses. It is called the "polymerase" domain because of strong evidences in favor of a role of the corresponding proteins in nucleotide (nt) polymerization during viral replication. It is defined by a GDD (the one-letter code for amino acids is used when dealing with consensus sequences) consensus surrounded by hydrophobic residues (Kamer and Argos, 1984) and preceded 20 to 50 aa upstream by a conserved sequence [S/T]GxxxTxxxN[S/T]. Other conserved motifs exist that restrict the homology to one or the other virus supergroup (reviewed in Hodgman and Zimmern, 1988).

A second domain is also well characterized and very ubiquitous. It is designated the nucleoside triphosphate (NTP)-binding site, based on direct and indirect evidences that the corresponding proteins are able to bind NTP (reviewed in Gorbalenya et al., 1988b). The consensus core of the domain is the sequence GxxGxGK[S/T]. It is followed at a distance that can vary from one virus group to the other by a degenerated consensus constituted of a D residue preceded by 3 hydrophobic residues. These two consenses together form part of larger domains spanning over 250 aa and 130 aa in the Sindbis-like and picorna-like supergroups respectively. These domains can be aligned throughout most members of each supergroup and reveal as many as 21 aa in the former and 45 aa in the latter supergroup, that are conserved, i.e. either invariant or having similar biochemical properties (Gorbalenya et al., 1988b). Surprisingly such a domain is lacking in some viruses such as barley yellow dwarf virus (a luteovirus; Miller et al., 1988), CarMV and CNV (Goldbach and Wellink, 1988). Thus, classification of these viruses in the Sindbis-like supergroup may need to be revised.

Other functional domains have been described in (+) RNA virus-coded proteins that are specific for a given supergroup or a given ensemble of virus families. One such domain is the proteinase domain found in picorna and picorna-like viruses. The proteolytic activity required by these viruses for the processing of their polyprotein is encoded in the polyprotein itself; it has been characterized as a cysteine proteinase different from all cellular cysteine proteinases known to date. The essential C residue is present in a consensus sequence Txx[E/G]xCG[S/G]x[L/V/I] followed 5 to 10 aa downstream by a second consensus [I/V]xGxH where the H residue could be part of the C- and H-containing active site of the proteinase (Argos et al., 1984; Gorbalenya et al., 1986; Domier et al., 1987).

Finally a domain has recently been characterized as specific for enzymes involved in DNA unwinding and may be called the helicase domain. It is present in all members of the Sindbis-like supergroup where it overlaps the NTP binding domain (Gorbalenya et al., 1988a; Hodgman, 1988).

A completely different kind of conserved domain is found in the capsid proteins of (+) RNA viruses possessing isometric particles. All the coat proteins involved in such particles and studied so far contain a similar structural motif described as an eight-stranded anti-parallel β-barrel, referred to here as the β-barrel domain (reviewed in Gibbs, 1987). This motif is not found in the coat protein of TMV which forms helicoïdal particles. The definition of a conserved structural motif for the coat proteins of filamentous or rod-shaped RNA viruses awaits further studies on these proteins.

The clearly established existence of conserved domains leads to several speculations. It could for example favor the hypothesis of divergent evolution according to which all protein sequences associated with a given function derive from a common ancestral sequence. The purpose here is rather to use the various domains as units of genetic information for viral genomes and analyze their recombination potential.

*2.2.2b. Different Organizations of Functional Domains.* The relative association of one domain with regard to the other can vary according to three parameters, 1) the nature and the number of the domains that are associated, 2) their organization in the genome, and 3) their mode of expression. Examples of these various combinations follow.

In the Sindbis-like supergroup, the same polymerase domain can be found in isometric particles (AlMV, BMV, CMV, Sindbis virus) as well as in helicoïdal particles (TMV). This domain is thus combined in the former case with the β-barrel domain specific for isometric particles and in the latter case (helicoïdal particles) with a domain specifying another type of coat protein. Similar combinations (Gibbs, 1987) are possible among the picorna-like viruses which possess either isometric (picornaviruses, comoviruses), or rod-shaped (potyviruses) virion particles.

Reciprocally one should stress that the same β-barrel motif can be found associated with Sindbis-type as well as with picorna-type functional domains.

In alphaviruses a domain coding for the glycoproteins engaged in the lipid envelope of the virion is also added to the capsid (Fuller and Argos, 1987). Recently evidence has been provided that recombination has brought together the capsid domain of one alphavirus species with the glycoprotein domain of another species and thus has given rise to a third alphavirus species (Hahn et al., 1988).

Interestingly, the conserved structural motif is present in one copy in the genome of the isometric viruses of the Sindbis-like supergroup and in three copies in the isometric viruses of the picorna-like viruses (Rossmann et al., 1985). These three β-barrel motifs are carried either by three different proteins (VP1, VP2 and VP3) in picornaviruses or by two proteins only (VP37 and VP23) in CPMV (Goldbach, 1986).

Other domains can be present twice on a genome. Two proteinase domains are found in some of the picorna-like viruses (proteinases 2A and 3C of poliovirus, protein 32K and proteinase 24K of CPMV). One of the two copies may have lost its original function and thus have diverged from the consensus (case of the 32K protein of CPMV; reviewed in Wellink and Van Kammen, 1988). Two NTP-binding domains are also found among some members of the Sindbis-like supergroup (BNYVV, BSMV). Here again one copy of the domain is less conserved than the other suggesting diversification through independent evolution following duplication (Goldbach and Wellink, 1988; Gorbalenya et al., 1988a).

Differences in genome organization with respect to the polymerase and the NTP-binding domains appear clearly in the Sindbis-like supergroup. In tripartite genomes (AlMV, BMV, CMV, TSV), these two domains are carried by two separated RNAs, whereas in monopartite genomes such as TMV or Sindbis virus, they are carried by the same RNA.

In TMV, the two domains remain associated on a single large polypeptide whereas in Sindbis virus they become dissociated by proteolytic cleavage. Moreover, sequence alignments reveal in Sindbis virus the insertion of an additional as yet non-assigned domain between the polymerase and the NTP-binding domains which are contiguous in TMV (Ahlquist et al., 1985).

Picornaviruses and CPMV also differ in the sense that the former have a monopartite and the latter a bipartite genome. However the relative organization of the functional domains remains very parallel between these two types of genome (Goldbach, 1986).

In monopartite genomes the capsid β-barrel domains can be found either at the 3' end (e.g. tymoviruses), at the 5' end (e.g. picornaviruses), or in internal regions of the genome (e.g. tombusviruses).

The mode of expression of a given domain may be different among the members of a same virus family. The polymerase domain of the Sindbis virus is a cleavage product of a polyprotein precursor that itself derives from the readthrough of a UGA termination codon located just upstream of the region coding for the polymerase domain. However in SFV, the polyprotein precursor for the polymerase domain is translated without interruption by any termination codon (reviewed for the other alphaviruses in Strauss et al., 1987).

A similar situation, although between two different virus families, has recently been observed when comparing TMV and TYMV. Readthrough of a termination codon is involved in the former virus to account for the translation of the polymerase domain whereas continuous translation of an uninterrupted open reading frame (ORF) in the latter case yields a polyprotein containing the polymerase domain (Morch et al., 1988).

*2.2.2c. The Hypothesis of "Modular Evolution".* Functional domains have been defined and are associated in various combinations throughout many virus families. This statement recalls the hypothesis of modular evolution proposed by Botstein (1980) on the basis of studies performed with the DNA bacteriophages λ and P22. As reported by Gibbs (1987), this hypothesis is defined by the "joint evolution of sets of functionally and genetically interchangeable elements (the domains or modules), each of which carries out a particular biological function". In other words (Hodgman and Zimmern, 1988), "the distinct virus groups are related by the association of shared common components with unique components (such as the transport proteins in plant viruses) that define the specific features of this group". In the case of retroviruses, the possibility is not excluded that rearrangements of modules have taken place during the DNA step of the viral cycle. However for the other RNA viruses neither DNA intermediates nor pseudogenes of viral genomes have been described. Thus, whatever the origin and evolution of a given virus family, the fact remains that the interpretation of all sequence analyses performed on contemporary viral proteins strongly suggests that recombination and gene duplication may occur at the RNA level in viral genomes.

## 2.2.3. Acquisition of Exogenous Genetic Material

Recombination has been considered so far between RNA genomes but exchanges may also occur between cellular and viral genetic material. However not all cellular sequences are known and divergence has occurred since their putative integration into the viral genome. Thus it is often very difficult to trace such sequences in a viral genome. Indirect evidence derived from comparative sequence analyses and more direct evidences based on very limited observations attempt to support the statement of recombination between cellular and viral sequences.

*2.2.3a. Homologies of Viral Domains with Cellular Proteins.* The comparison of the functional domains described above with cellular proteins has stressed striking homologies. The NTP-binding domain of (+) RNA viruses especially in the Sindbis-like supergroup can be aligned with a series of cellular prokaryotic and eukaryotic proteins that have in common the property of interacting with nucleic acids and unwinding helical turns (Hodgman, 1988, Gorbalenya *et al.*, 1988a; Lane, 1988).

The proteinase of picorna-like viruses, although claimed to be a new type of cysteine proteinase can be aligned with both cysteine and serine proteinases suggesting it could derive from a common ancestor of both cellular proteinases (Gorbalenya *et al.*, 1986).

Other homologies of virus-specific sequences with cellular sequences may still be revealed. For example the basic N-terminal region of several but not all capsid proteins of isometric viruses has been proposed to be related to histone or protamine sequences (Argos, 1981).

Although the hypothesis of convergent evolution of sequences is not excluded, these reports are to be taken as indications for the possible cellular origin of some viral sequences. Some direct observations may give weight to this hypothesis.

*2.2.3b. Direct Evidence of Transfer from Host to Virus.* A very well known system of genetic recombination between host and virus is the transducing ability of retroviruses such as the viral copies of oncogenes (v-*onc*) which originated from cellular copies of the same genes (c-*onc*) that were integrated into a retroviral genome (reviewed in Bishop, 1983). This type of recombination is likely to occur at the DNA level and thus will not be further developed here.

An interesting case is however that of some DI particles of Sindbis virus (Monroe and Schlesinger, 1983). Sequence analyses of these DI genomes established the presence at their 5' end of part of a cellular tRNA$^{Asp}$ (see 3.6.). Another example is found in influenza virus: transcription of the mRNAs of this (−) RNA virus is primed with capped oligonucleotides derived from cellular mRNAs (Plotch *et al.*, 1981). In both cases the extent of cellular sequences combined with the viral-related sequences is very small. These examples provide indications that during the virus cycle viral RNAs come in contact with cellular RNAs.

Further speculations on the possible involvement of cellular RNA sequences in the constitution of viral particles relies on more knowledge about RNA recombination mechanisms (see 4.).The possible existence of "cellular RNA genetic elements" has been postulated (Zimmern, 1982) that would be capable of being replicated, as are for example viroids, and of being integrated into viral RNAs via mechanisms that could be the reversal of splicing mechanisms.

## 3. EXAMPLES OF RECOMBINATION AMONG PLANT RNA VIRUSES

This section reviews the examples of (likely) recombination among plant viruses: the first is the only well-established example of recombination, whereas the other four correspond to very probable recombination events. The sixth part briefly reviews a few well-studied examples of recombination among animal RNA viruses and their DI RNAs.

## 3.1. RNA3 of Brome Mosaic Virus (BMV)

This is the only clear-cut example of exchange of genetic material among plant RNA viruses reported to date (Bujarski and Kaesberg, 1986; reviewed in Ahlquist et al., 1987).

The genome of BMV is composed of three distinct RNA molecules (designated RNA1, 2 and 3 in the order of their decreasing size) encapsidated in separate virions. All three components are necessary to induce detectable systemic infection in whole barley plants. The sequence of each RNA has been established. The 3' terminal 193 nt of each RNA are 97-99% homologous with the other two components, suggesting that their conservation is under considerable selection pressure. This 3' region is responsible for the tRNA-like properties of the viral RNA and it also directs the synthesis in vitro of the (-) RNA strand by the virus-specific RNA polymerase complex.

Advantage has been taken of these conserved regions to probe RNA recombination in plant cells. An engineered BMV RNA3 (designated m4 RNA3) has been constructed in which nt 81-100 from the 3' end of RNA3 have been deleted. This region corresponds to a stem and loop that is normally present at the 3' end of the RNAs of BMV; it is conserved in the RNAs of cowpea chlorotic mottle virus, but absent from the RNAs of broad bean mottle virus, two other bromoviruses. When barley plants are inoculated with wild-type (wt) BMV RNA1 and 2 and the deleted m4 RNA3, accumulation of progeny RNA is lower than if m4 RNA3 is replaced by wt RNA3. Interestingly, in systemic infection, m4 RNA3 is unstable: multiple pseudorevertants appear in which the deleted sequence has been restored due to recombination between m4 RNA3 and either wt RNA1 or RNA2. This is accompanied by a concomittant decrease in m4 RNA3.

A close analysis of the pseudorevertants has revealed that the stem and loop region has been regained by RNA3. Each revertant also bears one or more base changes 3' of the restored sequence and a few additional changes 5' of this sequence. These changes correspond to sequences present in either RNA1 or RNA2, so that it is possible to ascribe the pseudorevertants as resulting from a recombination event between RNA3 sequences upstream of the missing stem and loop, and 3' sequences of RNA1 or RNA2. Of the five pseudorevertants analyzed at the molecular level, three result from recombination within the highly conserved region of 193 nt. The other two result from recombination outside of this homologous region, demonstrating that recombination is not limited to homologous sequences (see 4.2.).

## 3.2. RNA2 of Tobacco Rattle Virus (TRV)

TRV contains a bipartite genome, RNA1 and RNA2. The viruses of this group are divided into three clusters, cluster 1 (strains of TPV), cluster 2 (strains of PEBV) and cluster 3 (strain CAM of TRV). Whereas only the 3' terminal third of the sequence of RNA1 is known, RNA2 has been totally sequenced for representatives of all three clusters. The length of RNA2 differs from strain to strain: it is 1799 nt long for strain CAM, 1905 nt for strain PSG of cluster 1, and 3389 nt for strain TCM, also of cluster 1.

The 3' termini of CAM-RNA2, PSG-RNA2 and TCM-RNA2 show 100% homology with the corresponding RNA1 over a length of 459, 497 and 1099 nt respectively, indicating a strong selection pressure in this region and suggesting recombination between the 3' region of RNA1 and RNA2. The most striking case is that of TCM-RNA2. Its 1099 nt-long RNA stretch homologous to RNA1 contains the C-terminal region of the 28.8 K (K = kilodalton) ORF of RNA1, followed by the 16 K ORF, the 3' proximal ORF of RNA1 (Cornelissen *et al.*, 1986; Angenent *et al.*, 1986).

## 3.3. tRNA-Like Regions in Tobamoviruses

Tobamoviruses contain a monopartite RNA genome whose 3' terminal region possesses tRNA-like properties. These viruses have been divided into two subgroups, on the basis of the location of the assembly origin (reviewed in Takamatsu *et al.*, 1983). The common (*Vulgare*) strain and the tomato strain (L strain) belong to subgroup 1 (assembly origin is 800-1000 nt from the 3' end), whereas the cowpea strain (Cc strain) and cucumber green mottle mosaic virus (CGMMV) belong to subgroup 2 (assembly origin 300-500 nt from the 3' end).

The complete sequence of the genome of the *Vulgare* and L strains, and the sequence of the 3' region of the genome of the Cc strain and CGMMV are known, allowing comparisons between the 3' non coding regions of these RNAs (reviewed in Morch and Haenni, 1987). Furthermore, the RNA of the *Vulgare* strain and of CGMMV accepts histidine, whereas that of the Cc strain accepts valine (reviewed in Joshi *et al.*, 1983b).This is an interesting situation because in all other cases of aminoacylation among plant RNA viruses, the RNA of the viruses belonging to a given taxonomic group accept the same aa. On the other hand, the RNA of tymoviruses such as TYMV accepts valine.

A search for sequence similarities has been made between the tRNA-like regions of the *Vulgare* and Cc strains, since they both belong to tobamoviruses but accept different aa, and between the tRNA-like regions of TYMV and the Cc strain, since they belong to different taxonomic groups but accept the same aa (Joshi *et al.*, 1983a). The overall tRNA-like region of TYMV and the Cc strain presents considerable sequence conservation, suggesting that among these conserved regions lie recognition sites for the Val-tRNA synthetase. In contrast, sequence similarity between the tRNA-like regions of the *Vulgare* and Cc strains is confined to the first 3' coterminal $\cong 42$ nt; this conserved region could be required for interaction with other tRNA-specific enzymes, and/or with the RNA polymerase.

The 3' terminal sequences of the CGMMV and L strains are extensively homologous to that of the *Vulgare* strain, consistant with the observation that the RNA of the *Vulgare* strain and of CGMMV accept histidine; it can thus be proposed with confidence that the 3' region of the L strain also accepts histidine, although this has not been tested.

The sequence similarities observed between the 3' region of the Cc strain and TYMV on one hand, and between the 3' region of the Cc strain and the other tobamoviruses on the other hand suggest that the Cc strain may have arisen by recombination between the coding body of a tobamovirus such as the *Vulgare* strain, and the tRNA-like non-coding 3' region of a tymovirus.

## 3.4. Satellite of Turnip Crinkle Virus (TCV)

Satellite RNAs are small molecules that require the company of a helper virus for their replication. Their presence in the plant can modify (intensify or reduce) the symptoms produced by the helper virus alone. Most satellites are encapsidated in helper virus particles, although in certain cases they are encapsidated in separate particles which they themselves encode. They vary in size (200-1700 nt), coding capacity, and form (linear or circular). There is little in common between satellites that accompany viruses of different groups. Moreover, there is little homology between satellites and their helper virus genomes, a feature that is said to distinguish satellites from DI particles (reviewed in  Murant and Mayo, 1982; Francki, 1985; Morch and Haenni, 1987).

However, an unexpected situation had recently been described for RNA C, a virulent satellite of TCV (Simon and Howell, 1986). Its 355 nt-long linear sequence is composed of 2 major domains of about equal lengths. The 3' domain presents 93% homology with the 3' non-coding region of TCV RNA, whereas the 5' domain is homologous to other satellites of TCV. Thus RNA C appears to be a composite molecule with elements of the TCV genome and of other TCV satellites. It thus shares features of DI RNAs and conventional satellites, and it seems possible that it may have arisen from recombination between TCV RNA and one of the other satellites of this virus.

### 3.5. DI RNA of Tomato Bushy Stunt Virus (TBSV)

To date, only one report has clearly established the existence of a DI particle in association with a plant virus (Hillman et al., 1987). Preparations of TBSV contain a 0.4 kilobase-long, genome-related and symptom-modulating species whose characteristics comply with those of DI RNAs. This DI RNA is a colinear deletion mutant of the parental virus and it is a mosaic of 7 juxtaposed RNA stretches ranging in size from 5 to 130 nt in the same relative order as in the parental virus.

### 3.6. Brief Survey Among Animal RNA Viruses and Their DI RNAs

Among picornaviruses, recombination was first described in poliovirus (Hirst, 1962; Ledinko, 1963) and soon thereafter (Pringle, 1965) in foot-and-mouth disease virus (FMDV). More recently evidence for recombination has been obtained for mouse hepatitis virus (MHV), a coronavirus (Lai et al., 1985).

Recombinants of the RNA genomes of poliovirus (Kirkegaard and Baltimore, 1986), FMDV (King et al., 1982; reviewed in King et al., 1987) and MHV (Keck et al., 1988) have been analyzed in detail by T1 fingerprinting and/or nt sequencing.

A recent example of possible natural recombination has been reported (Hahn et al., 1988) with respect to an alphavirus species that contains domains of two other alphavirus species (see 2.2.2b.).

Among animal viruses, DI particles have been found associated not only with minus-stranded RNA viruses such as VSV, influenza virus and Sendai virus (reviewed in Lazzarini et al., 1981), but also with positive-stranded RNA viruses such as the alphaviruses Sindbis virus and SFV (Stollar, 1980) and poliovirus (Kuge et al., 1986).

The structure of these DI RNAs generally involves complex recombinations of parental viral sequences (Lazzarini et al., 1981; Söderlund et al.,1981; Tsiang et al., 1988). This can best be illustrated by the mosaic DI RNA of influenza virus: it is composed of one region of segment 1 flanked by different regions of segment 3. In addition, the regions of segment 3 are borrowed alternatively from the 5' and the 3' part of that segment (Fields and Winter, 1982).

DI RNA formation can sometimes encompass the acquisition of non viral sequences. From this point of view, the example of several DI RNAs isolated from Sindbis virus-infected cells is striking: studies of these DI RNAs have revealed that they possess at their 5' termini the 67 nt deriving from the 3' end of a cellular tRNA$^{Asp}$ covalently attached to either nt 23 or 31 of the virion RNA (Monroe and Schlesinger, 1983, 1984). The exact significance of this surprising feature remains unclear since it does not seem to provide any selective advantage to this DI RNA population over those lacking the host sequence (Tsiang et al., 1985, 1988).

# 4. POSSIBLE MECHANISMS OF RECOMBINATION

The exact mechanisms of recombination are still the subject of speculation. However, two major mechanisms can be distinguished, splicing and the "copy-choice" mechanism.

## 4.1. Splicing and the Copy-Choice Mechanism

Splicing, as postulated at least for the formation of some plant satellite RNAs, requires the pre-existence of a precursor RNA in which excision and ligation at consensus sequences occurs (reviewed in Symons et al., 1985). Trans-splicing (Solnick, 1985; Konarska et al., 1985) could also account for recombination between viral genomes or between a viral genome and a cellular RNA (Zimmern, 1982). To date among viral genomes and DI RNAs, no consensus sequences for splicing events have been observed at the junction of the recombinant regions. For this reason, the copy-choice mechanism is the favored model for recombination.

In the copy-choice mechanism, the polymerase stops synthesis on its template strand and pursues synthesis by reinitiating elsewhere without releasing the nascent daughter strand (Lazzarini et al., 1981; Holland et al., 1982; Kolakofsky and Roux, 1987; Emerson and Schubert, 1987). The polymerase can reinitiate synthesis further on the same strand such as by slippage of the enzyme along the template, or it can switch template and reinitiate synthesis on another copy of the same RNA molecule or on a different RNA molecule (Lazzarini et al., 1981). The different RNA molecules can be anything: the nascent daughter strand itself leading for instance to "snap-back" particles (Lazzarini et al., 1981), another segment in viruses with a multipartite genome as might occur in the case of RNA3 of BMV (see 3.1.) and of RNA2 of TRV (see 3.2.), the RNA of another coinfecting virus as for the Cc strain in the tobamoviruses (see 3.3.), the RNA of the helper virus in the case of the satellite of TCV (see 3.4.), and finally a cellular RNA such as the tRNA$^{Asp}$ of certain DI RNAs of Sindbis virus (see 3.6.). More than one reinitiation event is required in some cases to account for the formation of a recombinant RNA, such as the DI RNA of TBSV (see 3.5.). In the case of the DI RNA of influenza virus (see 3.6.), it has been proposed that the enzyme could zigzag between the 5' and 3' ends of one or two copies of segment(s) 3, assuming these regions are close to one another; in addition, intersegment recombination with segment 1 would occur to complete formation of the DI RNA. In MHV-infected cells, discrete RNA species containing the MHV leader sequences as well as larger leader-containing RNA species are found (discussed in Makino et al., 1986a, 1986b). They might correspond to normal nascent intermediates of RNA replication, or to incomplete transcription products. The presence of such free RNA intermediates supports the copy-choice model of recombination.

## 4.2. Homologous Versus Non-Homologous Recombination

In an effort to define at the molecular level the elements that dictate recombination, attempts have been made to classify recombination events into homologous and non-homologous (reviewed in King et al., 1987; Hodgman and Zimmern, 1988).

Homologous recombination occurs in regions of homology of the protagonist RNAs. As a result, the region of homology is present in the progeny RNA, making the exact site of recombination difficult to assess. A general alignment of the parental molecules with a certain degree of base-pairing, or the stabilization of a structure on one and/or the other side of the immediate cross-over point may favor homologous recombination. This type of recombination appears to be responsible for at least some recombination events observed among picornaviruses (Kirkegaard and Baltimore, 1986; reviewed in King et al., 1987). Indeed, the only region of the genome where recombination has not been observed in FMDV is the region with the most poorly conserved sequence: sequence matches of 18 and 32 bases are found in two cross-over points (King et al., 1986). Homologous recombination is also postulated to be responsible for the generation of certain poliovirus-associated DI RNAs (Kuge et al., 1986).

However, since sequence homology is not a prerequisite for all recombination events, non-homologous recombination must also be invoked. The essential difference between homologous and non-homologous recombination rests in the minimum length of base-pairing required for a primer chain to be elongated, and it has been suggested that this may depend on the polymerase (King et al., 1987).

Non-homologous recombination is less well documented and has not been extensively studied. DI RNAs presumably result from this kind of recombination: few (if any) matching bases are detected at the cross-over point, the most common situation being one base match (Jennings et al., 1983), with a preference for a purine residue at the 3' end of the homologous region. The DI RNA of TBSV (see 3.5.) and the mosaic DI of influenza virus (see 3.6.) are likely examples of such non-homologous recombinations among plant and animal viruses respectively. Another example is found among certain DI RNAs of Sindbis virus (see 3.6.): only a single matching base can be distinguished at the cross-over point between the tRNA$^{Asp}$ and the DI RNA sequence. Non-homologous recombination could account for the situation observed in the tRNA-like region of the Cc strain of TMV (see 3.3.), and the satellite of TCV (see 3.4.). However, it is also possible that a region of sequence homology existed previously in the two parent genomes that permitted homologous recombination to occur; with time, point mutations and evolutionary pressure would have obliterated such a sequence in at least one parent so as to no longer be detectable. This example points to the caution that is warranted in designing models of recombination based on sequence homologies.

Whatever the mode of recombination, the copy-choice model postulated for the generation of DI RNAs supposes that re-initiation of synthesis can be directed efficiently by a single base-pair formed between a 3'-terminal nt (preferentially an A or a G) in the primer and a complementary base in the template.

It is worth returning to the situation encountered in the RNA3 or BMV (see 3.1.). Of the five recombinants analyzed, three have occurred within the 193 nt of the conserved region, whereas the other two occurred upstream, in a region devoid of sequence homology. Thus the RNA polymerase of BMV has presumably performed both homologous and non-homologous recombinations: the mode of recombination does not seem to be an inherent property of at least this polymerase, as has been proposed for others (King et al., 1987).

4.3. Dissecting Recombination

Two questions directly concern recombination: 1) the reasons for which the nascent polymerase-RNA complex leaves its original template, and 2) how it chooses its new template.

Pausing during elongation could cause the nascent polymerase-RNA complex to leave its template. This might occur because of a specific sequence in the template. Indeed, in the case of the mosaic DI RNA of influenza virus, it has been proposed (Fields and Winter, 1982) that uridine-rich regions in the template cause the polymerase to pause: if the polymerase-RNA complex were then to leave its original template, it might resume synthesis at another site. Secondary and tertiary structures within the template might also lead to pausing during elongation (reviewed in Lazzarini et al., 1981). In addition, one cannot exclude the possibility that other factors, such as proteins (capsid proteins ?) might interfere with RNA elongation.

Whatever the reason for which the nascent polymerase-RNA complex is detached from its original template, in homologous recombination it is fairly easy to visualize how the complex might base-pair to a new template, allowing RNA synthesis to proceed.

In non-homologous recombination, the parameters dictating resumption of synthesis are far more elusive. Since base-pairing is not a prerequisite in this form of recombination, other elements are most likely involved. A simple model that can be proposed is one in which the nascent polymerase-RNA complex would bind to the new template at a site for which the polymerase would have a high affinity. In this respect, it is interesting that in strains PSG and TCM of TRV (see 3.2.), just downstream of the junction point between RNA1 and RNA2, the sequence AUAAUUGUU occurs; this sequence resembles the 5'-termini of the genomic and subgenomic TRV RNAs and could be the site of internal initiation for the replicase on minus-strand RNA (Angenent *et al.*, 1986). It would thus appear logical for the polymerase to bind to this sequence. An additional possibility is that structure rather than sequence could constitute the target for binding of the polymerase, as proposed in the case of the DI RNAs of Sindbis virus (Tsiang *et al.*, 1988); indeed, the 5'-terminal region in these DI RNAs can undergo extensive variations while still maintaining the ability to be replicated by the viral-coded enzyme.

Pausing and reinitiation of synthesis at a new template site might be brought about by way of certain transient highly ordered structures that would bring potential cross-over regions in close proximity with one another. This could be brought about by looping-out of sequences that would consequently be skipped by the polymerase. A computer-aided search of poliovirus DI RNAs has revealed a secondary structure present in every deletion point in the parental genome that could be involved in DI RNA formation (Kuge *et al.*, 1986) In this model, the region(s) at which pausing and re-start would occur (either on the same or on a different RNA template) would be maintained together by "supporting" RNA sequences. Such supporting sequences could reside within poliovirus-specific RNA sequences themselves. This "supporting sequence-loop" model suggests that specific interactions exist between interruption and resumption sites. The model has been developed for poliovirus-associated DI RNAs, but it might be extended to explain the formation of DI RNAs of other viruses. Indeed, a similar computer-aided search has been made of rearranged sequences of DI RNAs of influenza virus and of Sindbis virus: supporting sequences for every rearranged site have been discovered in these mutated genomes (Kuge *et al.*, 1986).

## 5. CONCLUSIONS

One sould bear in mind that the recombinant RNAs we observe most likely only represent a fraction of those that are produced by Nature, since only the viable RNA progeny that has acquired a sufficient level of replication capacity and encapsidation is detectable.

For recombination to occur in Nature between distinct parental viral RNAs, it is obviously mandatory for the same host cell to allow both parents to replicate. This is probably not a difficult constraint for related viruses that in any event develop in the same host. On the other hand, this condition becomes more difficult to meet when the viruses are unrelated. One can postulate that recombination has occurred between the genomes of two plant viruses belonging to different families in the case of the Cc strain of TMV (see 3.3.), if indeed the genome of this virus results from recombination between a tobamovirus and a tymovirus. An investigation of the propagation hosts of tymoviruses suggests that this is not unlikely, since several tymoviruses replicate in tobacco plants or in peas (Blok *et al.*, 1987) that are also hosts of tobamoviruses.

A further level of complexity arises if one considers recombination between plant and animal viruses, since such an event also requires that both viruses replicate in the same host cell. A feature that is common to several virus families of both kingdoms is their propagation by insect vectors. It has been proposed (Goldbach, 1986) that insects may have harboured common ancestor viruses from which the present-day viruses derive. Indeed, the insect host range of plant and animal viruses overlaps, and certain insect RNA viruses also infect mammalian cells, such as black beetle virus (a nodamura virus), an insect virus with a divided genome, a common feature among plant viruses.

To complete the picture of the viruses discussed here, the demonstration of a plant RNA virus (single-stranded and of positive polarity) also capable of replicating in an insect or vice versa, is still lacking. A virus endowed with this capacity may be difficult to observe if the pressure of evolution were to have modified it to such an extent that it no longer replicates in one or the other host. Such a possibility is supported by studies with wound tumor virus: if this plant virus is propagated in its host plant for up to two years without passage through an insect vector, mutants appear that are no longer capable of being transmitted by the insect. Such mutants lack segments of their genome: the host range has thus been changed by changing the ecology of the virus (reviewed in Taylor and Hershey, 1987).

What advantage does recombination represent for RNA viruses? It is likely that recombination is functionally important, particularly for viruses with unsegmented genomes. No proof-reading mechanism exists for RNA genomes that might correct errors made by the RNA polymerase. However, "correction" by recombination rather than by proof-reading might circumvent this problem and could help to establish a population consensus. Recombination might also constitute a natural source of variants of a given virus, and from this point of view could play a similar role in virus evolution as reassortment of genome parts plays for viruses with a segmented genome. Finally, by enabling the virus to draw on host material, recombination offers a new dimension for viral variation.

ACKNOWLEDGMENTS   We thank F. Chapeville for his constant interest and his encouragements. J.C.B. is grateful to the "Ministère de la Recherche et de l'Enseignement Supérieur" for a fellowship. This work was financed in part by a grant from the "A.I.P. N° 87/4560 - Obtention Plantes Résistantes aux Virus". The Institut Jacques Monod is an "Institut Mixte, C.N.R.S. - Université Paris VII".

REFERENCES


Ahlquist, P., French, R., and Bujarski, J. J., 1987, Molecular studies of brome mosaic virus using infectious transcripts from cloned cDNA, Adv. Virus Res. 32:215-242.
Ahlquist, P., Strauss, E. G., Rice, C. M., Strauss, J. H., Haseloff, J., and Zimmern, D., 1985, Sindbis virus proteins nsP1 and nsP2 contain homology to nonstructural proteins from several RNA plant viruses, J. Virol. 53:536-542.
Air, G. M., and Laver, W. G., 1986, The molecular basis of antigenic variation in influenza virus, Adv. Virus Res. 31:53-102.
Angenent, G. C., Linthorst, H. J. M., Van Belkum, A. F., Cornelissen, B. J. C., and Bol, J. F., 1986, RNA 2 of tobacco rattle virus strain TCM encodes an unexpected gene, Nucl. Acids Res. 14:4673-4682.
Argos, P., 1981, Secondary structure prediction of plant virus coat proteins, Virology 110:55-62.

Argos, P., Kamer, G., Nicklin, M. J. H., and Wimmer, E., 1984, Similarity in gene organization and homology between proteins of animal picornaviruses and a plant comovirus suggest common ancestry of these virus families, *Nucl. Acids Res.* **12**:7251-7267.

Bishop, J.M., 1983, Cellular oncogenes and retroviruses, *Ann. Rev. Biochem.* **52**:301-354.

Blok, J., Gibbs, A., and Mackenzie, A., 1987, The classification of tymoviruses by cDNA-RNA hybridization and other measures of relatedness, *Arch. Virol.* **96**:225-240.

Botstein, D., 1980, A theory of modular evolution for bacteriophages, *Ann. N. Y. Acad. Sci.* **354**:484-491.

Bruening, G., 1977, Plant covirus systems: two-component systems, in: *"Comprehensive Virology"*, H. Fraenkel-Conrat, and R. R. Wagner, eds., Plenum Press, New York, vol. **11**, pp. 55-141.

Bujarski, J. J., and Kaesberg, P., 1986, Genetic recombination between RNA components of a multipartite plant virus, *Nature* **321**:528-531.

Cave, D. R., Hendrickson, F. M., and Huang, A. S., 1985, Defective interfering virus particles modulate virulence, *J.Virol.* **55**:366-373.

Cornelissen, B. J. C., Linthorst, H., J., M., Brederode, F., Th., and Bol, J. F., 1986, Analysis of the genome structure of tobacco rattle virus strain PSG, *Nucl. Acids Res.* **14**:2157-2169.

DePolo, N. J., Giachetti, C., and Holland, J. J., 1987, Continuing coevolution of virus and defective interfering particles and of viral genome sequences during undiluted passages:virus mutants exhibiting nearly complete resistance to formely dominant defective interfering particles, *J.Virol.* **61**:454-464.

Domier, L. L., Shaw, J. G., and Rhoads, R. E., 1987, Potyviral proteins share amino acid sequence homology with picorna-, como- and caulimoviral proteins, *Virology* **158**:20-27.

Emerson, S. U.and Schubert, M., 1987, Molecular basis of rhabdovirus replication, in *"The Molecular Basis of Viral Replication"*, R. Perez Bercoff, ed., Plenum Press, New York, London, pp. 255-276.

Fields, S., and Winter, G., 1982, Nucleotide sequences of influenza virus segments 1 and 3 reveal mosaic structure of a small viral RNA segment, *Cell* **28**:303-313.

Francki, R. I. B., 1985, Plant virus satellites, *Ann. Rev. Microbiol.* **39**:151-174.

Fuller, S. D., and Argos, P., 1987, Is Sindbis a simple picornavirus with an envelope?, *EMBO J.* **6**:1099-1105.

Gibbs, A., 1987, Molecular evolution of viruses; 'trees', 'clocks' and 'modules', *J. Cell. Sci. Suppl.* **7**:319-337.

Goldbach, R. W., 1986, Molecular evolution of plant RNA viruses, *Ann. Rev. Phytopathol.* **24**:289-310.

Goldbach, R., and Wellink, J., 1988, Evolution of plus-strand RNA viruses, *Intervirology*, submitted for publication.

Gorbalenya, A. E., Blinov, V. M., and Donchenko, A.P.,1986, Poliovirus-encoded proteinase 3C: a possible evolutionary link between cellular serine and cysteine proteinase families, *FEBS Lett.* **194**:253-257.

Gorbalenya, A. E., Koonin, E. V., Donchenko, A. P., and Blinov, V. M., 1988a, A novel superfamily of nucleoside triphosphate-binding motif containing proteins which are probably involved in duplex unwinding in DNA and RNA replication and recombination, *FEBS Lett.* **235**:16-24.

Gorbalenya, A. E., Blinov, V. M., Donchenko, A.P., and Koonin, E.V., 1988b, An NTP-binding motif is the most conserved sequence in a highly diverged monophyletic group of proteins involved in positive strand RNA viral replication, *J. Mol. Evol.*, in press.

Grantham, R., Gautier, C., Gouy, M., Jacobzone, M., and Mercier, R., 1981, Codon catalog usage is a genome strategy modulated for gene expressivity, *Nucl. Acids Res.* **9**:r43-r74.

Hahn, C. S., Lustig, S., Strauss, E. G., and Strauss, J. H., 1988, Western equine encephalitis virus is a recombinant virus, *Proc. Natl. Acad. Sci. USA*, **85**:5997-6001.

Hillman, B. I., Carrington, J. C., and Morris, T. J., 1987, A defective interfering RNA that contains a mosaic of a plant virus genome, *Cell* **51**:427-433.

Hirst, G.K., 1962, Genetic recombination with Newcastle disease virus, polioviruses and influenza, *Cold Spring Harbor Symp. Quant. Biol.* **27**:303-309.

Hiruki, C., 1987, The dianthoviruses: a distinct group of isometric plant viruses with bipartite genome, *Adv. Virus Res.* **33**:257-300.

Hodgman, T. C., 1988, A new superfamily of replicative proteins, *Nature* **333**:578.

Hodgman, T. C., and Zimmern, D., 1988, Evolution of RNA viruses, in: "RNA Genetics", J. Holland, E. Domingo, and P. Ahlquist, eds., CRC Press, Boca Raton, Fla., in press.

Holland, J. J., Kennedy, S. I. T., Semler, B. L., Jones, C. L., Roux, L. and Grabau, E. A., 1980, Defective interfering RNA viruses and host-cell response, in: "Comprehensive Virology", H. Fraenkel-Conrat and R. R. Wagner, eds., Plenum Press, New York, vol. **16**, pp. 137-192.

Holland, J., Spindler, K., Horodyski, F., Grabau, E., Nichol, S., and VandePol, S., 1982, Rapid evolution of RNA genomes, *Science* **215**:1577-1585.

Horiuchi, K., 1975, Genetic studies of RNA phages, in: "RNA Phages", N. D. Zinder, ed., Cold Spring Harbor Laboratory, New York, pp. 29-50.

Jennings, P.A., Finch, J.T., Winter, G., and Robertson, J.S., 1983, Does the higher order structure of the influenza virus ribonucleoprotein guide sequence rearrangements in influenza viral RNA?, *Cell* **34**:619-627.

Joshi, R. L., Joshi, S., Chapeville, F., and Haenni, A.L., 1983a, Primary and secondary structures of the tRNA-like regions of the genomes of plant RNA viruses, in: "Endocytobiology II", H. E. A. Schenk, and W. Schwemmler, eds., Walter de Gruyter, Berlin, pp. 57-68.

Joshi, S., Joshi, R.L., Haenni, A.L., and Chapeville, F., 1983b, tRNA-like structures in genomic RNAs of plant viruses, *Trends Biochem. Sci.* **8**:402-404.

Kamer, G., and Argos, P., 1984, Primary structural comparison of RNA-dependent polymerases from plant, animal and bacterial viruses, *Nucl. Acids Res.* **12**:7269-7282.

Keck, J. G., Matsushima, G .K., Makino, S., Fleming, J .O., Vannier, D. M., Stohlman, S. A., and Lai, M. M. C., 1988, In vivo RNA-RNA recombination of coronavirus in mouse brain, *J. Virol.* **62**:1810-1813.

King, A. M. Q., McCahon, D., Slade, W. R., and Newman, J. W. I., 1982, Recombination in RNA, *Cell* **29**:921-928.

King, A. M. Q., Ortlepp, S. A., Newman, J. W. I., and McCahon, D., 1987, Genetic recombination in RNA viruses, in: "The Molecular Biology of the Positive Strand RNA Viruses", D. J. Rowlands, M. A. Mayo, and B. W. J. Mahy, eds., Academic Press, London, pp. 129-152..

Kirkegaard, K., and Baltimore, D., 1986, The mechanism of RNA recombination in poliovirus, *Cell* **47**:433-443.

Kolakofsky,D., and Roux,L., 1987, The molecular biology of paramyxoviruses, in: "The Molecular Basis of Viral Replication", R. Perez Bercoff, ed., Plenum Press, New York, London, pp. 277-297.

Konarska, M. M., Padgett, R. A., and Sharp, P. A., 1985, *Trans* splicing of mRNA precursors in vitro, *Cell* **42**:165-171.

Kuge, S., Saito, I, and Nomoto, A., 1986, Primary structure of poliovirus defective-interfering particle genomes and possible generation mechanisms of the particles, *J.Mol.Biol.* **192**:473-487.

Lai, M. M. C., Baric, R. S., Makino, S., Keck, J. G., Egbert, J., Leibowitz, J. L., and Stohlman, S. A., 1985, Recombination between nonsegmented RNA genomes of murine coronaviruses, *J. Virol.* **56**:449–456.

Lane, D., 1988, Enlarged family of putative helicases, *Nature* **334**:478.

Lazzarini, R. A., Keene, J. D., and Schubert, M., 1981, The origins of defective interfering particles of the negative-strand RNA viruses, *Cell* **26**:145–154.

Ledinko, N., 1963, Genetic recombination with poliovirus type 1. Studies of crosses between a normal horse serum-resistant mutant and several guanidine-resistant mutants of the same strain, *Virology* **20**:107–119.

Makino, S., Keck, J. G., Stohlman, S. A., and Lai, M. M. C., 1986a, High-frequency RNA recombination of murine coronaviruses, *J. Virol.* **57**:729–737.

Makino, S., Stohlman, S. A., and Lai, M. M. C., 1986b, Leader sequences of murine coronavirus mRNAs can be freely reassorted: evidence for the role of free leader RNA in transcription, *Proc. Natl. Acad. Sci. USA* **83**:4204–4208.

Mayo, M. A., 1987, A comparison of the translation strategies used by bipartite genome, RNA plant viruses, in: *"The Molecular Biology of the Positive Strand RNA Viruses"*, D. J. Rowlands, M. A. Mayo, and B. W. J. Mahy, eds., Academic Press, London, pp. 177–205.

McCahon, D., and Slade, W. R., 1981, A sensitive method for the detection and isolation of recombinants of foot-and-mouth disease virus, *J. gen. Virol.* **53**:333–342.

Miller, W. A., Waterhouse, P. M., and Gerlach, W. L., 1988, Sequence and organization of barley yellow dwarf virus genomic RNA, *Nucl. Acids Res.* **16**:6097–6111.

Monroe, S. S., and Schlesinger, S., 1983, RNAs from two independently isolated defective interfering particles of Sindbis virus contain a cellular tRNA sequence at their 5' ends, *Proc. Natl. Acad. Sci. USA* **80**:3279–3283.

Monroe, S. S., and Schlesinger, S., 1984, Common and distinct regions of defective-interfering RNAs of Sindbis virus, *J. Virol.* **49**:825–872.

Morch, M. D., and Haenni, A.L., 1987, Organization of plant virus genomes that comprise a single RNA molecule, in: *"The Molecular Biology of the Positive Strand RNA Viruses"*, D. J. Rowlands, M. A. Mayo, and B. W. J. Mahy, eds., Academic Press, London, pp. 153–175.

Morch, M. D., Boyer, J. C., and Haenni, A. L., 1988, Overlapping open reading frames revealed by complete nucleotide sequencing of turnip yellow mosaic virus genomic RNA, *Nucl. Acids Res.* **16**:6157–6173.

Morch, M. D., Valle, R.P.C., and Haenni, A.L., 1987, Regulation of translation of viral mRNAs, in: *"The Molecular Basis of Viral Replication"*, R. Perez Bercoff, ed., Plenum Press, New York, London, pp. 113–159.

Murant, A. F., and Mayo, M. A., 1982, Satellites of plant viruses, *Ann. Rev. Phytopathol.* **20**:49–70.

Nishihara, T., Mills, D. R., and Kramer, F. R., 1983, Localization of the Qβ replicase recognition site in MDV-1 RNA, *J. Biochem.* **93**:669–674.

Perrault, J., 1981, Origin and replication of defective interfering particles, *Curr. Top. Microbiol. Immunol.* **93**:151–207.

Plotch, S. J., Bouloy, M., Ulmanen, I., and Krug, R. M., 1981, A unique cap ($m^7$GpppXm)-dependent influenza virion endonuclease cleaves capped RNAs to generate the primers that initiate viral RNA transcription, *Cell* **23**:847–858.

Pringle, C. R., 1965, Evidence of genetic recombination in foot-and-mouth disease virus, *Virology* **25**:48–54.

Reanney, D. C., 1982, The evolution of RNA viruses, *Ann. Rev. Microbiol.* **36**:47–73.

Rossmann, M. G., Arnold, E., Erickson, J. W., Frankenberger, E. A., Griffith, J. P., Hecht, H. J., Johnson, J. E., Kamer, G., Luo, M., Mosser, A. G., Rueckert, R. R., Sherry, B., and Vriend, G., 1985, Structure of a human common cold virus and functional relationship to other picornaviruses, *Nature* **317**:145-153.

Simon, A. E., and Howell, S. H., 1986, The virulent satellite RNA of turnip crinkle virus has a major domain homologous to the 3' end of the helper virus genome, *EMBO J.* **5**:3423-3428.

Söderlund, H., Keränen, S., Lehtovaara, P., Palva, I., Pettersson, R. F., and Kääriäinen, L.,1981, Structural complexity of defective-interfering RNAs of Semliki Forest virus as revealed by analysis of complementary DNA.*Nucl. Acids Res.* **9**:3403-3417.

Solnick, D., 1985, *Trans* splicing of mRNA precursors, *Cell* **42**:157-164.

Strauss, J. H., and Strauss, E. G., 1988, Evolution of RNA viruses, *Ann. Rev. Microbiol.* **42**: in press.

Strauss, J. H., Strauss, E. G., Hahn, C. S., Hahn, Y. S., Galler, R., Hardy, W. R., and Rice, C. M., 1987, Replication of alphaviruses and flaviviruses: proteolytic processing of polyproteins, in: *"Positive Strand RNA Viruses"*, M. A. Brinton, and R. R. Rueckert, eds., Alan R. Liss, Inc., New York, pp. 209-225.

Stollar, V., 1980, Defective interfering alphaviruses, in: *"The Togaviruses, Biology, Structure, Replication"*, R. W. Schlesinger, ed., Academic Press, New York, pp. 427-457.

Symons, R. H., Haseloff, J., Visvader, J. E., Keese, P., Murphy, P. J., Gordon, K. H. J., and Bruening, G., 1985, On the mechanism of replication of viroids, virusoids and satellite RNAs, in: *"Subviral Pathogens of Plants and Animals: Viroids and Prions"*, K. Maramorosch, and J. J. McKelvey, eds., Academic Press, New York, pp. 235-263.

Takamatsu, N., Ohno, T., Meshi, T., and Okada, Y., 1983, Molecular cloning and nucleotide sequence of the 30K and the coat protein cistron of TMV (tomato strain) genome, *Nucl. Acids Res.* **11**:3767-3778.

Taylor, M., W., and Hershey, H., V., 1987, Viruses: an overview, in: *"The Molecular Basis of Viral Replication"*, R. Perez Bercoff, ed., Plenum Press, New York, London, pp. 3-23.

Tsiang, M., Monroe, S.S., and Schlesinger, S., 1985, Studies of defective interfering RNAs of Sindbis virus with and without tRNA[Asp] sequences at their 5' termini, *J.Virol.* **54**:38-44.

Tsiang, M., Weiss, B. G., and Schlesinger, S., 1988, Effects of 5'-terminal modifications on the biological activity of defective interfering RNAs of Sindbis virus, *J.Virol.* **62**:47-53.

Van Duin, J., 1988, Single-stranded RNA bacteriophages, in: *"The Bacteriophages"*, R. Calendar, ed., Plenum Press, New York, vol. **1**, pp. 117-167.

Van Vloten-Doting, L., and Jaspars, E. M. J., 1977, Plant covirus systems: three-component systems, in: *"Comprehensive Virology"*, H. Fraenkel-Conrat, and R. R. Wagner, eds., Plenum Press, New York, vol. **11**, pp. 1-53.

Webster, R. G., Laver, W. G., Air, G. M., and Schild, G. C., 1982, Molecular mechanisms of variation in influenza viruses, *Nature* **296**:115-121.

Wellink, J., and Van Kammen, A., 1988, Proteases involved in the processing of viral polyproteins, *Arch. Virol.* **98**:1-26.

Zimmern, D., 1982, Do viroids and RNA viruses derive from a system that exchanges genetic information between eukaryotic cells?, *Trends Biochem. Sci.* **7**:205-207.

YEAST PRE-mRNA SPLICING MUTANTS

Usha Vijayraghavan and John Abelson

Division of Biology, 147-75
California Institute of Technology
Pasadena, California 91125

## INTRODUCTION

The removal of introns from the primary transcript occurs by RNA splicing. Three major types of RNAs; tRNA, rRNA and mRNA, are known to contain introns. There are two general questions concerning any splicing mechanism: first, how is the precise recognition and alignment of the splice junctions achieved in introns whose lengths vary from a few hundred nucleotides to several thousands of nucleotides, and second, what are the biochemical mechanisms of the cleavage and ligation reactions.

The specificity of the splice junction selection and the juxtaposition of the exons is achieved in Group I introns (e.g.,Tetrahymena rRNA, some yeast mitochondrial introns, choloroplast tRNA introns) and Group II introns (e.g. fungal mitochondrial introns) by the conserved secondary and tertiary structure of the intron. In group I splicing, the intron is excised in a RNA catalysed reaction. The catalytic RNA is the intron itself and splicing occurs in the presence of a guanosine cofactor. Splicing of Group II introns have also been shown to take place by self excision, and the excised intron is released as a branched lariat RNA, a structure also produced in nuclear pre-mRNA splicing (reviewed by Cech and Bass 1986).

The splice site selection in nuclear pre-mRNA is dictated only by short stretches of sequences at or near the splice junctions (Shapiro and Senapathy 1987). Splicing involves interaction of these cis-acting sequences with many cellular factors including proteins and small nuclear ribonucleoproteins (snRNPs) (reviewed in Sharp 1987; Maniatis and Reed 1987). The development of in vitro splicing systems from mammalian cell lines and from the yeast, Saccharomyces cerevisiae, has revealed a two step splicing reaction scheme for splicing (Fig. 1). First, cleavage at the 5' splice junction results in the formation of a lariat intermediate of intron-exon2 in which the first base of the intron is covalently linked by a 2'-5' phosphodiester bond to an internal adenosine near the 3' splice site. In the second step, the 3' splice junction is cleaved and the exons are ligated to give the mRNA and the lariat intron. Mechanistically, the reaction scheme is very similar to that of the self-splicing Group II introns; however, the nuclear pre-mRNA splicing machinery is more complex. In splicing

Pre-mRNA

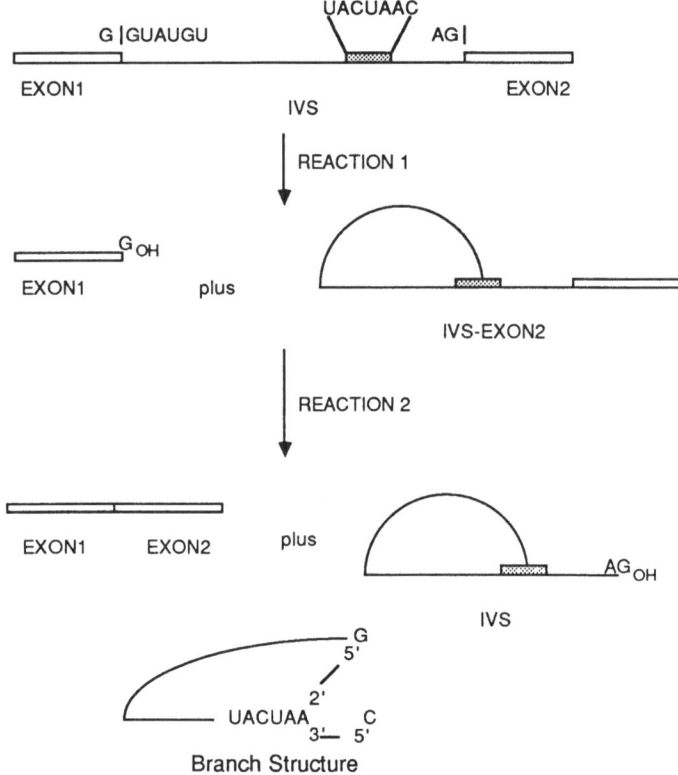

Branch Structure

Fig 1.    The pathway of pre-mRNA splicing reactions. The reactions
         are described in the text. The intron-exon junctions are
         represented by vertical bars. The structure of the yeast
         branch is shown.


reactions studied in vitro the pre-mRNA and the splicing intermediates
are associated with a large complex termed the spliceosome (Brody and
Abelson 1985 ; Grabowski et al., 1985; Frendeway and Keller 1985). The
spliceosome resembles the complexity of ribosomal subunits in that it
consists of several RNAs and proteins. It is also clear now that at
least five snRNPs (U1, U2, U4, U5 and U6) are required for splicing in
both yeast and mammalian systems. The participation of at least one
heterogeneous nuclear ribonuclear protein (hnRNP) is also indicated in
pre-mRNA splicing.

    Enumeration, identification and determination of the function of
spliceosomal components are essential for the complete understanding of
the splicing machinery and its operation. The extensive genetic
repertoire of the yeast Saccharomyces cerevisiae allows the isolation
and the characterization of mutants in the splicing pathway. The
analysis of such mutants and the isolation of the wild type gene product
of these loci is a powerful ally of the biochemical studies. A combined
genetic and biochemical approach has therefore been taken by several
groups working with yeast in their attempt to identify the factors
involved in the assembly and function of the spliceosome.

THE PRP MUTANTS OF YEAST

An important aspect in the genetic analysis of splicing in yeast has been the availability of several trans-acting mutants that affect pre-mRNA splicing. These mutants were identified as temperature sensitive (ts) lethals in the genes PRP2 through PRP11 in a general search for mutants affecting macromolecular biosynthesis in a bank of ts lethals (Hartwell 1967). These mutants originally called rna mutants are now being named prp for pre RNA processing mutants. Many of these mutants turned out to be involved in splicing because of a peculiarity of the yeast genome. The prp2-11 mutants showed a rapid cessation of net RNA accumulation upon shifting cultures to nonpermissive temperature. The mutants were subsequently shown to be defective in rRNA synthesis but not in tRNA or mRNA synthesis. The defect in rRNA synthesis was not in transcription but in the processing of the large pre-rRNA (Warner and Udem 1972). Interestingly, these mutations severely decreased the levels of most ribosomal protein mRNAs, and most non-ribosomal protein mRNAs were not affected. Subsequently a number of laboratories found that the genes coding for most ribosomal proteins contain introns. Intron-containing ribosomal transcripts accumulate at non-permissive temperature in these strains, while non-intron containing transcripts were unaffected (Rosbash et al., 1981). These data finally suggested that the defects in rRNA processing were due to defects in the splicing of precursors to ribosomal protein mRNAs. The apparent specificity of the defect was due to the preponderance of introns in genes coding for ribosomal proteins and the rarity of introns in non-ribosomal protein genes. Accumulation of pre-mRNA in other intron containing yeast transcripts for e.g. actin has also been observed (Teem et al., 1983). Although the deficiency in splicing of the prp mutants does not prove that they encode a part of the pre-mRNA splicing apparatus, these mutants were the best candidates for the genetic analysis of the components of the yeast nuclear splicing machinery (Warner 1987).

Evidence for the direct participation of many of the PRP 2-11 gene products in the splicing reaction came from in vitro biochemical work (Lustig et al., 1986). This was made possible by the use of a heat inactivation assay with the in vitro splicing system. Splicing activity in extracts from prp strains are heat sensitive in vitro. Seven of the nine complementation groups tested were heat inactivated. This loss of activity in different prp mutants was due to the inactivation of an exchangeable component as shown by in vitro complementation of pairs of inactivated extracts.

The PRP 3, 5, 7, 8, and 11 gene products are required early in the splicing pathway (Lin et al., 1987) as heat inactivated extracts do not form splicesomes (Fig 2). The prp4 extract does form an early splicing assembly intermediate (complex B, Fig. 2) after heat inactivation (J. Banroques and J. Abelson, unpublished results). The reason for involvement of many of the PRP gene products early in the splicing pathway is not understood. One reason could be that a large number of factors are required early in spliceosome assembly, or that there are shared components between the different steps of splicing. The PRP2 gene product acts after spliceosome assembly, as heat inactivated prp2 extracts accumulate spliceosomes (Lin et al., 1987; Cheng and Abelson 1987). In the splicing pathway this splicing complex immediately preceeds the reactive spliceosome (complex A1, Fig. 2). Spliceosomes formed in heat inactivated prp2 extracts are active intermediates in the splicing pathway, and, when isolated from glycerol gradients, the accumulated pre-mRNA can be chased to products. These experiments indicated that PRP2 is an extrinsic component of the spliceosome.

195

These experiments also support the presence of at least two other sets of extrinsic factors, b-n and c-n, required for the first and second reactions of splicing, respectively. Factor b-n has been shown to be a micrococcal nuclease resistant, heat stable protein(s) required with PRP2 for the first step of splicing.

Several of these PRP genes (PRP 2, 3, 4, 5, 8 and 11) have been cloned and their molecular characterization has been initiated. The characteristics are summarised in Table 1. The role that these genes play in pre-mRNA splicing is being investigated by the use of antibodies directed against the proteins. The antisera raised against PRP2, PRP3 and PRP11 products have been used to localize these proteins to the nuclei (Last and Woolford 1986 ; Chang et al. 1988). In the case of PRP11 immunoelectron microscopy has placed the protein in the periphery of the yeast nucleus.

Antibodies against PRP8-'β gal fusions have identified the protein as a exceptionally large 260 Kilodalton (kD) protein. A role for PRP8 in the splicing pathway has been shown, with experiments demonstrating loss of splicing activity after immunodepletion of PRP8 (Jackson et al., 1988). The PRP8 antibodies have been used to probe for a stable association of this protein with the yeast snRNAs. These antibodies can specifically precipitate U5 from yeast extracts indicating an association of the PRP8 probe in and U5 snRNA (Lossky et al., 1987). Preincubation of extracts with ATP resulted in immunoprecipitation of U6 and U4 along with U5. The co-precipitation of U4/U6 under these conditions may result from snRNP-snRNP interactions.

Recent experiments with PRP4 antibodies reinforce the view of a direct role in splicing for the PRP4 protein. There are two principal findings. First the PRP4 antisera in immunoprecipitation experiments precipitate yeast U5, U4 and U6 snRNAs. U6 snRNA is known to exist in a complex of proteins as U6 snRNP and also in a particle with U4 snRNA in it (Cheng and Abelson 1987). Thus, the co-precipitation of U4 and U6 is not unexpected. Second the PRP4 antibodies inhibit in vitro splicing reactions completely. These two results are corroboratory inhibition of splicing occurs after U2 binds the pre-mRNA (complex B, Fig 2) because anti PRP4 antibodies precipitate U4/U5/U6 snRNAs (J. Banroques and J. Abelson, unpublished observations).

In the case of PRP11, cloned DNA has been transcribed and translated in vitro to generate a $^{35}$S labeled PRP11 protein (Chang et al., 1988). This labeled protein complements in vitro the heat sensitivity of prp11 extracts. The complementation using the labeled protein has allowed tracing of PRP11 in splicing complexes. In extracts prepared from prp11 strain most of the protein enters a 30S complex whose formation is not dependent on the presence of pre-mRNA. A fraction of the labeled protein does associate with 40S spliceosomes; this association is dependent on the presence of splicing signals in the pre-mRNA. Interestingly, the protein sequence is predicted to have a Zn finger domain, previously shown to be a motif found in DNA and RNA binding proteins.

Interactions between the PRP gene products have been indicated both by characterization of extragenic supressors and by experiments showing that increased dosage of one PRP gene product can rescue the lethality of another. The first extragenic suppressor of RNA mutants to be isolated was SRN1 (Pearson et al., 1982). Mutants in SRN1 were isolated on the basis of supressing a prp2 prp6 double mutant. However, SRN1 does not display any allele specificity in its supression, as supression of several prp mutants was observed. These results indicate that the

196

TABLE 1.  MOLECULAR CHARACTERIZATION OF *PRP (RNA)* GENES

| | ts in vitro | Required for Spliceosome formation | GENE Cloned | GENE Essential | Gene Products | Antibody Experiments |
|---|---|---|---|---|---|---|
| *PRP (RNA)2* | Yes[a] | No | Yes[b,c,d] | Yes[b,c,d] | ~2.5 kb Transcript[b] ~100 kD protein | Nuclear[b] protein |
| *PRP (RNA)3* | Yes[a] | Yes[a] | Yes[b] | Yes[b] | ~1.5 kb Transcript[b] ~56 kD protein | Nuclear[b] protein |
| *PRP (RNA)4* | Yes[a] | Yes[a] | Yes[d] | Yes[d] | ~1.6 kb Transcript[d] ~52kD protein[f] | Inhibit splicing[f] Precipitate U5, U4, U6 snRNA |
| *PRP (RNA)5* | Yes[a] | Yes[a] | Yes[e] | ? | ~1.4 kb Transcript[e] ~47 kD protein | In progress[e] |
| *PRP (RNA)6* | No[a] | ? | No | ? | ? | ? |
| *PRP (RNA)7* | Yes[a] | Yes[a] | No | ? | ? | ? |
| *PRP (RNA)8* | Yes[a] | Yes[a] | Yes[g] | Yes[g] | ~7.4 kb Transcript[g] ~260 kD protein | Inhibit splicing[g] Precipitate U5snRNA |
| *PRP (RNA)9* | No[a] | ?[a] | No | ? | ? | ? |
| *PRP (RNA)10/11* | Yes[a] | Yes[a] | Yes[b,d] | Yes[b,d] | ~1.0 kb Transcript[b] ~30 kD protein[h] | Inhibit splicing[h] Nuclear protein |

a. Lin et al. 1987; Lustig et al. 1986
b. Last et al. 1984; Last et al. 1986
c. Lee et al. 1984
d. Soltyk et al. 1984
e. McFarland and Abelson unpublished
f. Barroques and Abelson unpublished
g. Jackson et al. 1988; Lossky et al. 1987
h. Chang et al. 1988

different mutants are related but does not indicate any specific interaction between the PRP gene products. Last et al (1987) have defined a specific interaction between PRP3 and PRP4, as extra plasmid-borne copies of PRP3 can rescue the ts lethality of three alleles of prp4. The supression of prp4 requires the presence of the mutant prp4 protein, indicating that PRP4 is not being bypassed. An extragenic supressor of prp2 (SRN2) has also be en obtained. SRN2 supressed six different alleles of prp2 but not other prp loci. SRN2 is itself an essential gene. The mode of action of SRN2 and PRP3 in supression of prp2 or prp4 respectively can be explained by different models. For example: SRN2 and prp2 are interacting proteins, and PRP3 and PRP4 interact, and the overproduction of one allows stabilization of the other by virtue of the interaction.

That the PRP3 and PRP4 gene products interact is also suggested by in vitro heat inactivation and complementation data where a weaker complementation than that obtained for other combinations was found (Lustig et al., 1986). These results suggest that they could be part of the same macromolecular complex, a hypothesis consistent with the evidence for the interaction between these gene products.

ISOLATION OF NEW MUTANTS AFFECTING PRE-mRNA PROCESSING

As is evident from the discussions of the prp2-11 mutants, the use of such an approach to isolation of gene products that play a role in the splicing in yeast has been tremendously fruitful. It is possible to isolate more mutants defective in pre-mRNA splicing. One approach is the generation of temperature or cold-sensitive banks of mutagenised strains, which can be screened for pre-mRNA processing defects. Analysis of conditional phenotypes other than temperature sensitivity would be useful to isolate mutants in some genes that do not readily mutate to give ts alleles.

Our effort to isolate new ts lethals, defining genes involved in pre-mRNA has resulted in the identification of several new complementation groups (U. Vijayraghavan M. Company and J. Abelson, manuscript in preparation). Five new complementation groups with three novel phenotypes not seen in the prp2 to prp11 mutants were isolated: in one case accumulation of lariat intermediate of splicing was observed and in another case accumulation of the released intron was seen, and in the third an accumulation of both pre-mRNA and intron is observed. The isolation of a ts mutations causing accumulation of the lariat intermediate provides a unique opportunity to study the components needed for the second step of splicing, cleavage at the 3' splice site and the ligation of exons. In vitro splicing using extracts prepared from one of these mutant strains can be heat inactivated, and inactivation results in accumulation of the lariat intermediate and exon1. Intact snRNPs are not required for complementation as micrococcal nuclease treated extracts complement inactivated extracts. Spliceosomes formed after heat inactivation contain as expected lariat intermediate and exon1. These intermediates can be chased to products upon addition of complementing fractions to the assembled mutant spliceosome. These chasing experiments have also defined a requirement for ATP in the second step of the splicing reaction.

The accumulation of the intron in one complementation group does not confer a temperature sensitive phenotype, and the isolation of the mutant as a ts was serendipitous. The cellular intron in this mutant exists in a large particle of about 40S in extracts prepared from mutant strains. Co-sedimentation of snRNPs with the intron containing particle suggests an association of the intron with the snRNPs. Tri-methyl G cap

and Sm specific antibodies that precipitate snRNPs, also precipitate
intron from whole cell extracts, supporting the view that the intron is

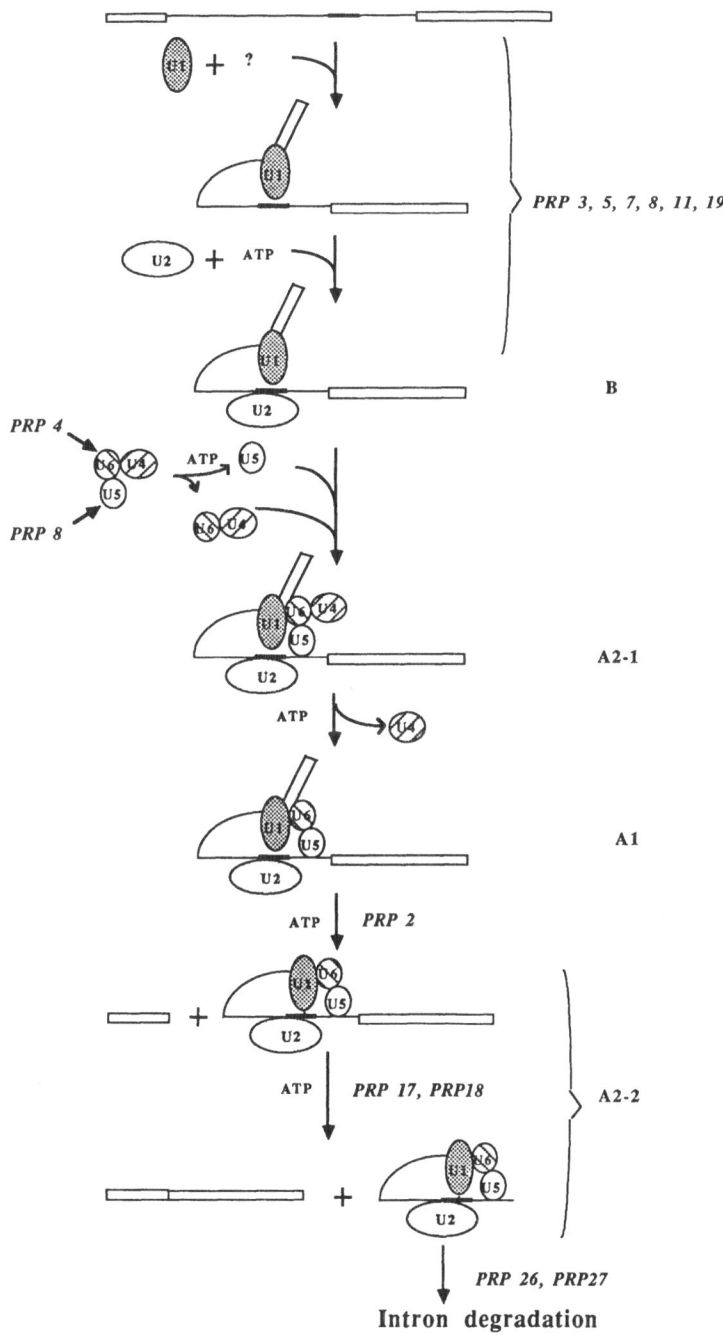

Fig 2. Proposed pathway for the assembly of the spliceosome. The
role of the PRP gene products in the assembly is indicated
where known.

associated with snRNPs.  The accumulation of the intron could be due to whe defect in a lariat debranching function, or due to a defect in the disassembly of the intron-containing particle after splicing has been completed.

A different approach to isolation of mutants in gene products involved in splicing is applicable to yeast for the study of macromolecular complexes.  This is the isolation of extragenic supressors to ts mutants in the prp2-11 groups.  These supressors could define interacting gene products, which are themselves a part of the splicing machinery.  In such an approach several pseudo-revertants to prp4 that allow growth at a temperature that is nonpermissive have been obtained.  These pseudo-revertants (srp 4) themselves confer a cold-sensitive phenotype to the cell and segregate as single nuclear recessive genes (J. Maddock and J. Woolford, per. comm.).  None of these pseudo-revertants of prp4 are allelic to known prp genes.  srp4-1, and srp4-2 supress prp4 on the basis of their isolation and also supress prp3 .  PRP4 and PRP3 have previously been shown to interact and the isolation of a supressor to both is not surprising.  The supressors do not work by substituting for the PRP4 or the PRP3 function a s supression by srp4  does not work with null alleles of PRP4 or PRP3.  srp4-1 and srp-2 are candidates for mutations in genes whose products directly interact with these PRP genes.  The use of cold-sensitive supressors of ts mutations, should bring the system around to a complete circle genetically with mutants in genes for some components supressing ts mutants in genes for other components.  Such analysis would also be extremely useful to map out interactions between these gene products.

A parallel genetic apprcach for identifying the components of the splicing machinery is the isolation of trans-acting supressors to point mutants in the pre-mRNA splicing signals that are essential for splicing.  A trans-acting supressor that restores splicing of a branch point mutation in a yeast intron has been isolated (Couto et al., 1987).  This allele-specific supressor behaves as a single Mendelian gene, with a dominant supression phenotype.  A recessive effect of accumulation of unspliced wild type pre-mRNA was also observed.  These two phenotypes argue for a direct role for the PRP16 locus in the splicing machinery.  A similar approach to isolate supressors to point mutations in essential regions of spliceosomal snRNAs should allow definition of components of snRNPs.

The isolation of trans-acting conditional mutants affecting pre-mRNA splicing, the isolation of supressors to these mutants and the isolation of supressors to mutants in the pre-mRNA should define many gene products of the splicing apparatus.  The analysis of the effects of conditional mutants biochemically should be possible when extracts prepared from these strains are inactivated in vitro.  Inactivation and complementation of these extracts would be useful in delineating the steps in splicing assembly pathway that they affect.  The analysis of defects in the assembly itself is now possible using improved methods of separation of splicing complexes on native gels, and the detection of binding of splicing components to pre-mRNA on solid supports.

The assembly of snRNPs in wild type extracts using in vitro synthesised snRNAs are now being studied.  Once the wild type pattern of assembly is known, study of the effect of inactivated mutant extracts or extracts from supressor strains on the assembly pattern of individual snRNPs and also their effects on multi-snRNP complexes (e.g. U4/U5/U6) can be attempted.  Thus, a variety of tools necessary for a concerted biochemical and genetic analysis of pre-mRNA splicing are available, and analyses toward this end are in progress.

ACKNOWLEDGEMENTS

We thank Drs. M. Aebi, D. Horowitz and G. Dalbadie-McFarland for discussions and careful reading of this manuscript. The work on yeast pre-mRNA splicing in J. Abelson's laboratory has been supported by National Institutes of Health grant GM32637.

REFERENCES

Brody, E., and Abelson, J., 1985, The "Spliceosome": Yeast pre-messenger RNA associates with a 40S complex in a splicing dependent reaction, Science 228:963-967.

Cech, T. R., and Bass, B. L., 1986, Biological catalysis by RNA, Ann. Rev. Biochem., 55:599-629.

Chang, T.-H., Clark, M. W., Lustig, A. J., Cusick, M. E., and Abelson, J., 1988, RNA11 protein is associated with yeast spliceosome and is localised in the periphery of the cell nucleus, Mol. Cell Biol., 8:2379-2393.

Cheng, S.-C., and Abelson, J., 1987, Spliceosome assembly in yeast. Genes Dev., 1:1014-1027.

Couto, J. R., Tamm, J. J., Parker, R., and Guthrie, C., 1987, A trans acting supressor restores splicing of a yeast intron with a branch point mutation, Genes Dev., 1:445-455.

Frendewey, D., and Keller, W., 1985, Stepwise assembly of a pre-mRNA splicing complex requires U-snRNPs and specific intron sequences, Cell, 42:355-367

Grabowski, P. J., Seiler, S. R., and Sharp, P. A., 1985, A multicomponent complex is involved in splicing of messenger RNA precursors, Cell, 42:345-353.

Hartwell, L. H., 1967, Macromolecular synthesis in temperature-sensitive mutants in yeast, J. Bact., 93:1662-1670.

Jackson, S. P., Lossky, M., and Beggs, J. D., 1988, Cloning of RNA8 gene of Saccharomyces cerevisiae. Detection of the RNA8 protein and demonstration that it is essential for nuclear pre-mRNA splicing, Mol. Cell Biol., 8:1067-1075.

Last, R. L., and Woolford, J. L., 1986, Identification and nuclear localization of yeast pre-mRNA processing components: RNA2 and RNA3 proteins, J. Cell Biol., 103:2103-2112.

Last, R. L., Stavenhagen, J. B., and Woolford, J. L., 1984, Isolation and characterization of the RNA2, RNA3 and RNA11 genes of Saccharomyces cerevisiae, Mol. Cell Biol., 4:2396-2405.

Last, R. L., Maddock, J. R., Woolford, J. L., 1987, Evidence for related functions of the RNA genes of Saccharomyces cerevisiae, Genetics, 117:619-631.

Lee, M. G., Young, R. A., and Beggs, J. D., 1984, Cloning of the RNA2 gene of Saccharomyces cerevisiae, EMBO J., 3:2825-2830.

Lin, R.-J., Lustig, A. J., and Abelson, J., 1987, Splicing of yeast nuclear pre-mRNA in vitro requires functional 40S spliceosomes and several extrinsic factors, Genes Dev., 1:7-18.

Lossky, M., Anderson, G. J., Jackson, S. P., and Beggs, J. D., 1987, Identification of yeast snRNP protein and the detection of snRNP-snRNP interactions, Cell, 51:1019-1026.

Lustig, A. J., Lin, R.-J., and Abelson, J., 1986, The yeast RNA gene products are essential for mRNA splicing in vitro, Cell, 47:953-963.

Maniatis, T., and Reed, R., 1987, The role of small nuclear ribonucleoproteins in pre-mRNA splicing, Nature, 325:673-678.

Pearson, N. J., Thornburn, D. C., and Haber, J. E., 1982, A supressor of temperature sensitive RNA mutations that affect general and specific messenger RNA processing in yeast, Mol. Cell Biol., w:571-577.

Rosbash, M., Harris, P. K. W., Woolford, J. L., and Teem, J. L., 1981, The effect of temperature sensitive RNA mutants on the transcription products from cloned ribosomal protein genes of yeast, Cell, 24:679-686.

Shapiro, M. B., Senapathy, P., 1987, RNA splice junctions of different classes of eukaryotes: sequence statistics and functional implications in gene expression, Nucl. Acids Res., 15:7155-7174.

Slotyk, A., Tropak, M., and Friesen, J. D., 1984, Isolation and characterization of the RNA2, RNA4 and RNA11 genes of Saccharomyces cerevisiae, J. Bact., 160:1093-1100.

Teem, J. L., Rodrigues, J. R., Tung, L., and Rosbash, M., 1983, The RNA2 mutation affects the processing of actin mRNA as well as ribosomal protein mRNAs, Mol. Gen. Genet., 192:101-103.

Warner, J. R., 1987, Applying genetics to the splicing problem, Genes Dev., 1:1-3.

Warner, J. R., and Udem, S. A., 1972, Temperature sensitive mutations affecting ribosome synthesis in Saccharomyces cerevisiae, J. Mol. Biol., 65:243-257.

# ALTERNATIVE SPLICING TO TISSUE   SPECIFIC SPLICING -

# AN EVOLUTIONARY PATHWAY?

Edward Brody,  Joëlle Marie,  Maria S. Goux-Pelletan  and
Béatrice Clouet d'Orval

Centre de Génétique Moléculaire, Batiment 24 - C.N.R.S.
91190   Gif-sur-Yvette, France

## INTRODUCTION

   Tissue   specific splicing of pre-messenger RNA is an important mechanism of regulating metazoan gene expression.  There are scores of examples of single species of pre-messenger RNA which are synthesized in various tissues or in one tissue at various stages of development, but which are spliced to give slightly different collections of exons in each tissue or at each stage of development. We wish to contrast such tissue specific splicing to alternative splicing: in the latter, various collections of exons are spliced from the same pre-messenger RNA, but these different splicing events take place in the same cell at the same time. Alternative splicing, then, generates diversity, but not necessarily regulation. In this article, we shall argue that alternative splicing was a characteristic of all primitive splicing systems. We shall further argue that during evolution, the precision of splicing was increased and that this led to two different results. One was constitutive splicing , in which one and only one 5' end of an intron is spliced to a unique 3' end of this intron. The other was tissue specific splicing, in which the original ambiguity of primitive splicing was harnessed to the demands of differentiating tissues in metazoans.

## RIGID VERSUS FLEXIBILE SPLICING SYSTEMS

   There are at present many different types of splicing systems. In mitochondria, chloroplasts , bacteriophages  T4 and SP01 , and a few nuclear  genes, splicing takes place via 2 transesterification reactions, the   first of which is catalysed by a guanosine ribonucleoside or ribonucleotide. Some of these introns are capable of  autocatalysing their splicing (1), while others require proteins, either coded in the intron (2, 3) or elsewhere (4, 5) for intron removal. These introns, called type I, have been intensively investigated their  mechanism  of splicing  and possible biological significance have been reviewed elsewhere (1). Other organelle introns use a different splicing pathway and have a different RNA structure ; these are the type II introns, some of which are self-spliceable (6). Type II introns self-splice using transesterification reactions which generate the same intermediates as those found in nuclear  pre-messenger RNA splicing (see Fig. 1). Type II introns are thought to resemble the evolutionary precursors of nuclear pre-messenger RNA splicing. It is assumed that self-splicing, both types I and II, pre-dated

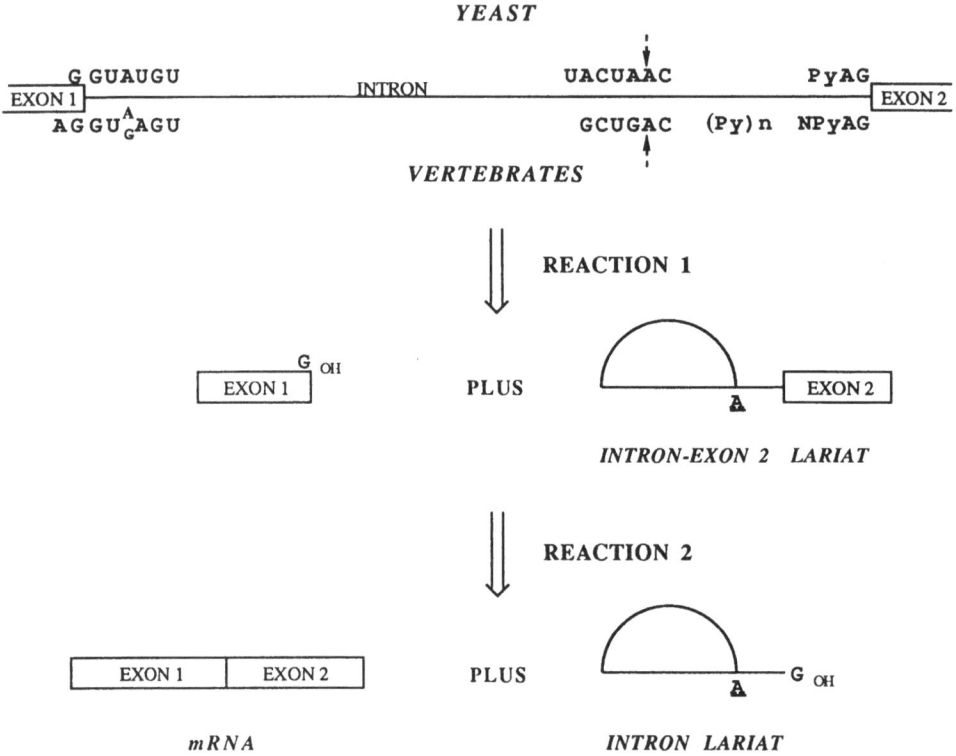

Fig. 1 : The Two Reactions of Nuclear Pre-Messenger RNA Splicing.

**A.** Two exons and one intron of a eukaryotic pre-mRNA are represented as substrate. Consensus sequences for S.cerevisiae (above) and vertebrates (below) are shown on an expanded scale. Reaction 1 yields cleavage between exon 1 and the intron with simultaneous formation of the first lariat product containing the intron and exon 2. The second reaction yields mRNA and the second lariat RNA which is the precisely excised intron.

**B.** The branch point structure of the lariat RNAs is shown. Note the double 2'-5' and 3'-5' phosphosdiester  bonds between adenosine and guanosine (2'-5') and between the same adenosine and the pyrimidine nucleotide which follows (3'-5'). This is the **A** which is indicated by an arrow in the  consensus branch point sequences.

protein assisted splicing, and perhaps even pre-dated protein synthesis. There is evidence that the open reading frames in group I introns can evolve independently of the ribozyme part of the intron sequence (7).

The overwhelming majority of nuclear splicing in eukaryotic organisms takes place on a multi-component complex called the spliceosome (8, 9, 10). The spliceosome consists of U1, U2, U5 and U4-U6 small nuclear RNP particles (although U4 release seems to be required for the first splicing reaction to take place (11, 12, 13)) plus some known and probably many unknown proteins. All of this complex apparatus seems to have evolved to aid what are basically the same two reactions shown in Figure 1 and which correspond to splicing of type II introns.

Nuclear splicing has been most thoroughly studied in a unicellular eukaryote S.cerevisiae, and a metazoan H. sapiens (primarily in the Hela cell line). Although the spliceosomes, splicing reactions, and sequences required for splicing in these two organisms resemble each other greatly , the differences in splicing between yeast and man are important. In man, and in fact, in all metazoans studied up to now, most genes contain many introns. Thus splicing is an obligate step in expression of almost all proteins. In yeast, introns are relatively rare, are limited in size , and when they do occur are positioned distinctly at the 5' end of genes. Another remarkable difference between the two systems is the apparent rigidity of the yeast system with respect to the requirement for nucleotide signals in introns : little deviation is seen from consensus sequences (Fig. 2). The absolute conservation of sequence around the branch point in yeast introns reflects, at least in part, a need to form Watson Crick base pairs with complementary sequences in the yeast U2 RNA (14). Although the mechanism of U2 RNP-branch point interaction seems to be similar in vertebrate pre-spliceosomes, the recognition may depend less stringently on complementary base pairing (15, 16). It is likely that various other interactions, each of which contributes a small amount to the final reaction, are responsible for U2 snRNP binding to the branch point in vertebrates.

U1 snRNP interacts via Watson-Crick base pairing with the 5' end of the intron, and this interaction is necessary (but not sufficient) both for spliceosome assembly and in vitro splicing (17, 18, 19,20). Again, the 5' ends of introns in S. cerevisiae are much more uniform than are the 5' ends of introns in vertebrates. Of 21 examples in S.cerevisiae 15 contain the consensus sequence GUAUGU, 5 deviate from this consensus by a change of 1 nucleotide, and the one example in which there is a 2 nucleotide deviation is a special case of regulation where the protein product of this gene is thought to interact at this sequence to inhibit splicing (21). In contrast, the 5' ends of vertebrate introns are much more variable (Fig.2).

We think it likely that the metazoan variability reflects the primitive situation more than the yeast stringency. S.cerevisiae is highly evolved because of the evolutionary pressure on it to divide rapidly, and it probably uses a greater portion of its DNA to code for proteins than do metazoan organisms. The few introns that have been retained in the yeast genome are uniform, although what the selective pressure driving this uniformity might have been is unknown.


## ACCURACY OF SPLICING

The accuracy of a biological process can only be defined by the biological purpose of that process. The accuracy of replication and of protein synthesis are respectively defined by in vivo mutation rates and by the misincorporation of amino acids into proteins. The accuracy of alternative and tissue specific splicing can only be defined within the context of their in vivo diversity. Troponin T mRNA can be said to be accurately spliced when the process generates the multiple species found in vertebrate fast skeletal muscle (22, 23). Before we can discuss accuracy in a reasonable way, we must know the in vivo splicing products for a given gene. This is not easy. Most in vivo splicing reactions in

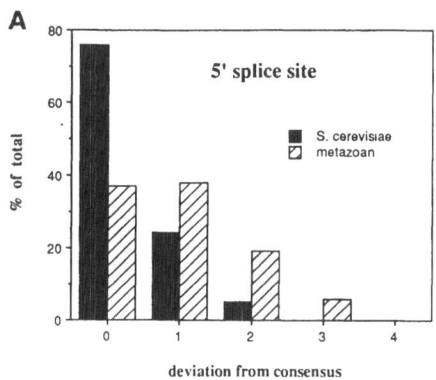

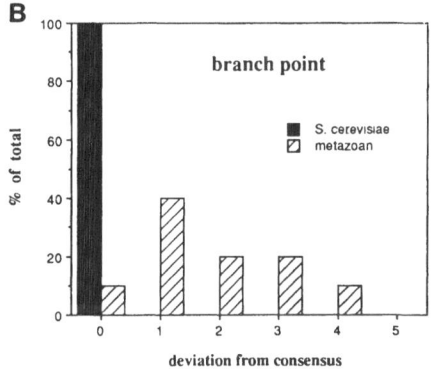

Fig. 2 : Variability of Conserved Sequences in Eukaryotic Introns.

**A.** 5' Splice site. Twenty-one examples of 5' splice sites from introns of S.cerevisiae were studied (61). The number of deviations from the consensus sequence GUAUGU was calculated for each site and the frequency of those which had O, 1, or 2 deviations was plotted. The same calculation was made with 119 examples of metazoan 5' splice sites (62). Most of the 119 splice sites were from vertebrates, but other metazoans were included. Introns from viruses were excluded. The metazoan consensus sequence was GUPuAGU, which, of course, is already less strict than the consensus sequence from S.cerevisiae.

**B.** Branch point. The same sort of analysis was done for branch points generated during lariat RNA formation. Twenty one examples were analysed for S.cerevisiae (61). Ten examples with mapped branch points from metazoans introns were used (59). No examples were taken from mutant introns, but had they been taken, it would have not changed the distribution.

vertebrates are defined by sequencing cDNA clones which have been reverse transcribed from either poly A selected or total cellular RNA. The direct nuclear products of splicing reactions may have been drastically altered and their abundances changed by the time mRNA molecules are selected for transport to the cytoplasm. Alternative or tissue specific splicing could be confused with differential stability, transport, and possibly polyadenylation. Pikielny and Rosbash (24) have already emphasized the importance of measuring both precursor and product to determine the efficiency of splicing, since splicing is not always rate-limiting for mRNA production. They have presented an example of an intron mutation which diminished the rate of splicing, but did not effect the level of mRNA in the (yeast) cell.

Even after these difficulties are resolved, we are left with the problem of why certain potential splice sites are chosen and others are ignored. Also, when in vitro splicing reactions give products different from those found in vivo, what is the source of this inaccuracy ? Again, the yeast splicing system seems more rigid than the vertebrate system. Because of the stringency of the sequence requirement, many mutations in consensus sequences of yeast introns eliminate splicing both in vivo and in vitro (25, 26, 27). Mutations that affect the first of the two splicing reactions (Fig. 1) also inhibit spliceosome formation (19, 27). When the second reaction is the stage at which inhibition occurs, mutant pre mRNAs do enter into spliceosomes (8, 11, 12, 19, 27). The evolution of this splicing system towards stringency has apparently selected a large free energy difference for spliceosome assembly between the consensus pre m RNA signals and signals with mutations (for many, but not all, of the consensus nucleotides).

This is important, since the greater the sequence information required to form a spliceosome, the more accuracy can depend on thermodynamic differences in binding. We emphasize at this point that conservation of sequence information in signals does not imply a particular mechanism. Although spliceosome formation requires Watson-Crick base pairing between 5' splice site sequences and U1 sn RNA, and also between branch point sequences and U2 sn RNA, this is not the way information is transferred for every conserved nucleotide. One example is particularly instructive. A G ---> A mutation in the fifth nucleotide of the conserved GUAUGU 5' splice site of yeast introns leads to aberrant cleavage upstream of the normal 5' splice site (29, 30). The wild type G can pair with the C at the fourth position of U1 sn RNA, whereas this base pair would be destroyed when the pre mRNA contains A at this position (31, 32). If the sn RNA is mutated to a U at the fourth position, an A:U base pair could form with the mutant pre mRNA . This construction leads to an increase in splicing efficiency, but not to an increase in accuracy (20). Thus, accuracy is transmitted by the G at this position, but by a mechanism other than simply base pairing. It is not clear of course, what the outcome would have been had the changes been a G -->C mutation in the 5' splice site and a C --->G mutation in the suppressing U1 sn RNA.

The emphasis on the rigid nature of yeast splicing does not mean that context plays no role in defining intron signals ; clearly , it does (33). That role is likely to be less important in discriminating between correct and incorrect splice junctions than it is in vertebrate splicing.

The picture is different in vertebrate introns. Because there is greater variability (flexibility) in the sequences that are used in correct 5' splice sites and branch point sequences, spliceosome formation must be able to take place in response to a wider distribution of sequences than was the case in yeast. It is perhaps interesting to note that the variability is high at both these signals. A priori there is no reason that this should be true. Perhaps accomodating a variety of U1 sn RNA -5' splice site interactions demands a similar flexibility at the other end of the intron where U2 sn RNP interacts with the branchpoint. A striking difference between yeast and vertebrate splicing is that although branch point mutations in vertebrates do affect discrimination between good and poor sites, this discrimination takes place after spliceosome formation (34). Thus, the possibility exists that the thermodynamic differences between binding correct and wrong sequences into an active complex are quite small. This, coupled with the requirement for

208

ATP hydrolysis at multiple steps in the spliceosome assembly process and in vitro reactions (for example 15, 35, 36, 37), makes one suspect that kinetic amplification may be necessary to assure accurate splicing in vertebrates (see 38, 39,40 for a discussion of various proofreading schemes). Reed and Maniatis (personal communication) have previously suggested that kinetic proofreading may be required to distinguish true from "cryptic" splice site selection in mammalian systems.

## RNA STRUCTURE AND SPLICING

Everyone seems to agree that RNA structure is of primary importance in nuclear pre-mRNA splicing, but the data gathered up to now do not give a coherent picture of how structure affects constitutive, alternative, or tissue specific splicing. Solnick (41) showed that long regions of base paired sequences allowed splice signals either in the stem or in the loop to be ignored by a Hela cell nuclear extract. This sort of structure permitted "jump-splicing" in vitro. But Solnick and Lee (42) demonstrated that such constructions, tested by transfection of Hela cells, showed no "jump-splicing" until stems were at least 41 base pairs long. They argued that in vivo, but not in vitro, potential RNA secondary structure was not permitted to form, and that an RNA-RNA helicase activity might be important for pre mRNA splicing. Nuclear RNA-RNA helicases have been described (43, 44), and at least one of the hnRNP C type proteins, which is necessary for in vitro splicing in Hela cell extracts (45), does melt RNA secondary structure (46). Moreover, Wollenzien et al (47) have used psoralen cross linking to probe intron secondary structure during in vitro splicing, and have found that much secondary structure is removed in an ATP hydrolysis dependent manner within the first 30 minutes of the reactions.

On the other hand, Munroe (48) has shown that anti-sense RNA can inhibit in vitro splicing. Inhibition was seen even when the 3' end of the probe was 82 nucleotides past the intron-exon 2 junction. In these studies, RNA-RNA duplex formation,as opposed to unwinding ,seemed to be stimulated by the Hela cell nuclear extract. It is of course too early to say what role antagonizing functions which both stimulate and unwind RNA secondary structure might play in fixing the pre mRNA into a correct conformation on the spliceosome surface.

RNA secondary and tertiary structure are thought to be important in specifying which exons are to be spliced or excluded in alternative and tissue specific splicing (49, 50,51). The rules are still not known . Weissman's group first pointed out (52) that in vitro splicing patterns could be inaccurate because pre-formed transcripts are added to extracts, whereas in the nucleus spliceosomes probably form along the pre-mRNA as splice sites appear during transcription. That such considerations must be important has recently been demonstrated in two independent ways. First of all, electron microscopic analysis of actively transcribing genes shows that stable RNP particles appear at 3' splice sites of introns essentially as soon as these RNA sites appear, and long before the rest of the pre-mRNA is synthesized (53) . Secondly, an apparent competition between secondary structure formation and spliceosome formation has been detected. Eperon et al (54) have "hidden" a 5' splice site in the stem of a stem-loop structure. When the sequence complementary to the 5' splice site is placed further and further downstream from the 5' splice site, there is a distance beyond which (about 60 base pairs) splicing is no longer inhibited. The interpretation given is that the time it takes for an inhibitory stem-loop structure to form in the RNA is of the same order of magnitude as the time it takes for an RNP or other sequestering function to bind to the 5' splice site. Thus the distance between the complementary structures or the transcription rate could tip the balance one way or the other. This intriguing result opens up the possibility of regulating alternative and tissue specific splicing by varying RNA polymerase elongation rates.

In sum, the role of RNA structure in splicing is beyond question, but precisely how it operates remains a mystery. Recent ideas about the evolution of RNA may help put these various, and sometimes conflicting, results into perspective. Splicing, like translation, requires transmission of linear information by long, multi-exonic RNA molecules. Both pre-mRNA splicing and translation require complex, energy consuming machines (the spliceosome and ribosome, respectively) to accomplish this task. The need for expending so much energy may be related to the evolutionary history of RNA. RNA structure has probably evolved in small catalytic molecules during most of evolutionary time. The evolutionary pressure on these molecules, such as self-splicing introns, was to present surfaces for catalysis of numerous reactions (see volume 52 of the Cold Spring Harbor Symposia on Quantitative Biology (1987) for reviews of RNA catalysis and its possible evolution). RNA structure, then, evolved primarily to transmit 3 dimensional information in small RNA molecules. Probably, the evolution of exons into multi-exonic genes was a late step in the overall evolution of cells. The need to transmit linear information on long RNA molecules must have been hampered by the complex 3 dimensional structures that had evolved up to that time. We think that the complicated machines mentioned above arose, in part at least, to solve this problem. Both splicing and translation use a lot of phosphate bond energy to make sure that RNA is read linearly, that is to avoid "jump" splicing and "jump" translation. However, in both splicing and translation some of the initial ambiguity has been retained in special cases in order to modulate these processes. The machines evolved to undo most secondary and tertiary structures that could mask sites, but at the same time structures were retained (or new ones evolved) to specifically modulate these two processes . Thus, RNA unwinding will occur, but some structures will direct alternative and tissue specific splicing on the one hand, and ribosome loading on the other. Even ribosome elongation could be affected, as recently demonstrated by the astounding finding that "jump"-translation occurs in the midlle of gene 60 mRNA from bacteriophage T4 (55). This is, we think, an analog of "jump"-splicing, and emphasize that ATP hydrolysis probably drives a general unfolding of potential RNA structure, but that special structures (not necessarily only Watson-Crick base pairs) have evolved to drive alternative and tissue-specific splicing.

## EXON DUPLICATION VERSUS GENE DUPLICATION

If tissue specific splicing evolved by using tissue specific factors to fix one or another alternative RNA configurations, one may wonder what advantage this had over the classical mode of generating isozymes by gene duplication . Breitbart et al (50) have already pointed out that tissue specific splicing is well suited to terminally differentiated tissues that have lost the capacity to replicate and that alternative splicing is potentially reversible. In terms of evolutionary choices, nonetheless, it is still not clear why some isozyme systems are generated by tissue specific splicing and others by gene duplication. Comparing the two sets of isozyme systems does not give an obvious answer. This was, we must add, also the case for the stereospecificity of dehydrogenase catalysed reduction of $NAD^+$ and $NADP^+$ before Benner and his colleagues (56) had the idea to group these enzymes by the redox potential of their presumed natural substrates. They then found that the enzymes fell into two distinct classes, each class corresponding to the utilization of the front or the back of the co-enzyme. We have not found the parameter which makes the isozyme list fall into two classes that correlate with the two aforementioned mechanisms of generating isozymes. There is, perhaps, one possibility, Reanney (57), following the analysis of Eigen and Schuster (56), has emphasized how the ability to faithfully replicate genetic information depends on mutation rates. Early genes, either RNA or DNA, without proof reading or mismatch repair, had to have a maximal size which was $10^2 - 10^6$ smaller than modern genes. Because of this, eukaryotic genes were less likely to have been faithfully duplicated than were eukaryotic exons. If exon duplication was a prerequiste for tissue specific splicing (see reference 50 for evidence on this point), then the classification may in fact be trivial. Isozyme production via tissue specific splicing may have arisen earlier than variation by gene duplication. The parameter which might help us arrange isozymes into 2 neat classes might simply be the time at which the isozymes first appeared

## CONCLUSION

We have attempted , in the context of this symposium, to present some qualitative ideas about tissue specific RNA splicing and its evolutionary past. Rather than focus on the details of constitutive splicing, which have been reviewed recently (59,60) , we have emphasized the probable role of ambiguous splicing in evolutionary history. It persists as alternative splicing and has had two derivatives. One, constitutive splicing, is best exemplified in Saccharomyces cerevisiae, where sequence information may define introns and splicing patterns via differential binding energies. The other, tissue specific splicing, has evolved by using tissue specific factors to fix alternative RNA structures. In vertebrates correct sites may be distinguishable from incorrect sites on the splicing apparatus only by using kinetic proofreading.

Acknowledgments

Work in our laboratory has been supported by the Centre National de la Recherche Scientifique (CNRS), Association pour la Recherche sur le Cancer (ARC), Institut national de la Santé et de la Recherche Médicale (INSERM), Association Française Contre les Myopathies (AFM), and the Ligue Nationale Française contre le Cancer.

## REFERENCES

1 - Cech, T.R. and Bass, B.L. (1986) Annu. Rev. Biochem. 55, 599-629.
2 - Lazowska, J., Jacq, C. and Slonimski, P.P. (1980) Cell 22 , 333-348.
3 - Dhawale, S., Hanson, D.K., Alexander, N.J., Perlman, P.S. and Mahler, H.R. (1981) Proc. Nat. Acad. Sci. USA 78, 1778-1782.
4 - Collins, R.A. and Lambowitz A.M. (1985) J. Mol. Biol. 184, 413-428.
5 - Herbert, C.J., Labouesse, M., Dujardin, G. and Slonimski, P.P. (1988) EMBO J. 7, 473-483.
6 - Peebles, C.L., Perlman, P.S., Mecklenburg, K.L., Petrillo, M.L., Tabor, J.H., Jarrell, K.A. and Cheng, H.C. (1986) Cell. 44, 213-223.
7 - Mota, E.M. and Collins, R.A. (1988) Nature 332, 654-656.
8 - Brody, E, and Abelson, J. (1985) Science 228, 963-967.
9 - Frendeway, D. and Keller, W. (1985) Cell 42, 355-367.
10 - Grabowski, P., Seiler, S.R. and Sharp, P.A. (1985) Cell 42 , 345-353.
11 - Pikielny, C.W., Rymond, B.C. and Rosbash, M. (1986) Nature 324, 341-345.
12 - Cheng, S.C. and Abelson, J. (1987) Genes Dev. 1, 1014-1027.
13 - Lamond, A.I., Konarska, M.M., Grabowski, P.J. and Sharp, P.A. (1988) Proc. Natl. Acad. Sci. USA 85, 411-415.
14 - Parker, R., Siliciano, P.G. and Guthrie, C. (1987) Cell 49, 229-239.
15 - Ruskin, B., Zamore, P.D. and Green, M.R. (1988) Cell 52, 207-219.
16 - Reed, R. and Maniatis, T. (1988) Genes Dev. 2, 1268-1276.
17 - Zhuang, Y. and Weiner, A.M. (1986) Cell 46, 827-835.
18 - Zillman, M., Zapp, M.L. and Berget, S.M. (1988) Mol. Cell. Biol. 8, 814-821.
19 - Ruby, S., and Abelson, J. (1988) Science, in press.
20 - Siliciano, P.G. and Guthrie, C. (1988) Genes Dev. 2, 1258-1267
21 - Dabeva, M.D. and Warner, J.R. (1987) J. Biol. Chem. 262, 16055-16059.

22 - Breitbart, R.E., Nguyen, H.T., Medford, R.M., Destreee, A.T., Mahdavi, V., and Nadal-Ginard, B. (1985) Cell 41, 67-82.

23 - Cooper, T.A. and Ordahl, C.P. (1985) J. Biol. Chem. 260, 11140-11148.

24 - Pikielny, C.W., and Rosbash, M. (1985) Cell 41, 119-126.

25 - Newman, A.J., Lin, R.J., Cheng, S.C. and Abelson, J. (1985 ) Cell 42, 335-344.

26 - Fouser, L.A. and Friesen, J.D. (1986) Cell 45, 81-93.

27 - Vijayraghavan, U., Parker, R., Tamm, J., Timura, Y., Rossi, J., Abelson, J. and Guthrie, C. (1986) EMBO J. 5, 1683-1695.

28 - Schneider, T.D., Stormo G.D., Gold, L. and Ehrenfeucht, A.,(1986) J. Mol. Biol. 188, 415-431.

29 - Jacquier, A., Rodriguez, J.R. and Rosbash, M. (1985) Cell 43, 423-430.

30 - Parker, R. and Guthrie, C. (1985) Cell 41, 107-118.

31 - Kretzner, L., Rymond, B.C., and Rosbash, M. (1987) Cell 50, 593-602.

32 - Siliciano, P.G., Jones, M.H. and Guthrie, C. (1987) Science 237, 1484-1487.

33 - Newman, A. (1987) Embo J. 6, 3833-3839.

34 - Reed, R. and Maniatis, T. (1988) Genes Del. 2, 1269-1276.

35 - Krämer, A. (1988) Genes Dev. 2, 1155-1167.

36 - Krainer, A.R. (1988) Nucleic Acids Res. 16, 9415-9429.

37 - Abmayr, S.M., Reed, R. and Maniatis, T. (1988) Proc. Natl. Acad Sci. USA 85, 7216-7220.

38 - Ninio, J. (1987) Cold Spring Harbor Symp. Quant. Biol. 52, 639-646.

39 - Hopfield, J.J. (1974) Proc. Natl. Acad. Sci. USA 71, 4135-4139.

40 - Hopfield, J.J. (1980) Proc. Natl. Acad. Sci. USA 77, 5248-5252.

41 - Solnick, D. (1985) Cell 43, 667-676.

42 - Solnick, D. and Lee, S.I. (1987) Mol. Cell. Biol. 7, 3194-3198.

43 - Rebagliati, M.R. and Melton, D.A. (1987) Cell 48, 599-605.

44 - Bass, B.L. and Weinraub, H. (1987) Cell 48, 607-613.

45 - Choi, Y.O., Grabowski, P.J., Sharp, P.A. and Dreyfuss, G. (1986) Science 231, 1534-1539.

46 - Kumar, A., Sierakowska, H. and Szer, W. (1987) J. Biol. Chem. 262, 17126-17137.

47 - Wollenzien, P.L., Goswami, P., Teare, J., Szeberenyi, J. and Goldenberg, C.J. (1987) Nucleic Acids Res. 15, 9279-9298.

48 - Munroe, S.H., (1988) EMBO J., 7, 2523-2532.

49 - Leff, S.E., Rosenfeld, M.G. and Evans, R.M. (1986) Ann. Rev. Biochem. 55, 1091-1117.

50 - Breitbart, R.C., Andreadis, A. and Nadal-Ginard, B. (1987) Ann Rev. Biochem. 56, 467-495.

51 - Libri, D., Lemonnier,M., Meinnel, T. and Fiszman, M.Y. (1989) J. Biol. Chem., in press.

52 - Aebi, M., Hornig, H., Padgett, R.A., Reiser, J. and Weissman, C. (1986) Cell. 47, 555-565.

53 - Beyer, A.L. and Osheim, Y.N. (1988) Genes Dev. 2, 754-765.

54 - Eperon, L.P., Graham, I.R., Griffiths, A.D. and Eperon, I.C. (1988) Cell, 54, 393-401.

55 - Huang, W.M., Ao,S.Z., Casjens, S., Orlandi, R., Zeikus, Weiss, R., Winge, D. and Fang, M. (1988) Science 239, 1005-1011.

56 - Benner, S.A., Alleman, R.K., Ellington, A.D., Ge, L., Glasfeld, A., Leanz, G.F., Krauch, T., MacPherson, L.J., Moroney, S., Piccirilli, J.A. and Weinhold, E. (1987) Cold Spring Harbor Symp. Quant. Biol. 52, 53-63.

57 - Reanney, D.C. (1987) Cold Spring Harbor Symp. Quant. Biol. 52, 751-757.

58 - Eigen, M. and Schuster, P. (1977) Naturwissenschaften 64, 541- 565.

59 - Green M.R. (1986) Ann. Rev. Genet. 20, 671-708.

60 - Padgett, R.A., Grabowski, P.J., Konarslka, M.M., Seiler, S. and Sharp, P.A. (1986) Ann. Rev. Biochem. 55, 1119-1150.

61 - Riedel, N., Wise, J.A., Swerdlow, H., Mak, A. and Guthrie, C. (1986) Proc. Natl. Acad. Sci. USA. 83, 8097-81O1.

62 - Mount, S.M. (1982) Nucleic. Acids. Res. 10, 459-472.

# WAS RNA THE FIRST GENETIC POLYMER?

Leslie E. Orgel

The Salk Institute for Biological Studies
P.O. Box 85800
San Diego, CA 92138

## INTRODUCTION

In recent years the hypothesis that an 'organism' based on RNA but lacking proteins is the ancestor of all living things has become increasingly popular (Cech, 1986; Darnell & Doolittle, 1986; Gilbert, 1986; Westheimer, 1986). The purpose of this article is to discuss the consequences of this hypothesis from the point of view of a chemist interested in the origins of life. How could such an organism have evolved? The nature of the 'metabolism' of the first RNA organisms is outside the scope of our discussion. We assume only that they made use of catalytic RNAs (ribozymes), one of which was capable of carrying out RNA replication.

The most conservative theory of the origin of RNA replication postulates that, from the first, replication depended entirely on the presence of preformed nucleotides or their components in the prebiotic soup. This theory must explain how the nucleotides were formed and polymerized in the absence of protein enzymes or ribozymes. The alternative theory assumes that some simpler replicating system evolved on the primitive earth and produced catalysts capable of synthesizing nucleotides (Joyce et al., 1987). We shall see that both theories face formidable obstacles.

There are at least 3 ways in which a polynucleotide might be assembled *de novo* or on a polynucleotide template:

(1) By the polymerization of nucleotides.
(2) By incorporation of bases into a preformed $3'$-$5'$-linked poly(ribose phosphate) backbone.
(3) By assembly of new nucleotides at the termini of pre-existing polynucleotides.

Routes (2) and (3) seem unlikely. We will consider only option (1). This takes us to the core of the problem - how could nucleotides have accumulated on the primitive earth?

One must recognize from the start that there is no possibility that the substrates of contemporary RNA synthesis, $\beta$-D-ribonucleotides, could have formed in a pure state. What is at issue is the synthesis of a racemic mixture containing several nucleotide-like molecules that includes a sufficient proportion of standard nucleotides to permit the *de novo* synthesis of an oligonucleotide template and its subsequent replication. The stereospecificity of the scheme for accumulating activated nucleotides and the resistance of the replication mechanism to inhibition by nucleotide analogues in the initial template or in the activated substrates (see below) are, therefore, interdependent factors that determine whether or not replication is possible.

# THE FORMOSE REACTION AND THE SYNTHESIS OF NUCLEOTIDES

The only prebiotic synthesis of ribose that has been reported is the polymerization of formaldehyde. Under a wide range of conditions, batch polymerization or polymerization in a flow reactor yields a complex mixture of straight chain and branched sugars (Harsch et al., 1983). Ribose is always a minor component of this mixture. Furthermore, all of the sugars decompose fairly quickly under the conditions of the reaction. It does not seem likely that the polymerization of formaldehyde under the standard conditions of the formose reaction could have been the source of the sugar moiety of the nucleotides.

This presents a challenging problem - is there some way of making the formose reaction more stereospecific under potentially prebiotic conditions? One can postulate a 'magic' catalytic mineral that does the trick (see below), but there is no precedent in synthetic organic chemistry. There may be some way of using a metal ion to form a ternary complex with glycolaldehyde and glyceraldehyde, intermediates in the formose reaction, in which a stereospecific condensation gives ribose directly. The extensive literature on these aldol reactions (Morgenlie, 1980; Harsch et al., 1984) does not make it clear why hexoses and tetroses would not be formed along with pentoses, nor is it clear why ribose would be the preferred product of the reaction rather than arabinose, xylose, lyxose or keto sugars. I have no doubt that a ribose derivative could be obtained in a reasonable state of purity from formaldehyde by an organic chemist. What remains to be shown is that there is a plausibly prebiotic route from formaldehyde to ribose.

Of course, there are modifications of the formose route that might be explored. The most promising is the direct synthesis of a phosphorylated ribose derivative from appropriately phosphorylated glyceraldehyde and glycolaldehyde derivatives. The self-condensation of glycolaldehyde phosphate yields threose and erythrose derivatives in roughly equal amounts (Fluharty & Ballou, 1959). However, under different conditions a metal ion might chelate to the phosphates and, in the case of the reaction between glycolaldehyde and glyceraldehyde derivatives, direct a stereospecific synthesis of the ribose moiety. Again, there are serious questions about the prebiotic relevance of such a reaction pathway. How could one get the correct phosphorylated aldehydes and avoid dihydroxyacetone phosphate or the unwanted glyceraldehyde phosphate isomer? What would prevent the synthesis of hexose-phosphates from two molecules of glyceraldehyde phosphate, etc.? Despite the difficulties, such potentially prebiotic syntheses as the polymerization of formaldehyde in the presence of excess trimetaphosphate are worth studying. Perhaps they would lead to just the right phosphate derivatives of glycolaldehyde and glyceraldehyde.

· No doubt an imaginative organic chemist could suggest many other possibilities for the stereospecific synthesis of ribose from formaldehyde or even of ribosides directly from nucleoside bases and formaldehyde. However, in every case one must ask whether it is realistic to propose such reaction sequences as the source of nucleosides on the primitive earth. What other possibilities are there? Could one find an alternative precursor to replace formaldehyde? Methane, methanol, formic acid, carbon dioxide, acetylene, ethylene, hydrogen cyanide all seem implausible starting points for a stereospecific ribose synthesis. There aren't too many other prebiotic carbon sources. Finally, one could appeal vaguely to the almost infinite number of microenvironments on the primitive earth and the hundreds of millions of years available for prebiotic synthesis. Surely one of them would have worked sooner or later! However, one should reflect that the almost infinite repetition, with minor variations, of a basically flawed organic synthesis is unlikely to prove rewarding.

If we ignore for the moment the ribose problem, and assume that the nucleoside bases and ribose or an ideal phosphorylated ribose derivative accumulated in high purity on the primitive earth, we must next explain the synthesis of ribonucleotides. No satisfactory prebiotic synthesis of the pyrimidine nucleosides or nucleotides has been achieved despite a good deal of effort (Sanchez & Orgel, 1970). The naturally occurring isomers of inosine, adenosine and guanosine have been obtained in 2-10% yield by direct heating of D-ribose and bases, but they are always contaminated with the $\alpha$-nucleotides and, in the case of adenine and guanine, with ribosylamine derivatives (Fuller et al., 1972). Non-enzymatic displacement of an $\alpha$-phosphate or pyrophosphate group from a ribose-1-phosphate derivative by the purine bases has been attempted, but without success (Sanchez, personal communication). It is hard to see how a $\beta$-ribonucleotide could have formed on the earth, even from pure ribose and the base, without substantial amounts of unwanted isomers.

## INHIBITORS OF TEMPLATE-DIRECTED SYNTHESIS

Preformed oligonucleotide templates, in certain cases, direct the synthesis of 3´,5-linked oligomers from activated mononucleotides (Orgel, 1987). Many problems must be solved before it will be possible to achieve exponential growth, but let us assume that by using pure activated nucleotide substrates we will be able to achieve true replication. How far could the replication process tolerate isomeric impurities added to the ribonucleotide pool?

An isomeric monomer can block chain growth efficiently and irreversibly only if it is converted to an activated derivative and is then incorporated, in a template-dependent process, at the growing end of an oligonucleotide chain. If growth occurs entirely in the 5´-3´ direction, the $\beta$-nucleotides derived from lyxose, xylose and arabinose should be inhibitors. Nucleotide derivatives of many hexose sugars might also be inhibitors, but this is less clear.

In the laboratory, it might be possible to fine-tune template-directed synthesis by choice of the activating group so that the ribonucleotides would react very efficiently and all potential inhibitors would incorporate much more slowly (Joyce et al., 1984a). However, it seems to me that it would require a remarkable coincidence for any environment on the primitive earth to provide a pure activating agent with the necessary selectivity.

We have discussed the problem of enantiomeric cross-inhibition in detail elsewhere (Joyce et al., 1984b). In brief, the L-isomer of a nucleotide in its *syn*-conformation has a three-dimensional structure similar to the D-isomer in its *anti*-conformation. This, we believe, accounts for the inhibition of the polymerization of activated D-guanosine-5´-phosphate derivatives on poly(C) by the corresponding activated L-derivatives. Since the prebiotic synthesis of only one enantiomer of a chiral nucleotide is impossible, this presents a very serious difficulty. One could postulate fine-tuning by choice of the appropriate activating agent so that once an all D- or all L-template was formed it would reject the mirror-image monomers, but again it is not obvious that this is plausible in the context of prebiotic chemistry.

While there may be a neat solution to one or more of the problems raised above, it does not seem too likely that any single set of conditions or any reasonable cycle of conditions would allow the establishment of self-replicating polynucleotides in an aqueous system. There are just too many arbitrary constraints to be met. The most plausible escape from this dilemma is to propose 'crypto-enzymes' that direct naturally non-specific organic chemistry along specific pathways. The only available 'crypto-enzymes' are minerals. Is it possible that some or all of the required monomer and polymer syntheses occur stereospecifically on mineral surfaces? Then an appeal to the large number of microenvironments available on the primitive earth, each with its own suite of minerals, might make the direct evolution of polynucleotide replication seem more reasonable. If one of the relevant mineral surfaces was optically active, it might also help overcome the problem of enantiomeric cross-inhibition, locally. Of course, there would be equal numbers of D- and L- faces so one would obtain a racemic mixture of all D- and all L-polymers rather than individual molecules containing equal numbers of D- and L-monomers.

Alternatively, if minerals did not catalyze the stereospecific synthesis of ribose, could they have adsorbed ribose selectively from the mixture of products obtained in the formose reaction? The presence of a pair of cis hydroxyl groups might, for example, provide the basis for separating ribose derivatives from the corresponding derivatives of most other sugars. This is certainly a possibility, but I am unaware of any literature that suggests that ribose or its derivatives differ significantly from other sugars and their derivatives in the strength with which they bind to mineral surfaces. Again, an appeal to the large number of different minerals that must have been present on the primitive earth might make the idea of selective adsorption more attractive.

I find it difficult to assess theories postulating catalysis or segregation by minerals in the absence of experimental evidence in their favor. Presumably the microenvironment in which life evolved must have been 'macro' by the scale of laboratory organic chemistry and it must have survived long enough to allow replication to become autonomous. It would be comforting to know that some abundant prebiotic mineral catalyzed at least one or two of the several required stereospecific reactions, or specifically adsorbed ribose or one of its derivatives. Until minerals with such properties are discovered, it seems sensible to explore possible indirect routes to polynucleotide replication.

It is perhaps worth emphasizing that, in the absence of membranes, any oligonucleotides or related polymers formed on the primitive earth would almost surely have been adsorbed on a mineral surface – there were lots of silicates and no siliconized test tubes! It seems very likely, therefore, that minerals played an important role in the origins of life. What is unclear is that minerals could have contributed to the stereospecificity of monomer synthesis or to the efficiency of template-directed synthesis. So far a number of experimental studies have failed to provide evidence for such contributions, although minerals have been shown to be useful in other ways.


## A PHILOSOPHICAL ASIDE

Perhaps there are very many potentially self-replicating systems of roughly the complexity of RNA. Then the chance of finding conditions under which one or another works would be much better than the arguments I have given above suggest. It could be argued that we have biased the odds against the evolution of replication by insisting on a series of coincidences that lead uniquely to $\beta$-D-ribonucleotides. Had the successful microenvironment been slightly different the genetic material might be based on glucose, say, and we would now be complaining about the difficulty of synthesizing glucose stereospecifically.

This is certainly a valid argument. However, I don't think it gets one very far. The fundamental difficulty is not that of making ribonucleotides specifically, but rather of making any single nucleotide-analogue specifically. It seems likely that any hypothetical 'RNA-like' replication scheme based on molecules as complex as nucleotides would have a group of analogue inhibitors, and could only work if a series of intrinsically non-specific reactions were made specific by some special feature of the prebiotic environment.


## 'SIMPLE' ANCESTORS OF RNA

The conservative theory that RNA was the first replicating molecule on the primitive earth may impose too severe constraints on the chemist's imagination. The alternative view, unfortunately, goes too far in the opposite direction. There is no limit to the variety of self-replicating systems that can be contemplated on paper. Virtually nothing is known about any such system - each would provide the subject matter for a large experimental program. Where should one start?

There seems little point in considering a novel family of monomers unless they can be synthesized under prebiotic conditions more easily than the ribonucleotides. The monomers must pair together in such a way as to permit template-directed synthesis based on molecular complementarity, and polymerization must proceed efficiently in aqueous solution. The synthesis of the monomers must not generate substantial amounts of analogues that would become incorporated into growing chains and act as terminators.

The above requirements are relatively well-defined. In addition it is desirable, but perhaps not essential, that a postulated first replicating system should bear some relation to the nucleic acid system. Cairns-Smith (1982) has argued that genetic 'take-over' of one replicating system by another completely unrelated system is possible. The first genetic system 'invents' a self-replicating system for its own (that is, the 'inventor's') purpose. The new system escapes control and becomes independent, possibly eliminating its 'inventor' in the process. There is nothing illogical about this argument; it is an essential feature of the 'replicating clay' hypothesis. Nonetheless the need for one organic replicating polymer to invent another unrelated replicating polymer is an ugly feature in any theory.

When we try to decide on a system for experimental investigation, we are faced, therefore, with a series of trade-offs. Should one take 'the phosphate problem' seriously and look for a self-replicating polymer that does not contain phosphate? Should we insist that the purine bases, or the pyrimidines, be present? Does one need four monomers from the start, or could one get by with two (or even with the two enantiomers of a single chiral monomer)? Each investigator must make his or her own choice, but I suspect that none will meet the essential constraints easily, no matter what they decide about the optional details. The problem of the origins of life still seems intractable.

I

HOH₂C, O, G — CH, CH₂

Structure I: $HOH_2C$ and $HOH_2C$ connected to CH, with O and CH₂ to G

II

$HOH_2C$ ... O ... G / CH₂ CH₂

III

O, CH₂, G / H₂C / CHOH / CH₂OH

IV

CH₂—G / O / H₂C / CHOH / H₂C / O / CH₂ ... G

V

$HOH_2C$ / CH—CH₂—G / $HOH_2C$

VI

$HOH_2C$ / CH₂—CH₂—G

VII

$HOH_2C$ / $HOH_2C$—C—CH₂—G / $HOH_2C$

VIII

CH₂—B / CH₂ / CH / $^-O_2C$ ... $NH_3^+$

In the least complicated models, each monomer must be one of a family of analogous trifunctional molecules. Two functionalities, for example, a phosphate and a hydroxyl or a carboxyl and an amine, are needed to permit polymerization. Some other part of the molecule is needed to differentiate between monomers, and to provide for complementary recognition. The fundamental problem is that trifunctional molecules are all fairly complicated; most prebiotic processes leading to one family of trifunctional molecules are likely to produce corresponding mono-, di- and tetrafunctional molecules, along with inhibitory analogues of the optimal trifunctional monomers. The only exception to this rule seems to be the α-amino acids, which are the predominant products of reasonable prebiotic reactions.

Let us look at two examples. Glycerol derivatives such as **I** have been proposed on a number of occasions as possible prebiotic monomers (Schwartz & Orgel, 1985; Joyce et al., 1987). But why wouldn't the riboside-like compound **I** be contaminated with a variety of isomers and analogues of which **II-IV** are typical? The acrolein-derived nucleoside analogue **V** could be formed by the addition first of a base, and then of two formaldehyde molecules to acrolein (Joyce et al., 1987). Why wouldn't they be contaminated with analogues **VI-VII**? The reader will soon convince himself that almost any choice of trifunctional monomers faces similar difficulties. The most plausible family of molecules that I can suggest to bypass the problem of analogue inhibition combine features of amino acids and nucleotides (**VIII**). Unfortunately, they don't seem at all likely to be formed under prebiotic conditions. Furthermore, we don't know if they are substrates for template-directed synthesis. Monomers of the type **VII** are also attractive in a number of respects (Miller, personal communication).

We have seen that it is easy to talk generally about monomers that can be synthesized more easily than ribose, but much less easy to come up with specific examples - the idea has been around for several years, but I know of no prebiotic synthesis of an alternative monomer. This might not be a serious argument if some potentially prebiotic polymers had really substantial advantages as substrates of template-directed synthesis, and had no compensating disadvantages. This does not seem to be the case.

Certainly some monomers have the substantial advantage that they might polymerize in aqueous solution more readily than ribotides. Pyrophosphate bonds, for example, are formed much more easily than phosphodiester bonds. Deoxynucleoside 3´,5´-bisphosphates and glycerol 1,3,bisphosphate derivatives polymerize readily in aqueous solution (Schwartz & Orgel, 1985). There is a very real possibility that some such compounds might be more tractable substrates for molecular replication than the nucleotides. Alternative substrates, if sufficiently flexible might also help to solve the problem of enantiomeric cross-inhibition (Joyce et al., 1987). However, theories that invoke some simpler polymer as a precursor of RNA face a very serious difficulty that is avoided in the conservative RNA model. How did the transition from the original genetic material to RNA occur?

219

$$^-O-\overset{\overset{O}{\parallel}}{P}-OCH_2$$

IX

$$HOH_2C \diagdown \quad CH-CH_2-A \diagup \quad HOH_2C$$

X

There are two possibilities. Perhaps the first genetic material was unrelated to RNA and invented RNA as a disjoint self-replicating system. Later, it was taken over by RNA, a la Cairns-Smith. I don't like this theory because it implies that little can be learned about the first replicating system from contemporary biochemistry. It makes the problem of the origin of life a problem in chemistry and planetology with little connection to biochemistry. However, personal preferences are not relevant, and anyone who believes in a totally novel organic replicating system as a precursor of RNA would be well-advised to test his or her ideas in the laboratory.

The other possibility is that there was a gradual transition. The original genetic polymer might incorporate a ribonucleotide into its structure in place of the dominant monomer without major structural change — a polyglycerol phosphate chain, for example, could incorporate an occasional ribonucleotide. With the passage of time, more and more ribonucleotides were incorporated until, finally, the polymers were made up entirely of ribonucleotides.

This brings us to the difficulty of all such theories. The only way in which ribonucleotides could become more readily available is through the catalytic activity of polymers synthesized from the simpler monomers. We must assume, therefore, that the simpler system could not only provide catalysts, but could provide catalysts that maintained their activity when the original monomers were substituted more or less at random by ribonucleotides. This doesn't seem likely, unless the original monomers were very like ribonucleotides. Furthermore, there doesn't seem a plausible way out of this difficulty by postulating that genetic and functional RNA molecules were formed in different ways and that the catalytic molecules completed the transition before the genetic molecules began to incorporate ribonucleotides, or vice-versa. One possible escape is to postulate that protein synthesis was invented before the transition to RNA. Although this idea is, at least superficially, in conflict with the postulate of an RNA world, it may still be worth considering the possibility that the very first self-replicating polymer 'invented' a possibly simplified form of protein synthesis.

## CYCLIZATION AS AN OBSTACLE TO REPLICATION

Recent experiments emphasize that the cyclization reactions of activated derivatives of nucleotide analogues often compete seriously with polymerization. The phosphor-imidazolide of I does not polymerize in aqueous solution, even in the presence of poly(C), but is converted exclusively to the cyclic phosphate (von Kiedrowski, personal communication). The same is likely to be true for any flexible acyclic nucleotide analogue that can cyclize to form a five-membered or six-membered ring.

The analogue IX is closely related to adenosine, but contains a six- rather than a five-membered sugar ring. Although the OH and $CH_2\text{-}OPO_3^{2-}$ groups are trans across the ring, cyclization is efficient and suppresses polymerization (Hill et al., 1988). It seems almost certain that cyclization will always be extensive unless the OH and $CH_2\text{-}OPO_3^{2-}$ groups are trans with respect to a five-membered ring. If so, ribose and arabinose are unique among pentose and hexose sugars in the resistance of their activated nucleoside-5´-phosphates to cyclization.

It is perhaps more surprising that activated analogues of the nucleoside bis-monophosphates also cyclize readily, since the cyclic derivatives that are formed contain a seven-membered ring. Cyclization always competes with polymerization (Schwartz et al., 1987), and in the case of the many open chain analogues, for example the bis-monophosphate of X, suppresses it almost completely (Tohidi & Orgel, unpublished).

220

Clearly cyclization reactions are a serious obstacle to polymerization for many simple analogues of nucleotides. They make it less plausible that analogues such as **I-VII** could have been the monomeric constituents of an early replicating system. On the other hand, cyclization provides a mechanism, admittedly inefficient, for excluding nucleoside analogues other than arabinosides from growing polymer chains; the analogues would cyclize before they could incorporate and act as chain terminators.

If the activated derivatives were polyphosphates, cyclization would lead to products with a much lower anionic charge than the starting materials. ATP, for example, carries 4 negative charges at neutral pH's, while 3′-5′-cyclic AMP carries only one. This might be crucial in segregating the appropriate activated monomers by adsorption on the surface of a mineral.

The arguments presented in this section are not conclusive. However, they do provide some encouragement for those who believe that RNA was the first replicating system. While cyclization does not ameliorate the problem of enantiomeric cross-inhibition, it might help to explain how chain-terminating nucleotide analogues, other than arabinosides, were excluded from growing RNA chains.

## REPLICATING CLAYS

According to Cairns-Smith (1982), the first genetic materials were clay particles. The nucleoprotein system or one of its precursors was 'invented' by replicating clays. If this is true, clay particles directed the stereospecific synthesis of the monomeric components of the first organic replicating system either directly or, indirectly, by assembling organic macromolecular catalysts (presumably proteins). This implies the existence of a family of clay particles each capable, on the one hand, of catalyzing a stereospecific organic reaction and, on the other, of reproducing sufficiently accurately that its descendents could carry out the same reaction. Is this plausible?

The active site of a stereospecific clay catalyst could not be much simpler than active sites of enzymes and ribozymes. Presumably it would require a 'cavity' between clay layers or at their edges around which at least 3-6 specific metal ions and oxide ions are arranged in precisely the correct orientation. Could a clay particle carrying such an active site replicate accurately?

Intuitively, it seems that features of clay structure such as the displacement of layers in a partially disordered particle or the statistical distribution of ions according to charge and size might be propagated from a pre-existing nucleus to new material deposited on it. Similarly, the gross structure of a spiral defect might be transmitted from ancestor to descendant. However, I can see no way in which the highly idiosyncratic structures of the cavities constituting active sites could replicate. There is no chemical mechanism for replication by lateral spread within an interlayer, while the two dimensional aluminosilicate layers would seem an insuperable barrier to the transmission of detailed structural information perpendicular to the layer direction. No mechanism has been suggested for the propagation of an 'edge' cavity. Furthermore, more than 20 years after the first presentation of the theory, there is no experimental evidence for stereospecific catalysis or information-conserving replication. I conclude that the clay theory is implausible in the absence of experimental evidence or convincing theoretical arguments for the replication of adequately complex aluminosilicate structures.

## CO-PRECIPITATION AND CO-CRYSTALLIZATION

Theories in which it is assumed that the first genetic material was an organic copolymer are attractive because they postulate a replication mechanism that is known to work. They suffer a compensating disadvantage that they require that an efficient dehydration-polymerization reaction occurred in aqueous solution. Cairns-Smith's clay theory does not suggest a replication mechanism that I find convincing. However, it does have one very attractive feature - it replaces polymerization by precipitation (or crystallization). Deposition of inorganic solids from aqueous solution must have occurred on a vast scale throughout geological time. Can one construct a hybrid theory in which information is stored in the spatial arrangement of organic moieties, but the organic molecules are locked into position as monomers within an inorganic matrix?

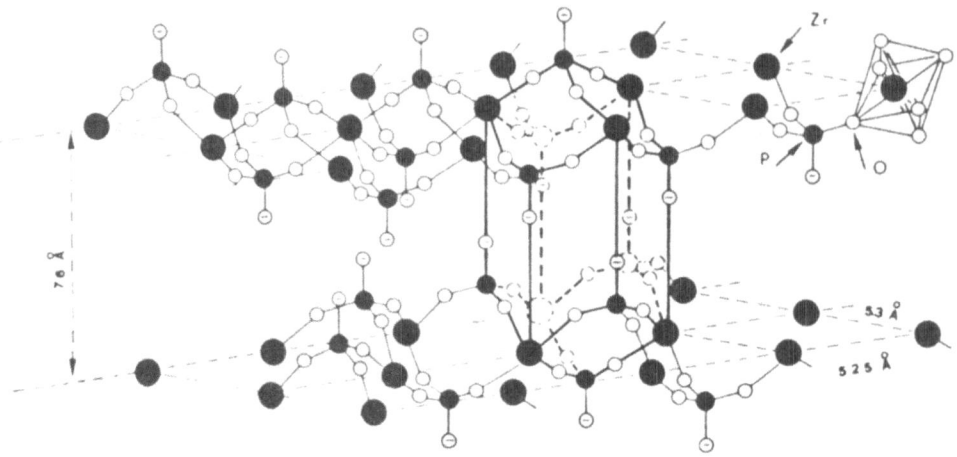

**Figure 1.** Idealized structure of $a = \text{ZrP}$ showing one of the zeolite cavities created by the arrangement of the layers.

There is one well-studied system which, although it cannot have any relevance to the sequence of reactions leading to the origins of life, has features that are of interest in the context of replication. Crystalline zirconium phosphate obtained from strongly acid solution has the empirical composition $\text{Zr}(\text{HPO}_4)_2$. The structure is made up of discrete layers (Figure 1). Each phosphate group provides 3 oxygen atoms that bind together the central $\text{Zr}^{4+}$ layer. The fourth oxygen lies far away from the $\text{Zr}^{4+}$ layer — the bond to this oxygen is perpendicular to the plane of the layer (Clearfield, 1984). Some or all of the $\text{PO}_4\text{H}^{2-}$ anions in this structure can be replaced by organic phosphonates ($\text{RPO}_3^{2-}$). The phosphorus-carbon bond is always perpendicular to the $\text{Zr}^{4+}$ plane. The interaction between the organic radicals then dominates the interlayer interaction, controlling the interlayer spacing (Figure 2) and the horizontal displacement of one layer relative to the next (Alberti et al., 1978).

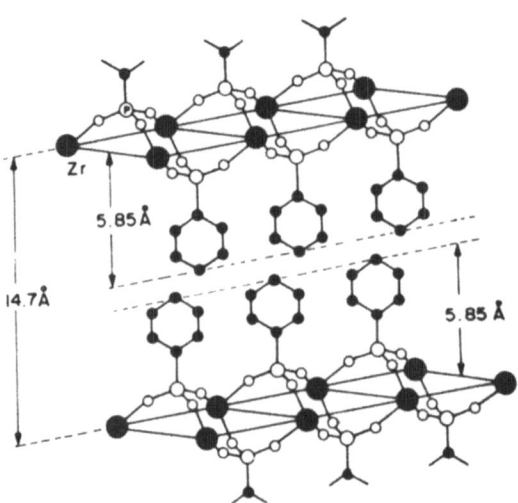

**Figure 2.** Idealized crystal structure of zirconium bis(benzenephosphonate).
●, Zr; O, P: o, 0; ●, C.

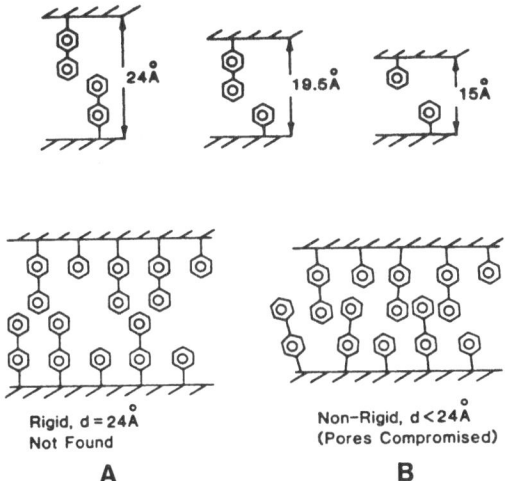

**Figure 3.** Alternative structures of an intermediate and end members of the series: (top) pure-phase biphenyl, ordered phenyl-biphenyl, and phenyl compounds; (bottom) situation for the propped structure (A) and situation for the bending of biphenyl groups, which compromises pores (B).

We cannot describe this fascinating chemistry in detail here. Instead, I have reproduced a diagram which is said to illustrate the structure obtained when $Zr^{4+}$ is crystallized with a mixture of phenylphosphonate and diphenylphosphonate (Dines & Griffith, 1983) (Figure 3). If this structure is correct, it suggests that a preformed layer directs the two dimensional arrangement of phenyl and diphenyl groups in a layer which is deposited on it. Could such a mechanism work with more plausibly prebiotic components in neutral aqueous solution? This possibility has been discussed in some detail elsewhere (Orgel, 1986).

We have not explored this model exhaustively. However, a substantial number of experiments have been completed. They fail to demonstrate any effect of preformed precipitates containing $Ca^{2+}$ and $PO_4^{3-}$ ions, together with nucleotides, or nucleoside-$5'$-triphosphates, on the precipitation of solutions containing $Ca^{2+}$, $PO_4^{3-}$ and complementary nucleotides. We were not inclined to pursue this model further!

CONCLUSIONS

There is at present no convincing theory that can account for the origin of replicating RNAs. There are several logically coherent scenarios, but each of them is obliged to assign to chemical entities properties that they seem unlikely to possess, or to postulate stereospecific synthetic pathways which seem implausible in a prebiotic context. But life did originate spontaneously, probably about 4 billion years ago on the primitive earth. Something has to give! A number of now familiar, prebiotic reaction sequences seem implausible *a priori*, for example the synthesis of adenine from ammonium cyanide. There must be a few more waiting to be discovered.

Obviously, I do not know the correct solution. It must be found in some combination of the following six possibilities. Most of them require that some chemical processes that seem *a priori* improbable did occur.

(1) There never was an RNA world. Self-replicating proteins came first or functional proteins evolved in parallel with RNA. These scenarios have been discussed repeatedly elsewhere. They have severe difficulties of their own.
(2) There never was an RNA world. A genetic polymer simpler than RNA invented protein synthesis. Maybe protein synthesis was perfected by RNA.
(3) RNA came first. There is, indeed, a reasonably stereospecific prebiotic synthesis of activated ribonucleotides, and an associated sufficiently enantiomer-resistant and isomer-resistant replication mechanism.
(4) There is at least one series of prebiotic monomers, simpler than the ribo-

223

nucleotides, that can generate replicating catalytic polymers. A transition from the simpler polymers to RNA, without loss of catalytic function, occurred on the primitive earth.

(5) There are stereospecific catalytic clays or catalytic organic-inorganic co-precipitates that can replicate. They 'invented' an organic replicating polymer.

(6) There is a different solution.

Experimental evidence in favor of any aspect of any of the above would be an important contribution to studies of the origins of life.

# REFERENCES

Alberti, G., Constantino, U., Allulli, S., and Tomassini, N., 1978, Crystalline Zr(R-PO$_3$)$_2$ and Zr(R-OPO$_3$)$_2$ compounds. (R = Organic Radical) A new class of materials having layered structure of the zirconium phosphate type. *J. Inorg. Nucl. Chem.*, 40:1113.

Cairns-Smith, A.G., 1982, "Genetic takeover and the mineral origins of life," Cambridge University Press, Cambridge (Great Britain).

Cech, T.R., 1986, A model for the RNA-catalyzed replication of RNA, *Proc. Natl. Acad. Sci. USA.*, 83:4360.

Clearfield, A., 1984, Inorganic ion exchanges with layered structures. *Ann. Rev. Mater. Sci.*, 205.

Darnell, J.E., and Doolittle, W.F., 1986, Speculations on the early course of evolution, *Proc. Natl. Acad. Sci. USA.*, 83:1271.

Dines, M.B., and Griffith, P.C., 1983, The mixed-component layered tetravalent metal phosphonate system, Th(O$_3$PPh)$_x$(O$_3$PC$_6$H$_4$Ph)$_{2-x}$. *Inorg. Chem.*, 22:567.

Fluharty, A.L., and Ballou, C.E., 1959, D-Threose 2,4-diphosphate inhibition of D--glyceraldehyde 3-phosphate dehydrogenase. *J. Biol Chem.*, 234:2517.

Fuller, W.D., Sanchez, R.A., and Orgel, L.E., 1972, Studies in prebiotic synthesis. VI. Synthesis of purine nucleosides. *J. Mol. Biol.*, 67:25-33.

Gilbert, W., 1986, The RNA World. *Nature*, 319:618.

Harsch, G., Harsch, M, Bauer, H., and Voelter, W., 1983, Produktverteilung und Mechanismus der Gesamtreaktion der Formose-Reaktion (Product Analysis and Mechanism of the Formose Reaction.) *Z. Naturforsch B*, 38B:1269; and references therein.

Harsch, G., Bauer, H., and Voelter, W., 1984, Kinetik, Katalyse und Mechanismus der Sekundarreaktion in der Schlussphase der Formose-Reaktion. *Liebigs Analen der Chemie*, 623; and references therein.

Hill, A.R., Jr., Nord, L.D., Orgel, L.E., and Robins R.K., 1988, Cyclization of nucleotide analogues as an obstacle to polymerization, *J. Mol. Evol.* 28:00.

Joyce, G.F., Inoue, T., and Orgel, L.E., 1984a, Non-enzymatic template-directed synthesis on RNA random copolymers: poly(C,U) templates. *J. Mol. Biol.*, 176:279.

Joyce, G.F., Visser, G.M., van Boeckel, C.A.A., van Boom, J.H., Orgel, L.E., and van Westrenen, J., 1984b, Chiral selection in poly(C)-directed synthesis of oligo(G). *Nature*, 310:602.

Joyce, G.J., Schwartz, A.W., Miller, S.L., and Orgel, L.E. 1987, A case for an ancestral genetic system involving simple analogues of the nucleotides. *Proc. Natl. Acad. Sci. USA*, 84:4398.

Miller, S.L., Personal communication.

Morgenlie, S., 1980, Gas Chromatography-Mass spectrometry of hexuloses and pentuloses as their O-isopropylidene derivatives: analysis of product mixtures from triose aldol-condensations. *Carbohydrate Research*, 80:215.

Orgel, L.E., 1986, Did template-directed nucleation precede molecular replication? *Origins of Life*, 17:27.

Orgel, L.E., 1987, Evolution of the genetic apparatus: a review, pp. 9-16, in "Cold Spring Harbor Symposia on Quantitative Biology", Vol. LII 'The Evolution of Catalytic Function', Cold Spring Harbor Laboratory, Cold Spring Harbor, New York.

Sanchez, R.A., and Orgel, L.E., 1970, Studies in prebiotic synthesis. V. Synthesis and photo-anomerization of pyrimidine nucleosides. *J. Mol. Biol.*, 47:531.

Sanchez, R.A., Personal communication.

Schwartz, A.W., and Orgel, L.E., 1985, Template-directed synthesis of novel, nucleic acid-like structures. *Science*, 228:585.

Schwartz, A.W., Visscher, J., Bakker, C.G., Niessen, J., 1987, Nucleic acid-like structures. II. Polynucleotide analogues as possible primitive precursors of nucleic acids, *Origins of Life*, 17:351.

Westheimer, F.H., 1986, Polyribonucleic acids as enzymes. *Nature*, 319:534.

# SEQUENCE SPACE, QUASISPECIES AND STATISTICAL GEOMETRY

# NEW CONCEPTS IN MOLECULAR BIOLOGY

Manfred Eigen and Ruthild Winkler-Oswatitsch*)

Max Planck Institut
für biophysikal. Chemie
Göttingen, Germany

## 1. Introduction

Let me start with two remarks of caution that seem appropriate to me when dealing with physical concepts in biology.

Physicists are accustomed to handle problems by abstraction, while biologists insist never to lose sight of the complexity of real living systems. My first remark addresses the physicists, the second the biologists.

A physicist looking at some biological object and describing his approach often starts in the following way: "Imagine an ideal spherical cow. In fact, the cow must neither be spherical, nor does it matter whether it really is a cow..."

I believe it was Max Delbrück, himself a theoretical physicist, who told his fellow physicists that the road of physics to biology has to start from biology rather than from physics. Life is a peculiar state of matter that is not easily acceptable to equations that have been worked out for other material systems. Many physicists, of course, realized that quite soon; some of them even became vitalists who thought that an entirely new physics is required in order to deal with the problems of biology. What Max Delbrück really has taught us is that you have to start working in biology in order to learn to ask the right physical questions of biology.

Now to the other side, the biologists! After I characterized a theoretical physicist I should try the same for a (molecular) biologist. He certainly is someone who is better in blotting than in plotting, someone who likes to go "southern", "western", or "northern", but – despite glasnost – still discriminates "eastern". And he likes abbreviations, such as SDS, PCR, PFGE or YACS, and be sure, those abbreviations mean neither political parties nor programs. In the early time of EMBO I organized – together with Jeffries Wyman and John Kendrew – a meeting at Konstanz to which we invited molecular biologists from all over the world in order to discuss possible programs to be included into the new European Laboratory (EMBL) that was soon to become reality. The president of the newly founded university of Konstanz gave us a reception. Before beginning his welcome address he took Sydney Brenner and me aside and asked us: "What does EMBO really mean? I know that you are some

---

*)Part 1 to 4 presented by Manfred Eigen; part 5 presented by Ruthild Winkler-Oswatitsch

sort of biologists and I do have some associations with the word JUMBO, but what the hell does EMBO mean? No dictionary could tell me."

Among biologists there is a common belief (or should I say fear) that physicists, by applying their theories and models, might find an alternative way to solve the biologist's problems. Such a view is simply wrong. There is no alternative to find a biological structure or mechanism other than painstakingly studying it in all detail using any experimental approach that is possible. Physical theories and models fulfill an entirely different purpose; they aim at an understanding, that is to say at an integration of the biological facts into the congruous framework of physics. In order to establish the principles that govern reality one has to abstract from reality – this is true both in theoretical and experimental work – in order to test the validity of principles under clearly defined boundary conditions which will guide further thinking. In this respect Einstein was right stating that only theory can tell us which experiment ought to be done and he meant this type of scrutinizing experiment, not merely descriptive exploration.

## 2. Sequence Space

We consider a gene. In the present biological reality a gene is a piece of DNA or RNA. However, I do not mean here the chemical structure, i.e. the arrangement of all its atoms in space. What I mean is its semantic nature, so to speak the abstraction of a gene. The biological significance of the gene lies in its sequence, i.e. the sequence of quaternary digits that defines an inheritable message. Evolution proceeded by progressively substituting these digits, and hence evolution has created a pattern that is recorded in the topological structure of related genes. Evolution may be viewed as a trajectory describing the change of sequences along a route in some appropriate space.

How does such a space have to be constructed? The problem is analogous to the problem of representing messages in informatics. In fact, it is identical with it, if we keep in mind that nature's alphabet is quaternary rather than binary. The space we are going to construct in order to deal with sequence relationships should have the following properties:

i) Each of the $4^\nu$ possible sequences of lengths $\nu$ have a unique locus, i.e. a defined point in the space.

ii) All points are arranged in such a way that all kinship relations are correctly represented by some distances.

We call that space a sequence space. The concept was first introduced for binary sequences by R.W. Hamming (1).

The two conditions mentioned can only be satisfied in a multidimensional space. In order to imagine such a space we need some abstract rule of construction. We start from a binary sequence, e.g. a nucleic acid in which we first distinguish only purines (R) from pyrimidines (Y), and look at only one position. It may be either R or Y. Hence two points represent the sequence space of one position, and the two points are connected by a line that indicates their mutual replacement called "transversion". Adding a second position means adding a second line and connecting those points which can be mutually transformed. Thereby we obtain a square. Adding a third position again requires doubling of the former diagram (i.e. the square) and connecting corresponding points yielding a cube. We now have learned the algorithm. Going from dimension $\nu$ to $\nu + 1$ means doubling the former diagram and connecting neighbouring points, i.e. points that have the distance of one. Hence in a $\nu$-dimensional hypercube there are $\nu$ lines starting from every point (representing a given sequence) and connecting it with its $\nu$ possible one-error mutants. In such a geometrical representation, the measure of separation, or metric, counts distances like street blocks and is called a Hamming metric.

We now use the binary sequence space as a logical basis for generalization.(2) Assignment of a quaternary symbol to any position may proceed in two successive binary steps. In the first step we assign the base class (R or Y) only, while in the second step we decide which of the two R's or Y's is involved. Hence a quaternary sequence space may be constructed by assigning to each corner point in the $\nu$-dimensional hypercube (representing the RY sequence) a subspace, that itself is a hypercube of dimension $\nu$. We thereby arrive at a point space of dimension $2\nu$ in which each of the $2^{2\nu}$ points identifies one of the $4^{\nu}$ possible sequences.

We may ask what is the advantage of retreating to such an abstract concept. The answer is that this is the only way to correctly describe the evolutionary adaptation of messages that are contained in a sequential form. Intuitively we encounter three features associated with this concept of sequence space:

i) The "volume" is enormous. It comprises $4^{\nu}$ points each of which is uniquely assigned to one of the possible sequences of length $\nu$. (As an example consider our universe as a volume comprising about $10^{107}$ Å$^3$. Thus we could map our universe Ångstroem by Ångstroem into a sequence space that corresponds to a sequence length of only $\nu = 180$, i.e. $4^{180} = 10^{107}$.)

ii) Despite the large volume all (detour or loop-free) distances remain smaller than $2\nu$. This does not help in a search process that is entirely based on trial and error, but it is of enormous advantage in any walk biased by gradients.

iii) The large connectivity – each point via mutational jumps of length k is connected with $\binom{\nu}{k}$ neighbouring points – provides access to any point through a large multitude of alternative routes. This is of advantage if the route leads through a rugged fitness landscape, preventing evolution from getting stuck on a minor fitness hill.


## 3. Quasispecies

An important point is that all RNA or DNA sequences – via complementary base pairing – are inherently self-reproductive. Inherent self-reproduction is the basis of the conservation as well as of the production of information. In order to represent information it is not sufficient that a molecule be comprised of a finite number of classes of monomers (defining symbols) that can be linked to polymeric chains (representing messages). Nucleic acid share this property with other macromolecules such as proteins. The essential prerequisite of being a message is that it is readable. Only through RNA and DNA have the molecules learned to read and thereby store information. Proteins lack this inherent chemical property that nucleic acids possess. Exclusive mutual complementarity of the four nucleobases is not only important for reading, conserving and processing information, it is also responsible for producing genetic information. Natural selection is a direct consequence of self-reproduction.(3)

Each of the sequences has its individual lifetime, rate and fidelity of reproduction which might be combined into a parameter called selective value. This parameter has to be positive for a sequence in order for it to be eligible for selection. The selective values appear as auto-catalytic (i.e. diagonal) terms in the system of rate equations that describes the population dynamics of individual sequences. At equilibrium the diagonal coefficients are always negative. Therefore the system has to be far from equilibrium. Each point in sequence space is characterized by such a diagonal coefficient the magnitude of which defines an "altitude" in the value landscape. Whatever the true functional value of a gene or its product is, it must feed back to its selective value. By its inherent autocatalytic nature the system seeks the highest peak in this value landscape.

On the other hand, a gene sequence is not solely produced by selfcopying, it also may come about through erroneous copying of closely related sequences. Hence there are also non-diagonal terms in the system of rate equations that are related to the diagonal coefficients via an error matrix. The ratio of the probability of falsely copying a symbol $(1 - q)$ to the probability of correct reproduction $q$ is an important parameter in this matrix. This mutational coupling ensures that it is not the "fittest" individual sequence, but rather the most efficient mutant clan – we call it a molecular quasi-species – that is the true target of selection.

In particular, the theory of the quasi-species (4) (5) shows that selection can be understood as some kind of condensation in sequence space. Phenomenally, Darwinian selection means that all sequences are localized at a given point in sequence space. This view is now modified. "Fittest" is not the property of a single sequence or point but rather that of a whole mutant clan, represented by a condensed cloud of points in sequence space. This cloud is by no means symmetrically arranged around the sequence with the highest (local) selection value. Natural selection simply means condensation of a gas-like distribution into such a cloud. There is an error threshold $(1 - q)_{max}$ which must not be surpassed. If it is, this cloud will dissolve and – like an ideal gas – expand into the vast volume of sequence space. Selection behaves physically like a phase transition and can be described by a theory analogous to a 2-dimensional Ising model.(6)

The detailed structure of the mutant cloud depends critically on the value topography in sequence space. Essentially, only peaks and ridges of the value landscape are populated and along those ridges the cloud has protrusions that reach far into sequence space. Since peaks are more likely to occur along such ridges, there is an efficient internal self-guiding of the evolutionary process along ridges to the highest peaks. This self-guidance is highly efficient because of the large connectivity in high-dimensional space. The final outcome is an acceleration of the evolutionary process by many orders of magnitude and a behavior which has little in common with the interplay of trial and error assumed in classical Darwinian theory.(4)

In vitro experiments with enzyme-assisted self-reproduction of RNA molecules confirm this new view of natural selection.(7)(8)(9)(10) With these insights we have built machines in which molecular evolution can be guided under optimal conditions.(2) Analysis of viral systems (phage $Q_\beta$, influenza A, foot and mouth desease and vesicular stomatities virus (11)) shows that natural evolution adheres to the same principles, operating near the error threshold under conditions of maximum flexibility. Retroviruses (including HIV) with their high intrinsic error rates also exemplify typical quasi-species behaviour.

## 4. Statistical Geometry in Sequence Space

Another application of the concept of sequence space has been recently proposed in connection with comparative sequence analysis. In the following I summarize the method which has been described in detail elsewhere (12).

Two points in binary sequence space are separated by a Hamming distance that counts the number of positions with different occupation. All classical methods of comparative sequence analysis are based on overall distances with the aim of reconstructing a topology of divergence. The pitfalls of this method lie in the non-uniform probabilities of mutation *and* fixation which may vary from position to position, as well as from species to species. Moreover, parallel and reverse mutation introduce some noise, blurring the true topology of divergence. Two sequences define a distance. Three sequences then define a tripod, the three segments of which can be matched with the 3 distances (i.e. 3 equations for 3 unknowns). A tripodal node for quaternary sequences, however, may be fictitious if two different substitutions occur at the same position. Four sequences must not necessarily fit any true dendrogram. There are 6 distances that have to be matched with 5 distance segments. Adding a fourth sequence to a tripod requires the assignment to one of three possible four-taxa trees, none of which needs to fit exactly. Tree construction then becomes a method of optimal assignment for which various procedures have been proposed.(13)(14)(15)(16)

With the concept of sequence space, a new approach is possible that circumvents the difficulties described. Projecting four sequences into sequence space, rather than constructing mere distance diagrams, provides an objective localization of nodes in a characteristic diagram.

Those characteristic diagrams, depending on the specification of the kind of substitution, involve boxes or polyhedra with protrusions. Determining the characteristic geometries of any of the $\binom{n}{4}$ possible quartets of sequences with a set of n sequences (e.g. $n = 4 : \binom{n}{4} = 91390$) leads to statistically well defined averages for every segment in the particular geometry. Therefore the method was called "statistical geometry in sequence space". Its advantage lies in four facts:

i) The segments in the characteristic geometry refer to true divergence in sequence space and, unlike overall distances, do not sum up, and thereby mutually compensate divergences of various parts of the sequences.

ii) All segments have to refer consistently to the same state of divergence which brings about some internal calibration that is absent in absolute distances.

iii) Averages are taken from a large number of quartets, and therefore, are statistically sound.

iv) The various segments respond with varying sensitivity to different stages of divergence. This is especially important for highly randomized kinship relations, where distances are very insensitive because they are close to the limiting values of accidental fit.

Characteristic geometries can be elucidated for three cases.

First one may look at transversions (i.e. changes between base classes R and Y) only. One is dealing then with binary sequences (R,Y) and one may specify 7 distance segments for any quartet of sequences: 4 protrusions, where one of the four sequences differs from the rest, (e.g. $A \neq B = C = D$) and three pairwise coincidences (e.g. $A = B \neq C = D$), which may be represented as three dimensions of a box. The 4 protrusions may be subsumed under $d_1$, (i.e. all positions with "three of a kind"), and the box dimensions similarly form the class $d_2$ (i.e. all positions with "two pairs"). The difference between sequence length $\nu$ and $d_1 + d_2$ yields $d_0$ (i.e. all positions with "four of a kind"). The characteristic geometry is a box with four protrusions starting from four mutually diagonal corners of the box.

If the concept is expanded to include a specification of the base class, we obtain more complicated geometries. In addition to the classes $d_0, d_1$ and $d_2$, we obtain $d_3$, subsuming all positions with "one pair", and $d_4$, referring to those positions at which all four nucleotides differ (i.e. "no pair"). While the degeneracies for $d_0, d_1$ and $d_2$ were 1, 4 and 3, there are 6 possible arrangements in $d_3$ and 1 for $d_4$. The 6 segments of $d_3$ can be combined into an irregular tetrahedron (or, if averages are taken, into a regular tetrahedron). Each edge of this tetrahedron is to be prolonged by $d_4$ and the whole figure to be combined with the former box arrangement for $d_1$ and $d_2$. In the final figure the 4 triangles of the tetrahedron (expressed by $d_3$ and $d_4$) are separated in the four corners by the box dimensions (expressed by $d_2$) from which the four protrusions (expressed by $d_1$) start.

A further complication may be introduced by looking separately at transversions and transitions (i.e. changes within base classes). Instead of 5 classes, $d_0$ to $d_4$, we now obtain 8 classes $d_{ik}$, where $i$ specifies the 5 distance segments in quaternary (AUGC) space and $k$ refers to the 3 distance segments in binary (RY) space.

Average geometries according to all 3 specifications have been determined for a large number of tRNA sequences. An application of the sequence space concept will be summarized below by Ruthild Winkler-Oswatitsch.(17)

5. How Old is the Genetic Code?

Roughly one thousand tRNA sequences are known today. There are two ways of ordering the wealth of data. First, one can specify a tRNA by its anticodon family and establish its phylogeny. Second, one can focus on a given species family and produce an alignment of its individual tRNAs.

The following concept is suggestive. The two forms of divergence that we find, tree-like phylogeny of an individual tRNA and bundle-like spread of tRNAs within a given species, project to different origins. All phylogenies are rooted in the first cells that differentiated into the precursors of the present kingdoms; however, the family spread within a species is rooted at the origin of the genetic code, or more accurately at the time of fixation of the various tRNA adaptors by which the amino acids became assigned to their codons. All that is required to establish the first case is the temporal assignment of some early tripod; in the second case a congruent point of convergence for all family spreads has to be identified.

Thus we are dealing with two types of distances. They may allow the establishment of temporal limits, depending on some calibration of evolutionary rates. If it is possible to assign precursor sequences to early phylogenetic reference points we can compare the noise accumulation due to nucleotide substitution in tRNA families at two notably different times of divergence.

The concept sounds straightforward. However, a difficulty lies in the non-uniformity of substitution probabilities at various positions. Vertical analysis reveals that among the 52 reference positions (where base pairs are counted as one position) 21 are practically constant, 21 have diverged moderately, while 10 appear to be highly diverged. Statistical geometry therefore was applied separately to the two classes of variable positions while the practically constant positions – obviously biased by present functional requirements – have been exempted from the analysis.

Construction of the phylogenetic trees and reconstruction of precursor species families refering to early nodes have been carried out according to a maximum parsimony method worked out by A. Dress and coworkers (18).

The method of statistical geometry in sequence space – introduced above – is optimally suited for quantitative analysis of divergence and noise accumulation in various phases of evolution.

We consider the 3 distance segments $d_0$, $d_1$ and $d_2$ that refer to box models of quartet combinations in RY space.

The experimental values of these segments, obtained for 14 species families (including eubacteria, archaebacteria, phages, chloroplasts, mitochondria and various eukaryotes) were compared with a simulation of the three distances as a function of the relative mutation distance (the equivalent of time for constant substitution rates).

The experimental distance segments refer consistently to identical relative mutation distances (i.e. identical times of divergence). All values referring to the 21 moderately diverged positions show an equal degree of divergence (based on relative distance segments) for eubacteria, archaebacteria, chloroplasts and eukaryotes, while mitochondria and T-phages have accumulated about 50% more noise. Precursor species families that refer to early nodes, show a much smaller noise level. Their relative divergence is only (depending on the position of the node) 1/3 to 1/2 of that in present species families (i.e. eubacteria etc.). Those positions that have been classified as highly divergent are quite different. They are nearly totally randomized, in present families as well as in precursor families. On the other hand, in phylogenetic diversification they have not changed much more than the moderately variable positions.

A corresponding analysis has been carried out for all distance segments that are obtained if the four bases are specified and if different rates of substitution are assigned to transitions and transversions.

Simulations that best fit the experimental data use a ratio of transition to (individual) transversion rates of 2.5. Again, moderately and hypervariable positions are clearly distinguished as in binary space.

We may conclude: Individual and master sequences of tRNA reflect kinship relations that are consistent with generally accepted evolutionary patterns. Their family divergences allow an extension of studies into a pre-phylogenetic time range.

Moderately variable positions of tRNA sequences appear to be the main source of evolutionary information. They are homologous in all master sequences of present and precursor families, as far as base classes R and Y are concerned. Statistical geometry reveals that their evolution was divergent; early precursors appear less diverged and blurred than present families. The ratios of divergences of early precursors and present families are about $0.35 \pm 0.1$, while the ratios of phylogenetic divergence to present family divergence are near $0.65 \pm 0.1$.

These ratios refer to divergence distances. They could be converted to time ratios only if substitution rates were time independent. In the very early phases of evolution error rates, as well as acceptability of changes, must have been larger than in present organisms, where structures are optimized and error-rates are minimized by virtue of error-correcting devices. Optimization of the biosynthetic machinery may have been nearly completed at the time of early phylogenetic differentiation, but it must have been quite crude when this machinery originated. Hence ratios of divergence distances that refer to such early periods can only represent upper limits, which strengthens our argument that the genetic code must be appreciably younger than our planet. If early phylogenetic reference nodes are dated between 2 and 3 billion years ago the code cannot be older than $3.8 \pm 0.6$ billion years.

REFERENCES

1) R.W. Hamming, Bell Syst. Tech. J. 29, 147 (1950).

2) M. Eigen, in "Evolution of Catalytic Function", Cold Spring Harbor Symposia on Quant. Biol. Vol. LII, 307 (1987).

3) M. Eigen, "Selforganization of Matter and the Evolution of Biological Macromolecules" Naturwiss. 58, 465 (1971).

4) M. Eigen, J. McCaskill and P. Schuster, J. Phys. Chem. November (1988).

5) M. Eigen, J. McCaskill and P. Schuster, "The Molecular Quasi-Species" in Advances in Chemical Physics (1988).

6) I. Leuthäusser, J. Chem. Phys. 84, 1884 (1986).

7) C.K. Biebricher and M. Eigen, in "RNA Genetics" Vol. I., 1, CRC Press, Boca Raton, Florida (1988).

8) C.K. Biebricher, M. Eigen and W.C. Gardiner, Biochemistry 22, 2544 (1983).

9) C.K. Biebricher, M. Eigen and W.C. Gardiner, Biochemistry 23, 3186 (1984).

10) C.K. Biebricher, M. Eigen and W.C. Gardiner, Biochemistry 24, 6550 (1985).

11) M. Eigen and C.K. Biebricher, in "RNA Genetics" Vol. III., 211, CRC Press, Boca Raton, Florida (1988).

12) M. Eigen, R. Winkler-Oswatitsch and A. Dress, "Statistical Geometry in Sequence Space: A Method of Quantitative Comparative Sequence Analysis" Proc. Natl. Acad. USA 85, 5913 (1988).

13) W.M. Fitch and E. Margoliash, "Construction of Phylogenetic Trees: A Method Based on Mutational Distances..." Science 155, 279 (1967).

14) W.M. Fitch, "Toward Defining the Course of Evolution: Minimum Change for a Specific Tree Topology" Syst. Zool. 27, 401 (1971).

15) J.A. Lake, "A Rate Independent Technique for Analysis of Nucleic Acid Sequences: Evolutionary Parsimony" Mol. Biol. Evol. 4, 167 (1987).

16) G.J. Olsen, in "Evolution of Catalytic Function", Cold Spring Harbor Symposia on Quant. Biol., Vol. LII, 825 (1987).

17) M. Eigen, B. Lindemann, M. Tietze, Ruthild Winkler-Oswatitsch, A. Dress and A. von Haeseler, Science (1989).

18) H.J. Bandelt and A. Dress, Adv. Appl. Math. $\underline{7}$, 309 (1986).

CHRONICLE OF THE SUMMER SCHOOLS

ON MOLECULAR BIOLOGY


(SPETSAI SUMMER SCHOOLS)


on the occasion

of the 20th anniversary

**MOLECULAR BIOLOGY OF THE CELL**
July 4 - 16, 1966

**Organizers**    M.Grunberg-Manago(Paris)(chairman),    A.E.Evangelopoulos
(Athens), B.Pullman (Paris), L.Zervas (Athens).

Lecturers and Lectures

**B.N.Ames** (Bethesda),   **H.Boedtcker** (Cambridge,MA),   **M.Bretscher** (Cambridge),
**E.Canellakis** (Yale),   **B.F.C.Clark** (Cambridge),   **M.Cohn** (Univ.Pennsylvania),
**F.H.C.Crick** (Cambridge),    **P.Doty**   (Cambridge,MA.),    **A.Garen**   (Yale),
**M.Grunberg-Manago** (Paris),   **I.C.Gunsalus** (Urbana),   **R.Langridge** (Boston),
**U.Z.Littauer** (Rehovot),   **A.M.Michelson** (Paris),   **R.Monier** (Marseille),
**A.Peterkofsky** (Bethesda),   **D.Photaki** (Athens),   **B.Pullman** (Paris),   **A.Rich**
(Mass.),    **J.Richardson**   (Paris),    **M.F.Singer**   (Bethesda),    **J.D.Watson**
(Cambridge,MA.), **H.G.Zachau** (Cologne).

Lectures : **Main aspects of the chemistry of heredity**

1.  Problems of the structure of nucleic acids and proteins:  chemistry of
peptides;  some  aspects  of  the  electronic  structure  of  macromolecules;
secondary structure of model polynucleotides;  RNA (transfer, messenger,
ribosomal);  DNA structure and properties of polypeptides and proteins in
solution.
2.  Mechanism of polynucleotide synthesis:  DNA-and RNA-polymerase;  viral
RNA  replicase;   polynucleotide  phosphorylase;   methylases,  role  of
methylated bases.
3. Mechanism of protein synthesis - Genetic Code - Suppressors.
4. Mechanism of enzyme synthesis regulation: regulatory units in bacteria;
biochemical mechanism of regulation in bacteria;  biological importance of
nucleases.
5. Metabolism of the nerve cell.


Participants
J.N.Abelson (Cambridge,MA),  J.Anderson (Cambridge),  A.Argoudelis (Upjohn
Co.),   G.Augusti-Tocco (Naples),   S.Avrameas (Villejuif),   Argyrakis
(Athens),   E.Baulieu (Paris),   A.Beltchev (Paris),   T.Berman (Israel),
K.Bhargava (S.Diego),  M.Billeter (Zürich),  S.A.Bonanou (London), L.Borda
(S.Diego),  Y.Brans (Louvain),  H.Bremermann (Berkeley),  P.Bretscher
(Cambridge),   W.Brill   (Mass.,USA),   K.Cammack   (Salisbury),   R.Cape
(Montreal),  P.J.Chapman (Hull), H.Chiarucci (France), P.Claverie (Paris),
L.Cole (San Francisco),  N.S.Constansas (Athens),  J.M.Cox (Cambridge),
L.Christodoulou (Athens),  S.B.Dahlberg (Chicago),  M.Davidson (Glasgow),
A.M.De-Recondo (Villejuif),   G.Dietz (Paris),   L.Dimitrievic (Paris),
R.Djurtoft (Copenhagen),  P.Douzou (Paris),  S.Dube (Cambridge, U.K.),
J.P.Ebel (Strasbourg),  F.Eckstein (Göttingen), R.Esnault (Paris), D.P.Fan
(Cambridge),   J.Ferris   (S.Diego),   A.Flemming   (Athens),   P.Friedman
(Columbia),   W.Flygare (Urbana),   A.Garen (Yale),   J.Garnier (Orsay),
M.Garstens  (Washington),   S.G.Georgopoulos  (Athens),   N.Glandsdorff
(Brussels),   J.Georgatsos (Thessaloniki),   R.Gomes (Rio de Janeiro),
P.Gaaloul (Paris),   H.Goodman (Cambridge),   P.Granboulan (Gif/Yvette),
D.E.Griffiths (Warwick),  W.Guschlbauer (Saclay),  J.Heedegaard (Paris),
C.Helene   (Paris),   J.Hindley   (Bristol),   G.Holme   (Montreal),
A.Jacquemin-Sablon (Villejuif),   H.C.Kaerner (Heidelberg),   P.Karatzas
(Athens),  C.Kittel (Berkeley, USA), K.Kleppe (Oslo), C.Klutchko (Paris),
C.Krimbas (Athens),  M.Kogut (London), A.Kovoor (Paris), G.Kreil (Vienna),
B.Labouesse (Orsay), J.Labouesse (Orsay), N.Ledinko (Bennington), Y.Le Gal
(Paris),  J.B.Leigh (Cambridge),  C.Letendre (Paris), F.Levin (Villejuif),

G.Levis (Athens), A.Lovlie (Oslo), D.McKenzie (London), A.McMullen (Buffalo), P.Magee (Gif-sur-Yvette), P.Mandel (Strasbourg), J.Mantzos (Athens), L.Marcaud (Paris), L.Margaritis (Athens), D.B.Millar (Bethesda), Y.Moule (Villejuif), C.A.Niavis (Athens), G.Pagoulatos (Paris), G.Pantazis (Athens), C.Paoletti (Villejuif), A.Parmeggiani (Göttingen), J.Paul (Glasgow), A.Peterkofsky (Bethesda), I.Photaki (Athens), L.Pica (Naples), D.Pluznik (Rehovot), F.Pochon (Paris), E.Preddio (Bronx), J.Quertier (Brussels), D.H.Rammler (Palo Alto), H.Robertson (New York), R.Rownd (Paris), C.Saccone (Bari), I.Schechter (Rehovoth), H.K.Scheit (Göttingen), M.Schott (Paris), L.Shapiro (Bronx), E.Signer (Cambridge, Mass.) R.Simard (Villejuif), D.Stathakos (Athens), M.Stewart (Glasgow), I.Svensson (Uppsala), J.Taylor (Athens), T.Uchida (Tokyo), K.Vakirtzi-Lemonias (Athens), M.van Montagu (Gent), S.Varrone (Naples), U.Wagner (Rehovoth), J.Weil (Strasbourg), R.Weil (Freiburg), U.Wintersberger (Vienna), M.Wintzerith (Strasbourg), I.Yannas (Cambridge, Mass.), C.Zioudrou (Athens).


## SPETSAI 1967

**Protein - Nucleic Acids Interactions**
July 7 - 20, 1967

PLANNED BUT CANCELLED


## SPETSAI 1969 (4th Advanced Study Institute)

### NUCLEIC ACID AND PROTEIN INTERACTIONS
July 6 - July 19

**Organizers** F.H.C.Crick (Cambridge) (chairman), M.S.Bretscher (Cambridge), B.F.C.Clark (Cambridge), A.E.Evangelopoulos (Athens).

### Lecturers and Lectures

**P.Berg** (Stanford): Activating enzyme - tRNA system. Oncogenic viruses.
**M.S. Bretscher** (Cambridge): Protein synthesis - chain initiation-chain termination.
**B.F.C.Clark** (Cambridge): Classification and composition of cellular nucleic acid species.
**F.H.C.Crick** (Cambridge): An introduction to embryology.
**A.E.Evangelopoulos** (Athens): On enzyme, anti-enzyme and substrate interaction.
**M.Grunberg-Manago** (Paris): Polynucleotide phosphorylase. Reconstituion of ribosomes.
**U.Henning** (Tübingen): Colinearity and protein structure. Control of enzyme function.
**R. Knippers** (Konstanz): Structure of viruses.
**A.M. Michelson** (London): Components of nucleic acids; structure and function.
**L.E.Orgel** (La Jolla): Prebiotic chemistry.
**D.C.Phillips** (Oxford): X-ray analysis of protein crystals: methods of isomorphous replacement and anomalous scattering; resolution and interpretation of image; direct methods of analysis and extension to larger molecules and molecular systems.
**A.Rich** (Boston): The structure of nucleotides and polynucleotides.
**R.Russel**: Control of gene expression.
**M.Singer** (Bethesda): DNA-dependent DNA polymerase, DNA-dependent RNA polymerase; RNA-dependent RNA polymerase, DNA ligase. Characterization and function of DNases.

**J.D. Smith** (Cambridge): Missense and chain terminating mutants and the mechanism of their suppression. Host controlled modification and restriction of DNA.
**A.Tissières** (Geneva): Introduction to protein synthesis. Ribosome structure – RNAs and proteins.
**H.G.Zachau** (München): tRNA structure; minor nucleosides; multiplicity of tRNA recognition sites.

## Participants

B.Acharia (Paris), G. and J.Akoyunoglou (Athens), H.Anderson (Copenhagen), C.Argoudelis (Urbana), P.Avner (Coventry), D.Bartosik (Shrewsbury), P.M.Bayley (Oxford), K.Beaucamp (Tutzing), S.Berry (Paris), P.Besmer (Zürich), S.Bohun (New York), C.Bordier (Geneve), A.P.Bretscher (Cambridge), C.J.Bruton (Cambridge) C.R. and L.Cantor (New York), A.Caratzas (Athens), A.Cashmore (Cambridge), S.Chang (Cambridge), P.Cole (New Haven), R.Cotter (London), R. Cox (London), T.Delovitch (Montreal), C.Dimitropoulos (Athens), R.M.Dittgen (Heidelberg), S.Dolfini (Milan), G.A.Donovan (Texas), A.Efstratiadis (Athens), H.P.Erikson (Cambridge), A.Fantoni (Rome), H.Feldmann (München), J.E.Flatgaard (Tübingen), M.Fried (London), M.L.Gefter (Cambridge), K.Geider (Heidelberg), J.G.Georgatsos (Thessaloniki), R.Giegé (Strasbourg), P.Gilbert (Cambridge), A.N.Granitsas (Thessaloniki), D.Grierson (Edinburg), R.Hehlmann (München), P.Henson (Oxford), M.Herzberg (Rehovot), J.Hindley (Bristol), N.Hogg (Oxford), M.L.Hooper (Cambridge), T.Igo-Kemenes (München), M.Ikeda (Copenhagen), M.Issidoridis (Athens), D.R.James (London) W. Kabsch (Heidelberg), J.Kallos (Montreal), T.Kalogerakos (Athens), A.Kalogeropoulos (Athens), A.Kappas (Athens), G.Keith (Strasbourg), D.Kettlewell (Bristol), M.A.Koblinsky (New York), T.A.Krulwich (New York), P.M.Leighton (Edinburgh), C.Lemonias (Athens), G.Leonidopoulou (Athens), K.Letnansky (Vienna), M.Levis (Athens), M.Levitt (Cambridge), R.Lohrmann (La Jolla), C.Jacobs-Lorena (Cambridge M), D.B.Malcolm (Edingburgh), J.Mantzos (Athens), F.H.Martin (Cambridge), F. Mazza (Rome), A.H.Mehler (Milwaukee), L.Miller (Oxford), P.S.Miller (Illinois), G.J.Mitchison (Cambridge), P.J.Oriel (Midland), J.Papadimitiou (Thessaloniki), F.Papetin (Tübingen), I.Patramanis (Athens), D.Paulin (Paris), J.M. Pesando (New York), P.Philippsen (München), G. Pieczenik (New York), R. Porter (London), P.M Pithova (Paris), E.H.Prizant (London), M.P. Rathbone (Hamilton), B.J.Reger (Oak Ridge), D.Rickwood, (Birmingham), P.Rigby (Cambridge), P.Ringrose (Cambridge), G.Rosenbaum (Heidelberg), J.P.Rossier (Brussels), F.Salvatore (Naples), M.L.Sartirana (Milan), M.Schoentjes (Liege),G. Schulz (Heidelberg), U.Schwarz (Tübingen), D.Serman (Zagreb), D.Scoutas (Urbana), P.R. Sharma (Geneva), A.E.Sippel (Würzburg), A.Smith (Cambridge), K.Smith (Cambridge), V.Ssymank (Göttingen), D. Stpejan (Paris), R.Taglang (Gif-sur-Yvette), J. O.Thomas (Cambridge), J.W.Thorner (Boston), J.Tsibris (Urbana), A.Vaheri (Helsinki), V.Vomvoyanni (Athens), V.Vondreijs (Prague), H.P.Vosberg (Giessen), J.N. Vournakis (Cambridge), R.C. Warrington (Vancouver), N.L.Webb (Cambridge), S.Weil (Cambridge), F.R. Williams (Gif-sur-Yvette), S.Zadrazil (Prague).

## SPETSAI 1970
July 19 – July 31

**Organizers** H.G.Zachau (München)(chairman), F.H.C.Crick (Cambridge), A.E.Evangelopoulos (Athens), H.Feldmann (München), P.H.Hofschneider (München).

Planned : SPETSAI 1970

July 19 – July 31

Organizers : H. G. Zachau, F. H. C. Crick, A. E. Evangelopoulos, H. Feldmann, P. H. Hofschneider.

*Nature, Vol. 223, September 13, 1969*   1186

CORRESPONDENCE

Molecular Biology at Spetsai

Sir, – We the undersigned, are a group of non–Greek scientists of the Fourth NATO Advanced Study Institute of Molecular Biology, held in July 1969 in Spetsai, Greece. Many of our colleagues refused to attend this meeting because of their objection to the military government now ruling Greece. Because of their position, we wish to inform them, and the private citizens of our countries of origin, of our attitude towards this meeting.

First, we believe that the holding of a scientific meeting of this sort in Greece is a distinct advantage to scientists of that country. To refuse to attend facilitates the isolation of the Greek intellectual community, thereby cutting them off from the free interchange of information and ideas. For these reasons we believe that such meetings should continue to be held in Greece. We believe that they are of value in demonstrating the international nature of science and the desirability of a free and full exchange of views. It is with these beliefs that we went to Spetsai ; our visit in no way reflects approval of the Greek government.

Second, attendance at such meetings should be dependent upon certain assurances being given concerning the financial and political independence of the meeting and the unrestricted admission of participants. In this case, Dr. Francis Crick, the chairman of the Organizing Committee of the Study Institute, sought the assurance by the Greek government that it did not intend to make any political propaganda in connexion with the meeting. Its representative stated that the Greek government appreciated the strictly scientific nature of the meeting. Dr. Crick also obtained the assurance of the freedom of admission for students from Eastern European countries resulting in the granting of visas to them for the duration of the course.

Finally, we believe that the exact sources of finance for such meetings should be declared in advance. We consider it important that this school should in no way depend financially upon a government which we consider to be oppressive. We feel strongly that the attitude of such of our home countries as, singly and collectively, support the Greek military government is deplorable.

Yours faithfully

| | | |
|---|---|---|
| H. Andersen | P. Henson | F. Salvatore |
| P. R. Avner | M. Herzberg | M. L. Sartirana |
| P. M. Bayley | W. Kabsch | M. Schoentjes |
| C. Bordier | D. Kettlewell | A. E. Sippel |
| S. Chang | M. A. Koblinsky | V. Ssymank |
| R. Cotter | T. A. Krulwich | J. W. Thorner |
| R. M. Dittgen | D. B. Malcolm | H–P. Vosberg |
| S. Dolfini | D. Paulin | J. N. Vournakis |
| A. Fantoni | J. M. Pesando | S. Weil |
| P. Gilbert | G. Pieczenik | F. R. Williams |
| | G. Rosenbaum | |

PLANNED BUT CANCELLED

ERICE 1971

**MOLECULAR AND DEVELOPMENTAL BIOLOGY**
July 27 - August 9

**Organizers:** H.G.Zachau (München) (chairman), F.H.C.Crick (Cambridge), M.Crippa (Naples), H.Feldmann (München), P.H.Hofschneider (München), A.Monroy (Naples).

**Lecturers and Lectures**

**P.Berg** (Stanford): Tumor viruses. Suppression.
**M.L.Birnstiel** (Zürich): Molecular aspects of chromosome organization in eukaryotes, chromomeres and genes. News about hybridization.
**J.Brachet** (Rhôde-St-Génèse): Nuclear cytoplasmic interactions in animal and plant cells. Extranuclear DNA.
**M.S. Bretscher** (Cambridge): Cell membranes.
**F.H.C.Crick** (Cambridge): Patterns of the insect cuticle.
**M.Crippa** (Naples): Oogenesis. Gene amplification.
**F.Gros** (Paris): Regulatory mechanisms at the transcriptional level in prokaryotic systems.
**M.Grunberg-Manago** (Paris): Protein synthesis. Reverse transcriptase.
**U.Henning** (Tübingen): Membrane bound enzymes in bacteria (dehydrogenases, active transport, reconstruction experiments in model systems). Bacterial morphogenesis.
**P.H.Hofschneider** (München): DNA dependent DNA replication. RNA dependent DNA replication.
**R.Levi-Montalcini** (Rome): Growth control mechanisms of the sympathetic nervous system. Patterns of nerve-growth in vitro in long-term cultures of invertebrate nervous tissue.
**P.A.Marks** (New York): Erythroid cell differentiation and control of hemoglobin biosynthesis.
**A.Monroy** (Naples): Ultrastructure and molecular analysis of fertilization and early development.
**A.Moscona** (Chicago): Morphogenetic cell interactions.
**M.Nomura** (Aarhus): Ribosome structure, function and assembly; general presentation of colicines and new experiments.
**L.Orgel** (London): Chemical evolution, model systems of biological evolution.
**K.Rajewski** (Cologne): Outline of the cellular immune system, cell interactions in immune response.
**M.Revel** (Rehovot): Control mechanisms in the translation of mRNA, bacterial amd mammalian systems, brief notes on control by interferon.
**F.Sanger** (Cambridge): Sequencing methods for RNA. Structure of bacteriophage RNA and functional implications.
**M.Siniscalco** (Leiden): Mammalian cell hybrids.
**G.Tomkins** (San Francisco): Principles of regulation in higher organisms, enzyme induction, hormone action.
**H.G.Zachau** (München): Recognition of tRNAs by aminoacyl tRNA synthetases.
**W.Zillig** (München): Structure and function of bacterial DNA dependent RNA polymerases. Regulation of bacterial transcription as studied in vitro, transcription in eukaryotes.

**Participants**
T.Ahern (Sussex), G.M.Air (Cambridge), F.Auricchio (Naples), B.R.Balda (München), N.G.Baptist (Cambridge), J.D.Baxter (San Francisco), K.Beaucamp (München), W.Bender (Cambridge), P.R.Bishop (New York), W.Bonner (Oxford), M.S.Bretscher (Cambridge), M.Campo (Edinburgh), L.Casola (Naples), M.Cazillis (Villejuif), S.E.Chang (Cambridge), B.F.C.Clark (Cambridge),

M.J.Clemens (London), G.Cognetti (Palermo), M.J.Coll (Madrid), J.C. Da Costamaia (Paris), J.E.Dahlberg (Madison), D.Dina (Naples), J.E.Edström (Stockholm), H.Falter (München), B.Fedecka-Brunner (Nogent-sur-Marne), L.Felicetti (Rome), M.S.Fischer (Madison), F.Fittler (München), H.Friedman (Cambridge), J.A.Fuchs (Stockholm), P.Fuchs (Ness-Ziona), J.H.Galey (London), M.Geisert (Mainz), D.M.Glover (London), G.Giudice )Palermo(, G.Goldin (Johannesburgh), H.Grosjean (Rodes-St.-Genese), I.C.Gunsalus (Urbana), W.Henning (Tübingen), G.Högenauer (Vienna), T.Hovi (Helsinki), L.A.Isaksson (Umeå), E.Jacob (München), M.Jaquet (Paris), B.R.Jordan (Marseille), E.Jost (Berlin), G.Keith (Strasbourg), R.Konings (München), K.Koschel (Würzburg), G.Kraus (Stöckheim), B.Küppers (Göttingen), T.Landau (Rehovot), A.Larsson (Stockholm), B.Lebleu (Paris), G.Lecatsas (London), R.Levi-Montalcini (Rome), J.Loeb (Villejuif), W.Loh (Erlangen), A.Lövlie (Oslo), F.Mangia (Rome), P.A.Marks (New York), E.Mattoccia (Rome), D.McConnell (Dublin), M.E.Mirault (Lausanne), B.Moav (Tel Aviv), K.Mölling (Tübingen), J.Morrow (Naples), I.Nardi (Pisa), W.Neupert (München), J.S.Normark (Umea), R.Nucci (Naples), E.Parisi (Naples), G.Peters (Edinburgh), B.de Petrocellis (Naples), M.Petrova (München), A.J.Pittard (Paris), J.H.va de Pol (Utrecht), Ponzy Lu (Göttingen), A.Quintero Ruiz (Paris), J.S.Raisman (Naples), M.Revel (Rehovot), M.Rosbash (Cambridge Mass.), L.Rymo (Göteborg), D.V.Santi (Santa Barbara), W.Schiebel (München), D.Schneider (Göttingen), R.Schwarz (Giessen), D.Solter (Zagreb), U.Sparren (Oslo), C.Taddei (Naples), U.Theze (Paris), R.Thiebe (München), V.D.Vaquier (La Jolla), V.Vomvoyanni (Athens), K.Wanner (München), R.M.Warn (Oxford), J.Wartiovaara (Helsinki), J.Weymann (Ludwigshafen).

SPETSAI 1972

## ORGANISATION AND FUNCTION OF SUPRAMOLECULAR STRUCTURES
August 16 - August 29

**Organizers**  L.E.Orgel (La Jolla)(chairman), M.S.Bretscher (Cambridge), B.F.C.Clark (Cambridge), A.E.Evangelopoulos (Athens).

### Lecturers and Lectures

**M.L.Birnstiel** (Edinburgh): DNA of higher organisms.
**S.Brenner** (Cambridge): Genetic analysis in eukaryotes.
**M.S.Bretscher** (Cambridge): Arrangement of lipids and proteins in biological membranes: the red blood cell plasma membrane. Cell surfaces and contact inhibition.
**J.Cairns** (Cold Spring Harbor): DNA synthesis in bacteria.
**B.F.C.Clark** (Cambridge): Bacterial protein synthesis: elongation and termination.
**F.H.C.Crick** (Cambridge): Chromosomal models.
**M.Grunberg-Manago** (Paris): Reverse transcriptase. Initiation of protein biosynthesis in prokaryotes.
**H.Huxley** (Cambridge): The organisation of the protein components in muscle. Structural changes in muscel during activity.
**R.Kornberg** (Cambridge): Arrangement of lipids and proteins in biological membranes: bilayers.
**B.Müller-Hill** (Cologne): Control mechanism in bacteria: lac and ara systems. Protein-nucleic acid recognition.
**D.Northcote** (Cambridge): Cellular supramolecular structure of chromosomes. Motile systems in cell structure.
**L.E.Orgel** (La Jolla): Cellular ageing.
**K.Rajewsky** (Cologne): Cell-cell recognition and antibody formation: differentiation of the cell surface in T and B lymphocytes; B cell activation.

240

**J.Roberts** (Harvard): RNA synthesis in bacteria. Bacteriophage lambda.
**R.E.Stephens** (Waltham Mass.): Primitive mobility and the role of actin-like filaments. The biochemistry of microtubular systems.
**G.Tomkins** (San Francisco): Macromolecular synthesis and control mechanisms in eukaryotes. General growth control in mammalian cells.
**H.G.Zachau** (München): Components of chromosomes and chromatin; structure and function.

## Participants

J.H.Akoyonoglou (Athens), J.R.Arrand (Bristol), F.J.Barrantes (Buenos Aires), R.Barzilai (Jerusalem), R.Berezney (Freiburg), G.Bonatsos (Athens), N.Brewin (Cambridge), C.Bruton (Cambridge), P.Babu (Cambridge), J.Capleton (New York), A.Charitou (Athens), A.Cihak (Prague), N.Constantinidou (Athens), P.A.Costello (Aberdeen), G.Deho (Milan), C.Dimitropoulos (Athens), J.E.Donelson (Cambridge), J.M.Dothie (Cambridge), M.Epiphaniou (Athens), D.E.Evangelidis (Patras), H.Feldmann (München), S.A.Fuhrmann (La Jolla), C.Gabrielidis (Athens), D.Galanopoulou (Athens), U.Gehring (San Francisco), C.Georgopoulos (Geneva), E.J.Griffiths (Coventry), M.D.Griswold (Middleton Wisc.), F.Hamilton (Larqs Scotl.), B.A.Hamkalo (Oak Ridge), J.Humbert (Lausanne), B.Jarry (Marseille), R.Joho (Zürich), T.Kalogerakos (Bures sur Yvette), A.Kalogeropoulos (Athens), J.N.Karli (Athens), K.Katsiris (Athens), T.P.Keneklis (Lausanne), S.Kim (Cambridge), A.Kindelis (Illinois), C.Klein (San Francisco), K.Klein (Cologne), K.K.Kotinis (Thessaloniki), H.Kröger (Berlin), P.E.Kyriakopolou (Athens), J.Ladner (Cambridge), P.Laggner (Graz), S.Lavi (Rehovot), J.J.Lawrence (Grenoble), L.H.Lazarus (Jerusalem), H.Leffler (Göteborg), J.R.Lillehaug (Bergen), C.Lossow (München), D.Loukopoulos (Athens), Ponzy Lu (Göttingen), A.Maelicke (Göttingen), N.G.Makris (Thessaloniki), E.A.Mamalaki (Athens), H.Manor (Rehovot), F.Marx (Heidelberg), J.P.Mather (La Jolla), C.G.Mesologites (Athens), K.Mölling (Berlin), H.Murer (Zürich), P.McAthey (Newcastle-upon-Tyne), D.J.McConnell (Dublin), C.Norris (London), M.Ombach (Warsaw), A.Orgel (La Jolla), O.Ozier (Orsay), E.Papdopoulou (Thessaloniki), G.Papgeorgiou (Athens), P.R.Parham (Middlesex), J.Patzer (Warsaw), G.Pepe (Bari), F.D.Petrogiannis (Athens), P.Philippsen (München), A.Pihl (Oslo), G.Pirro (Modena), A. Platel (Gif-sur-Yvette), V.Popovic (Zemun), P.M.Price (New York), C.L.Prives (Rehovot), P.Puigdomenech-Rosell (Barcelona), M.Renz (Tübingen), P.W.J.Rigby (Cambridge), M.O.Savas (Helsinki), H.Savaki (Athens), P.J.Seely (Henley-on-Thames), T.Sensky (London), D.Shields (London), G.R.Smith (Geneve), N.Smolar (Uppsala), M.A.Sodd (Washington), H.Söderlund (Helsinki), E.Solomon (Paris), J.Speirs (Edinburgh), R.Sperling (Cambridge), E.Stamataki (Athens), S.A.Stewart (Aberdeen), R.E.Streek (München), M.Suh (Vienna), Q.S.Tahin (Kibutz Kabri), C.Tavouxoglou (Athens), R.S.Taylor (Cambridge), A.Therwath (Lausanne), W.J.Todd (Colorado), C.Tsamandanis (Patras), H.Tsirimonaki (Athens), I.von Zabern (Heidelberg), L.P.G.Wakelin (Cambridge), J.E.Wheeler (Tuscon Ariz.), T.H.Yeo (Newcastle-upon-Tyne), H.M.Zacharis (Göteborg).

## SPETSAI 1973

## MOLECULAR AND DEVELOPMENTAL BIOLOGY
July 19 - July 31

**Organizers**  M.Grunberg-Manago (Paris)(chairman), F.Chapeville (Paris), A.E.Evangelopoulos (Athens).

## Lecturers

**J.M.Ashworth** (Leicester), **S.Benzer** (Pasadena), **G.Bernardi** (Paris), **M.S.Bretscher** (Cambridge), **P.Chambon** (Strasbourg), **F.Chapeville** (Paris), **B.F.C.Clark** (Cambridge), **P.Ebel** (Strasbourg), **A.Evangelopoulos** (Athens), **M.Gefter** (Cambridge, Mass.), **W.Gilbert** (Cambridge, Mass.), **F.Gros** (Paris), **J.B.Gurdon** (Cambridge), **H.Kornberg** (Leicester), **C.G.Kurland** (Uppsala), **M.Grunberg-Manago** (Paris), **P.A.Marks** (New York), **A.M.Michelson** (Paris), **R.Monier** (Villejuif), **L.Sachs** (Rehovot), **R.Shulman** (New Jersey), **M.Siniscalco** (Naples), **A.Smith** (London), **J.A.Steitz** (New Haven), **Stöffler** (Berlin), **A.Weiss** (London), **H.G.Zachau** (Munich).

## Lecture Topics

1. Aspects of structure, biosynthesis, and control mechanisms (nucleic acids, proteins, ribosomes, chromosomes).
2. Cell transformation and differentiation: membranes, oncogenic viruses, and egg development.
3. Evolution of oxygen activating systems.
4. Genetic approach to behavioural problems.

## Participants

N.L.Anderson (Cambridge), F.Andronico (Naples), P.Argoudelis (Kalamazoo), S.Artavanis (Cambridge), R.Astier (Bordeaux), B.S.Baliga (Cambridge, MA), M.Barbacid (Madrid), K.Barnoux (Paris), J.Baxter (San Francisco), J.P.Beck (Paris), K.Beckingham Smith (London), F.R.Bem (Athens), G.Bennet (Lafayette), O.Bernard (Paris), E.Bikoff (Cambridge,MA), G.J.Brahenhoff (Amsterdam), St.Bram (Paris), R.Buckingham (Paris), M.Buckingham (Paris), K.Burridge (Cambridge), G.Carrara (Rome), L.Carrasco (Madrid), D.Christodoulou (Thessaloniki), E.Collatz (Berlin), A.Coutnho (Stockholm), A.P.Czermilofsky (Vienna), Devauk (Paris), C.Dimitropoulos (Athens), G.Dimitriadis (Patras), B.S.d'Udine (Leiden), Y.Engelborghs (Leuven), G.Falcoff (Paris), L.Ferguson (Oxford), W.Fitschen (Cape Town), N.P.Franks (London), Z.Frot Coutaz (Villeurbanne), R.Gebert (New York), P.Giorgi (Newcastle upon Tyne), R.Goth (Tübingen), G.Guarneros (Geneva), T.Haeptle (Poznan), J.P.Henry (Paris), I.Hirsch (Prague), B.Hohn (Basle), G.Huszar (Cambridge,MA), M.T.Imaizumi (Lausanne), V.Ingram (Cambridge,MA), J.Isaakidou (Athens), G.Jullien (Athens), T.Kalogerakos (Athens), A.Varnava-Kalogerakos (Athens), M.Kanje (Goteborg), J.G.Kaplan (Ottawa), A.Karoylias (Edinburgh), E.Katz (Jerusalem), G.Kaufmann (Rehovot), J.Keidling (Copenhagen), A.Kohn (Tel-Aviv), A.Kortsaris (Salonica), K.Katsizis (Athens), K.Kotinis (Salonica), T.Kozelj (Ljubljana), M.Kress (Villejuif), C.Lacombe (Poitiers), C.Lane (Cambridge), E.Lapan (New Haven), M.C.Lechner (Oeiras), M.Lex (Liverpool), E.Long (Geneva), A.Loyter (Jerusalem), R.Martin (Stockholm), O.Michelsen (Copenhagen), S.Mitra (Gothenburg), G.Moore (Berlin), Y.Mory (Rehovot), S.Olsnes (Oslo), M.Patillon (Paris), B.Pearse (Cambridge), S.Perzynski (Warsaw), M.Petrova (Munich), C.D. Platsoucas (Patras), P.Pohjanpelto (Helsinki), V.Popovic (Zemun), H.Prinz (Göttingen), F.Rieger (Paris), J.M.Rossignol (Villejuif), G.M.Rubin (Cambridge), L.Sadzinska (Gliwice), H.C.Schaller (Tübingen), R.Schindler (Berlin), C.Schneider (Saclay), H.P.Seeburg (Tübingen), C.Self (Tübingen), T.Sensky (London), V.Serra (Naples), G.St.Sligar (Urbana), Th.Sotiroudis (Athens), G.Sperk (Vienna), D.Steinemann (Tübingen), R.Tanguay (Stockholm), D.Thiele (Saclay), S.Tzartos (Athens), C.Vaquero (Paris), L.Varesio (Torino), M.Wable (Berlin), G.Walter (Poitiers), R.Whalen (Paris), A.Wolff (Washington), J.Wood (Coventry), S.Zagorska (Warsaw), D.Zouzias (New York).

ERICE 1974

MOLECULAR AND DEVELOPMENTAL BIOLOGY
August 1 - August 14

**Organizers:** H.G.Zachau (München)(chairman), H.Feldmann (München), G.Giudice (Palermo), A.Monroy (Naples).

## Lecturers and Lectures

**E.K.F.Bautz** (Heidelberg): RNA polymerases of prokaryotes and eukaryotes.
**H.D.Berendes** (Nijmegen): Structure and function of polytene chromosomes.
**B.Bernardi** (Paris): The organization of the eukaryotic genome.
**M.L.Birnstiel** (Zürich): The organization of the eukaryotic genome, molecular hybridization, multiple gene copies, intermediate repetitive DNA.
**F.Bonhoeffer** (Tübingen): Replication of the DNA in prokaryotes and eukaryotes.
**E.M.Bradbury** (Portsmouth): Chromatin structure, histones, and control of mitosis.
**A.Gierer** (Tübingen): Hydra as experimental model for morphogenesis; theory of pattern formation.
**F.Gros** (Paris): Biochemistry of muscle differentiation.
**U.Z.Littauer** (Rehovot): Differentiation of neuroblastoma cells and their surface membrane.
**B.Mach** (Geneva): The genetic control of antibody diversity; the mRNA for immunoglobulin cains. Hybridization studies for gene dosage.
**A.Monroy** (Naples): Essentials of early development.
**A.A.Moscona** (Chicago): Morphogenetic interactions of embryonic cells.
**K.Murray** (Edinburgh): Nucleotide sequence determination in DNA. Restriction and modification enzymes as new tools in nucleic acid biochemistry.
**L.E.Orgel** (La Jolla): Prebiotic chemistry. Ageing.
**J.Paul** (Glasgow): Chromatin, non-histone proteins and chromosome structure.
**N.Ringertz** (Stockholm): Cell fusion and cytoplasmic control of nuclear activity.
**F.Ritossa** (Bari): Basic genetic techniques in Drosophila and the genetists' view of the eukaryotic chromosome.
**M.Siniscalco** (New York): Somatic cell genetics.

## Participants

C.Alonso (Madrid), R.Appels (Stockholm), S.Austin (Brighton), N.Avdalovic (Zagreb), P.R.Avner (Paris), B.R.Balda (München), H.Blüthmann (Tübingen), R.Beier (München), J.R.Brocklehurst (London), H.Bünemann (Braunschweig), M.Bustin (Rehovot), R.Caizzi (Bari), A.Cihak (Prague), I.Claeys (Geneva), P.M.Clissold (London), G.Cognetti (Palermo), H.Cooper (Manchester), J.C.Courvalin (Kremlin-Bicetre), A.Cupello (Göteborg), J.C.Daniel (Chicago), O.H.J.Destree (Amsterdam), J.C.J.Eeken (Nijmegen), G.Eggertson (Reykjavik), W.A.Elmer (Atlanta), F.L.Enea (New York), A.E.Evangelopoulos (Athens), R.S.Feldberg (Waltham Mass.), K.Foerst (Berlin), J.Forstova (Prague), F.Gissinger (Strasbourg, E.Grabczewska (Warsaw), E.Haakansson (Uppsala), K.Harbers (Montreal), E.Harms (Berlin), P.Harrison (Glasgow), I.Hartmann-Goldstein (Sheffield), D.Helland (Bergen), J.Houben (Nijmegen), M.Huet (Paris), T.Igo-Kemenes (München), R.Jargus-Smith (London), B.Johnson (Aarhus), J.K.Keski-Oja (Helsinki), J.F.Koninkx (Nijmegen), G.Kreysing (Göttingen), M.Kröger (Göttingen), S.Kühn (Freiburg), P.Laget (Angers), M.Laval (Villejuif), M.R.Levine (London), K.Lindahl-Kiessling (Uppsala), S.Longacre (San Francisco), U.Lönn (Stckholm), F.Machicao (München), R.Mandel (Waltham Mass.), P.Manduca (Naples), M.Manson (Edinburgh), J.Marvaldi (Marseille), H.R.Matthews (Portsmouth), H.Mayer

(Braunschweig), D.McConnell (Dublin), M.Meisler (Buffalo), I.Melchers (Köln), R.Meneghini (Stanford), G.Milanesi (Pavia), C.A.Morrison (Berlin), P.Mounts (Edinburgh), P.Müller (Braunschweig), B.Murray (London), G.Parry (London), B.Perles (Villejuif), R.L.Pictet (San Francisco), G.Pirro (München), J.W.Pollard (London), D.Rekosh (London), D.Riesner (Hannover), J.A.Roper (Sheffield), D.Rungger (Geneva), J.Satava (Szeged), F.Scalenghe (Bari), N.Schechter (Rehovot), M.F.G.Schmidt (Giessen), M.Schwärzler (Bern), S.Seaver (Strasbourg), G.Siwert (Berlin), R.Solomon (Rehovot), C.Spadafora (Naples), M.Stefanini (Pavia), M.Steinmetz (München), P.Suau (Barcelona), I.Sures (Marburg), A.Tolun (Uppsala), F.Turnowsky (Vienna), A.Ullrich (Heidelberg), E.Ullu (Rome), R.L.van Etten (Lafayette Ind.), R.Voss (New York), K.Zänker (München), E.J.Zöllner (Mainz), F.Zucco (Naples).

## SPETSAI 1975

### THE MOLECULAR BIOLOGY OF ANIMAL VIRUSES
August 21 - September 3

**Organizers:** B.F.C.Clark (Aarhus)(chairman), M.S.Bretscher (Cambridge), A.E.Evangelopoulos (Athens), A.E.Smith (London).

### Lecturers

**D.Baltimore** (Cambridge, Mass.), **J.Cairns** (London), **F.Cuzin** (Nice), **S.Fazekas de St.Groth** (Basle), **F.Gros** (Paris), **M.Grunberg-Manago** (Paris), **A.Huang** (Cambridge, Mass.), **N.O.Kjeldgaard** (Aarhus), **R.Kornberg** (Cambridge, Mass.), **R.Monier** (Villejuif), **L.Philipson** (Uppsala), **P.Reichard** (Stockholm), **L.Sachs** (Rehovot), **K.Simons** (Heidelberg), **J.Tooze** (Heidelberg), **K.Weber** (Göttingen), **C.Weissmann** (Zürich), **R.Williamson** (Glasgow).

### Lecture Topics

1. Basic molecular biology, with emphasis on the comparison of prokaryotes and eukaryotes.
2. Structure and life cycle of bacteriophage (lambda and Qβ).
3. Structure, life cycle and biological effects of animal viruses (polio, DNA and RNA tumor viruses).

### Participants
S.Adams (Bethesda), K.Alitalo (Helsinki), J.Almond (Cambridge), F.Angelatou (Athens), E.Barbarese (Quebec), P.Baudy (Paris), P.Beaudry (Paris), D.Becker (Giessen), J.Bell (London), E.Blomquist (Uppsala), T.K.Bradshaw (Coventry), F.Brunel (Glasgow), W.Büsen (Tübingen), C.E.Castro (Lubbock), J.Chroboczek (Warsaw), F.H.Crick (Cambridge), F.Cochran (New York), H.-H.Dahl (Aarhus), L.Dimitrijevic (Paris), P.D.Dixon (Cambridge), K.Dunker (New York), T.Dzieglielewski (Poznan), A.Epiphaniou (Athens), G.Evangelatos (Attiki), F.Fernandez-Madrid (Detroit), W.Ferreira (Lisboa), I.Fiser (Vienna), D.Frisby (High Wycombe), E.Gallori (Firenze), K.Gausing (Aarhus), J.Gee (Oxford), S.Glass (Cambridge), T.Haertle (Poznan), R.Hauptmann (Vienna), M.Hunter (Cambridge), S.Inglis (Cambridge), G.Isaksson-Forsen (Lund), U.Jellinghaus (Heidelberg), K.Johansson (Uppsala), R.Kaempfer (Jerusalem), M.Kaerlein (Würzburg), K.Kaltoft (Aarhus), M.Kaparianos (Patras), C.Karahalios (Attiki), K.Katsiris (Athens), J.C.S.Kim (Michigan), A.Klein (Tel-Aviv), B.Koller (Martinsried), P.Kontomichalou (Athens), H.Kornberg ( Leicester), K.Kotinis (Thessaloniki), H.Kroath (Graz), C.Krüger (Berlin), D.Labuda (Poznan), H.Langbeheim (Rehovot), E.Larsson (Uppsala), A.Lazar (Jerusalem), Z.Lev (Haifa), L.Liebes (Detroit), T.Lund (Aarhus), H.Lövdahl

(Bergen), E.Mansur de Oliveira (Rio de Janeiro), G.Margomenout-Leonidopoulou (Athens), C.Martinet (Orsay), L.Martinussen (Aarhus), S.Mastronicoli (Athens), N.Matsokis (Patras), R.Michalides (Springfield), W.Min Jou (Gent), A.Morabito (Torino), N.Moschonas (Athens), A.Neer (Haifa), N.Oikonomakos (Athens), D.Orphanoudaki (Athens), S.Ovrebo (Bergen), M.Papamichail (Athens), R.Patient (Birmingham), A.Pawson (London), B.Pearse (Cambridge), M. Petkevich (Oxford), P.F.Pignatti (Paris), P.Piper (Aarhus), B.Ponder (London), H.Potuzak (Vienna), D.Rekosh (London), W.Rhode (Giessen), C.Ross (Aberdeen), L.Roux (Geneva), W.G.Röwekamp (Heidelberg), L.Samson (London), D.Sheiness (Edinburgh), S.Scherneck (Berlin), D.Schiffmann (Würzburg), S.Sevall (Lubbock), B.Sjöberg (Göteborg), C.Smith (New York), N.Sonenberg (Rehovot), T.Sotiroudis (Athens), D.Stathakos (Aghia Paraskevi), H.G.Suarez (Villejuif), C.Syrett (Beckenham), D.Talbot (Sutton), D.Tapper (Winthrop, Mass.), V.Tate (High Wycombe), M.N.Thang (Paris), S.Tzartos (Athens), A.Ullrich (Heidelberg), J.Vacqier (California), H.van Ormondt (Leiden), M.Vasseur (Paris), W.Wolf (Tutzing).


## SPETSAI 1976

### MOLECULAR INTERACTIONS INVOLVED IN THE MORPHOGENESIS OF CELLULAR ORGANELLES AND IN CELLULAR RECOGNITION
August 23 - September 3

**Organizers** M.Grunberg-Manago (Paris)(chairman), F.Gros (Paris), A.E.Evangelopoulos (Athens).


### Lecturers and Lectures

**G.Bernardi** (Paris): General discussion on chromatin.
**P.Douzou** (Paris): Dynamics of intermolecular forces in system of genetic translation.
**J.P.Ebel** (Strasbourg): Interaction of ribosomal RNAs with ribosomal proteins.
**M.Feldman** (Rehovot): Interactions in immunology: (i) lymphocytes recognition of cell surface antigen, (ii) interactions between lymphocytes and tumor cells.
**F.Gros** (Paris): Gene expression and recognition during myogenesis.
**M.Grunberg-Manago** (Paris): The role of metal ions and protein factors in ribosomal subunit association.
**I.C.Gunsalus** (Urbana): Component organization of peripheral systems.
**E.Kellenberger** (Basle): Phage head assembly.
**A.Klug** (Cambridge): Assembly of tobacco mosaic virus. Structure of viruses and other macromulecular assemblies.
**M.Lazdunski** (Nice): Organization of the axonal membrane and molecular aspects of nerve condition.
**U.Z.Littauer** (Rehovot): Differentiation of neuroblastoma cells and their surface membrane.
**V.Luzatti** (Gif-sur-Yvette): Structure of biological membranes.
**A.M.Michelson** (Paris): Component organization of a bioluminescent organelle. **J.H.Miller** (Geneva): DNA interaction with altered lac operon repressor.
**A.A.Moscona** (Chicago): The role of cell recognition in mophogenesis and differcntiation.
**R.N.Perham** (Cambridge): Quarternary structure in biological macromolecules. Subunit interactions in multienzyme complexes.

**S.N.Timasheff** (Waltham Mass.): Self-association of tubulin: monomer, microtubules, and other structures.

**G.Warren** (Cambridge): How membrane structure is designed to regulate the function of membrane proteins.

**K.Weber** (Göttingen): Microtubules and microfilament structures in vivo.

**C.W.Wu** (New York): Molecular interactions involved in gene transcription. (i) The key-enzyme: DNA dependent RNA polymerase, (ii) regulatory proteins: lac repressor and cAMP receptor.

**J.Wyman** (Rome): Assembly of giant proteins: (i) general relations regarding the control of molecular assembly, (ii) examples.

**H.G.Zachau** (München): Chromatin.

## Participants

R.Arche (Madrid), L.Backman (Umea), D.Bennet (London), M.Bellard (Strasbourg), M.Bevan (Cambridge), A.Bienvenue (Paris), B.Blazy (Toulouse), T.C.Boghansen (Copenhagen), R.Bori (Madrid), M.Bradley (New York), A.van Broekhoven (Antwerpen), M.Buckingham (Paris), R.Buckingham (Paris), M.J.Butler (Oxford), Carlier (Fresnes), S.Cary (Urbana), M.Castagna (Villejuif), L.Dibbelt (München), M.Dorizzi (Paris), S.Dowsett (New York), R.L.Drake (Cambridge Mass.), R.van Driel (Basel), I.Economidi (Thessaloniki), L.Eisenstein (Urbana), Escudero (Madrid), F.Esposito (Pisa), M.Ferrer (Toulouse), Filipic (Ljubljana), A.Flavell (London), E.Fries (Heidelberg), A.Frisch (Jerusalem), G.Gasperi (Pavia), M.J.Gething (London), R.E.Glass (Edinburgh), M.Goppelt (Hannover), I.Gozes (Rehovot), M.Greil (München), G.Gross (Heidelberg), P.Hansen (Copenhagen), E.Hakansson (Uppsala), J.Heedegaard (Copenhagen), C.Henderson (Cambridge), R.Jacob (Tübingen), T.Kalogerakos (Paris), T.Karemifillos (Thessaloniki), M.Kalimi (New York), C.Katsiri-Evangelopoulos (Athens), G.Kessler (Rehovot), Klareskog (Uppsala), C.Komitopoulou (Athens), S.Kougianou (Athens), A.Kovoor (Paris), A.Krol (Strasbourg), C.Krüger (Berlin), A.Lambeir (Louvain), M.Lasser (Rehovot), A.Lax (London), M.Sousa Lechner (Oeiras), E.M.Lilius (Turku), M.Lindsay (Basle), O.Leoncini (Tübingen), J.Lever (London), J.Lotem (Rehovot), F.Lustig (Göteborg), M. da Conceicao Duque Magalhaes (Oeiras), A.Malacrida (Pavia), J.Mallet (Paris), K.Marx (Edinburgh), N.McGregor (Oxford), G.McMaster (Lausanne), S.Metafora (Naples), M.Meurnier-Rotival (Paris), T.Moss (Portsmouth), K.Müller (Graz), D.Nachkov (Sofia), C.R.S.Naylor (Oxford), T.Nishihara (New York), Onorato (Basle), S.Paglin (Rehovot), A.K.Pandey (New York), C.Pangalos (Athens), I.Papasotiriou (Thessaloniki), Pardo (Marseille), M.Paturneau-Jouas (Mont Saint Aignan), H.U.Petersen (Paris), T.Pfeuty (Fontenay-aux Roses), G.Philipps (Strasbourg), D.Phillipides (Cambridge), Pichon (Marseille), R.Pranab (Calcutta), Y.Prives (Rehovot), L.Rabbani (New York), A.Rafalski (Poznan), M.Rafalzki (Poznan), D.Raulston (Houston), T.Reid (New Haven), E.Reisler (Rehovot), L.S.Rodriguez (Madrid), M.Rossi (Naples), G.Roizes (Montpellier), Rubin (Uppsala), E.Runger-Brandle (Geneva), L.Rylander (Stockholm), N.Sandler (Jerusalem), P.Sau (Portsmouth), A.H.Sarris (New Haven), J.Schaefer (Strasbourg), D.Schmid (Tübingen), J.Schurch-Rathgeb (Basle), S.Schwartz (Santa Cruz), W.Schwarz (München), J.Skorve (Bergen), S.Sligar (Urbana), D.Spandidos (Toronto), J.Stanley (Strasbourg), M.Steinmann (Tübingen), M.Steinmetz (München), L.Swennen (Wilrijk), H.Szain (Göteborg), A.Vincent (Paris), C.Vola (Marseille), Wasylik (Strasbourg), J.C.Weill (Paris), D.Wachter (Zürich), J.Wehland (Göttingen), E.Whitehead (Rome), P.Wilairat (Bangkok), P.Wright (Torino), F.Y.H.Wu (New York), H.Yamanaka (Nagoya), E.Yefenof (Stockholm), K.Zanker (München), A.Ziemiecki (Heidelberg), H.P.Zingsheim (Göttingen).

GENOME ORGANIZATION AND FUNCTION
August 21 – September 3

**Organizers**  H.G.Zachau (München)(chairman),  A.E.Evangelopoulos (Athens),
H.Feldmann (München).

## Lecturers and Lectures

**G.Bernardi** (Paris):  Density gradient centrifugation as a method for
studying the eukaryotic genome. The mitochondrial genome of yeast.
**E.M.Bradbury** (Portsmouth):  Histone interaction and histone complexes.
Histone modifications and the cell cycle.
**C.R.Cantor** (New York):  Fluorescence techniques.  Ribosome structure.
Eukaryotic RNA polymerase and problems of chromatic transcription.
**B.F.C.Clark** (Aarhus):  Polypeptide chain elongation;  polypeptide chain
termination.
**M.Eigen** (Göttingen): How does information originate?
**D.J.Finnegan** (Edinburgh): Studies on Drosophila.
**M.Grunberg-Manago** (Paris): Initiation of protein synthesis ans translation
of phage and viral mRNA in prokaryotic and eukaryotic systems.
**A.Klug** (Cambridge):  The protein disk of tobacco mosaic virus and
specificity in the RNA recognition and packaging.  Physical studies on the
structure of chromatin.
**K.Murray** (Edinburgh):  Making use of phage lambda in recombinant DNA
research. DNA restriction enzymes and their uses.
**N.Murray** (Edinburgh): Structure and biology of phage lambda. Expression of
prokaryotic genes inserted into lambda and comparative aspects of lambda
and plasmids.
**L.Orgel** (La Jolla):  Cellular ageing.  An outsiders' view of the
neutralist-selection controversy.
**J.Paul** (Glasgow):  The complexity of nuclear and cytoplasmic RNA.  Control
of erythroid maturation in the Friend cell.
**P.Reichard** (Stockholm): DNA Replication.
**A.Rich**  (Cambridge  Mass.):  Biosynthesis  of  protein  hormones.
Protein-nucleic acid interaction.
**W.Schaffner** (Zürich):  Repeated genes: the histone gene cluster of the sea
urchin Psammechinus miliaris.  Surrogate genetics,  or how to learn more
about regulatory sequences in eukaryotes.
**P.Starlinger** (Cologne):  The  structure  of  the  bacterial  chromosome.
Transposable DNA elements.
**J.R.Tata** (London):  Hormones and gene expression.  Biological systems for
studying gene expression during development.

## Participants
A.Akam (Oxford),  M.Allan (Glasgow),  F.Amalric (Toulouse), A.D.Argoudelis
(Kalama Mich.),  N.A.van Arkel (Eindhoven),  F.Azorin (Barcelona),  A.Bär
(München),  B.R.Balda  (München),  A.Barta (Vienna),  I.B.Barthelmess
(Hannover), J.D.Baxter (San Francisco), K.Beaucamp (Tutzing), A.S.Berkower
(New York),  J.C.Boothroid (Edinburgh),  C.D.Boyd (Tygerberg SA), I.Bozzoni
(Zürich), R.Brambilla (Basle), T.Bryngelsson (Lund), J.Burckhard (Zürich),
H.Bustan (Jerusalem),  G.F.Crouse (Stanford),  J.R.Daban (Barcelona),
L.J.Degennaro (San  Francisco),  A.Depicker (Gent),  C.Dierks-Ventling
(Basle), I.Doxiadis (München), W.Dunnick (Cambridge), F.C.Eden (Bethesda),
B.Edvardsson (Uppsala),  E.Egyhazi (Stockholm),  J.Engberg (Copenhagen),
J.D.Fabricant (Paris), M.G.Farace (Rome),  W.Filipowicz (Warsaw),  K.Fink
(Copenhagen),  N.Furgac (Göttingen),  F.Gannon (Strasbourg),  A.C.Garapin
(Paris),  M.Geiser (Basel), E.Gilboa (Rehovot), R.Gjerset (Paris), A.P.van
Gool (Louvain),  H.M.Goodman (San Francisco),  J.Hackstein (Tübingen),
H.Hameister (Freiburg),  H.Hartmann (Braunschweig),  H.Hauser (Konstanz),

K.Henco (Darmstadt), H.A.Henriksson (Göteborg), C.Hentschel (London), R.Herrmann (Düsseldorf), H.van Heuverswyn (Gent), R.P.Hjelm (Portsmouth), T.Hofstätter (Tübingen), M.Horowitz (Rehovot), M.Ryniewicz (Warsaw), L.S.L. Hsu (London), U.Hübscher (Zürich), T.Igo-Kemenes (München), H.Jakubowski (Poznan), K.E.Jörstad (Bergen), J.L.Jocarno (Berlin), B.L.Kellas (London), S.Kidd (London), J.R.Kinghorn (Bern), S.Klenow (Copenhagen), T.Klopotowski (Warsaw), H.Klump (Eugene Ore.), A.Konieczny (Poznan), I.Kucan (Zagreb), Z.Kucan (Zagreb), W.Kuhn (Würzburg), J.J.Lawrence (Grenoble), B.Leber (Tübingen), B.Levy (Calgary), S.Löfdahl (Uppsala), H.Martin-Bertram (Karlsruhe), H.R.Matthews (Portsmouth), R.Mohun (London), U.Müller (Heidelberg), C.C.de Nava (Mexico), M.A.Nelson (Naples), M.Neuberger (London), M.C.Nguyen Huu (Berlin), K.M. O'Hare (Edinburgh), G.N.Pavlakis (New York), P.Pennequin (Bicetre), G.Peter (München), J.G.L.Petersen (Copenhagen), M.Philipp (Baltimore), M.Pirre (Villejuif), J.Pierre-Hebert (Villejuif), L.J.Polder (Groningen), A.Pühler (Erlangen), A.Quintero-Ruiz (Mexico), H.P.Ramjoue (Basle), A.Reisfeld (Rehovot), B.Robert (Paris), E.Rocha Perez (Barcelona), M.Salden (Nijmegen), C.Schäfer-Nielsen (Copenhagen), G.Scherer (Freiburg), A.I.Scovassi (Paris), T.Seebeck (Bern), B.H.Sells (St. Johns Can.), P.J.Southern (Edinburgh), C.Spadafora (Strasbourg), M.Steinmetz (München), P.G.Stockley (Cambridge), F.Thomas (Zürich), C.D.Triantaphyllidis (Thessaloniki), A.Ullrich (San Francisco), J.Vielkind (Giessen), J.Vuust (Aarhus), E.F.Wagner (Innsbruck), E.Wawra (Vienna), M.Wickens (Stanford), W.R.Willems (Giessen), K.R.Willison (Cambridge), L.Willmitzer (Braunschweig), R.Winkler (Göttingen), B.Wittig (Berlin), T.Zarvalis (Thessalonoki).

## SPETSAI 1978

### GENE EXPRESSION IN UNIFECTED AND VIRALLY INFECTED CELLS
August 28 - September 3

**Organizers:** B.F.C.Clark (Aarhus) (chairman), A.E.Evangelopoulos (Athens), N.O.Kjeldgaard (Aarhus), L. Philipson (Uppsala).

### Lecturers and Lectures

**D.Baltimore** (Cambridge MA): Small animal RNA viruses.
**M.Buckingham** (Paris): Myogenesis as a model system for the study of terminal differentiation.
**H.Bujard** (Heidelberg): The transcriptional process and its regulation in prokaryotic systems.
**J.E.Celis** (Aarhus): Search for assays to assess tumorigenicity at the cellular level in non-virally transformed cells. Microinjection of suppressor tRNAs into somatic cell mutants.
**B.F.C.Clark** (Aarhus): Polypeptide chain elongation and termination.
**S.G.Clarkson** (Zürich): Organisation of DNA sequences. Microinjection. Transcriptional and post-transcriptional controls.
**B.Daneholt** (Stockholm): Balbiani ring 2 - a model system for analysis of a specific gene and its activity in a eukaryotic cell.
**H.Feldmann** (München): Organisation of DNA sequences.
**W.Gehring** (Basle): Genome organisation and transposable elements in Drosophila. The heat shock genes of Drosophila melanogaster.
**H.Goodman** (San Francisco): Recombinant DNA.
**M.Grunberg-Manago** (Paris): Initiation and regulation of protein synthesis in prokaryotic and eukaryotic system.
**P.R.Harrison** (Glasgow): The friend cell as a model for cell differentiation.

**A.S.Huang** (Boston): Complex RNA viruses.
**F.Kafatos** (Harvard): The chorion of insects as a model system for the study of cell differentiation and molecular evolution.
**N.O.Kjeldgaard** (Aarhus): On the control of RNA synthesis in bacteria.
**A.Klug** (Cambridge): Structure and superstructure of chromatin.
**M. van Montagu** (Gent): RNA phages. Plant tumors.
**N.E.Murray** (Edinburgh): Bacteriophage lambda. Recombinant DNA.
**L.Philipson** (Uppsala): Transcription and processing of RNA.
**A.E.Smith** (London): Small DNA tumor viruses. Control of viral protein synthesis in eukaryotes.
**N.Wilkie** (Glasgow): The organisation and expression of genes in Herpes simplex virus. Herpes simplex virus and transformation.
**M.Yaniv** (Paris): Chromatin structure and gene activity. Replication and transcription of viral chromosomes.
**J.Zeuthen** (Aarhus): Higher levels in the organisation of chromatin. Application of cell fusion experiments for studies on genetic regulation. Somatic cell hybrids with lymphoid cells. The Epstein-Barr-Virus and its interaction with host cells.


## Participants

E.P.Amann (Berlin), P.Andersen (Aarhus), K.Andersson (Stockholm), A.Arellano Castillo (Braunschweig), C.Aulelah (Tübingen), C.Baglioni (New York), S.Barlati (Pavia), K.E.Batley (Vienna), C.Benicourt (Paris), C.Benoist (Strasbourg), I.Bikel (Boston), C.Brüschke (Freiburg), M.Busslinger (Zürich), M.S.Campo (Edinburgh), M.Caravatti (Zürich), C.Casimir (Glasgow), R.Castro (Madrid), Y.C.Chong (Cambridge), G.Clinton (Boston), P.Clissold (London), A.Cohen (Paris), S.Corbett (Portsmouth), W.Crumpton (Coventry), P.Daubas (Paris), J.Devine (London), N.Din (Copenhagen), L.Dixon (Edinburgh), P.Donner (Berlin), J.L.Drocourt, W.Earnshaw (Cambridge), M.G.Farace (Rome), V.Gamulin (Zagreb), F.Gautier (Braunschweig), S.D.Georgatsos (Athens), G.Giannopoulos (Patras), L.Gissmann, H.Goodman (San Francisco), E.Gounaris (Thessaloniki), O.Grau (La Plata), I.Grummt (München), A.Guialis (Athens), L.U.Güresci (Izmir), M.L.Hammarskjöld (Stockholm), M.Harson (London), R.Harvey (London), P.Hellung-Larsen (Copenhagen), W.G.Hesselink (Nijmegen), B.Hoffmann-Liebermann (Rehovot), M.Holsters, J.Holtlund (Oslo), B.M.Honda (Cambridge), J.Imbert (Marseille), M.Innis (Nutley NJ.), T.C.James (London), D.Johnson (Edinburgh), F.Johnston (Calgary), M.D.Jones (Aarhus), K.Kaiser (Edinburgh), E.Katsoris (Patras), P.Katsoris (Patras), ;.Kessel (Kiel), B.Knibiehler (Marseille), C.Kotinis (Thessaloniki), E.Krajewska (Warsaw), T.Kristensen (Oslo), C.Küger (Berlin), B.E.Lachmi (Ness Ziona), C.Lane (London), J.Langenakens (St.Genesius-Rode), A.Lazarou (Athens), A.Levine (Princeton), C.P.Lichtenstein (Cambridge), D.Liebermann (Rehovot), J.Luka (Stockholm), L.Lundberg (Göteborg), A.Maresca (Naples), B.Marian (Vienna), E.Markert (Lyngby), S.Maxwell (Paris), P.Melius (Auburn Ala.), A.Mellor (London), G.Meneguzzi (Nice), S.Michel (Berlin), A.Minty (Paris), A.Misra (Kahashwaranagar India), D.Morello (Paris), F.Müller (Zürich), P.O'Farrell (San Francisco), M.Olsson (Uppsala), D.Palmer (London), R.Pangiotidou (London), E.Paucha (London), G.Pavlakis (New York), A.Person (Paris), P.Pierandrei Amaldi (Rome), C.B.Post (San Diego), L.Potts (Coventry), R.Ralph (Auckland), T.Reiter (München), C.Reuveni (Windsor Ont.), W.Reynolds (Davis Cal.), R.Sammons (Birmingham), S.G.Sanders (Glasgow), S.Saragosti (Paris), E.Schwarz (Freiburg), R.Sharma (New Delhi), P.Shaw (New York), H.Shure (Rehovot), A.Smith (Cambridge), D.Spandidos (Toronto), N.K.Spurr (London), E.Szczesna (Warsaw), U.Szybiak (Poznan), N.Tsadaroglou (Athens), I.Ulmanen (Helsinki), I.S.Villadsen (Lyngby), D.V.Volsky (Jerusalem), E.Vuorio (Turku), S.Waddingham (Ann Arbor Mich.), G.Wagner (Uppsala), B.Wasylyk (Strasbourg), J.Weber (New York), S.West (Newcastle upon Tyne), I.Weygand (Zagreb), W.Wolf (Tutzing), W.Zagorski (Warsaw), T.Zarvalis (Thessaloniki).

SPETSAI 1979

PROTEIN - NUCLEIC ACID RECOGNITION INTERACTIONS
August 20 - September 2

**Organizers** M.Grunberg-Manago (Paris)(chairman), M.E.Buckingham (Paris), A.E.Evangelopoulos (Athens).


**Lecturers and Lectures**

**H.Bujard** (Heidelberg): Promoter polymerase interaction: physical mapping. Kinetic studies on function of RNA polymerase.
**C.Cantor** (New York): Cross-linking studies: structural sites on ribosomes. Fluorescence techniques: applications to nucleosomes.
**P.Chambon** (Strasbourg): Active chromatin. Ovalbumin genes.
**B.F.C.Clark** (Aarhus): tRNA interaction with elongation factor. tRNA interaction with ribosome.
**B.S.Cooperman**: Ribosomal sites for antibiotic binding. Affinity labeling with application to ribosomes.
**H.Delius** (Basle): Heteroduplex mapping. Visualization of nucleic acid and protein complexes.
**J.P.Ebel** (Strasbourg): Ribosomal RNA and protein interactions. tRNA interaction with aminoacyl tRNA synthetase.
**W.Gilbert** (Cambridge Mass.): Sequencing methods for DNA and RNA.
**D.Glover** (Oxford): Organization of the Drosophila genome. rRNA genes in Drosophila.
**F.Gros** (Paris): Multiple gene families of the contractile proteins. Hu protein.
**M.Grunberg-Manago** (Paris): Components and sequence of initiation steps in prokaryotes - localization on ribosomes of initiation factors. Mechanism of selection of mRNA by ribosomes.
**C.Hélène** (Paris): Functional groups in proteins and nucleic acids. Mechanism of quenching of aromatic aminoacid fluorescence in protein nucleic acid complexes.
**J.Hershey** (Davis Cal.): The mechanism of action of eukaryotic initiation factors. Translational control mechanisms in mammalian cells. Radio-immune assays, preparation and use of antibodies.
**P.H.von Hippel** (New York): Structural features potentially useful in protein- nucleic acid interactions. Thermodynamic parameters and molecular properties of the lac repressor operator inducer DNA system. Interactions of nucleic acids and helix destabilizing proteins.
**A.Klug** (Cambridge): Nucleosome structure and its relation to higher order foldings. Interactions between TMV RNA and capsid protein.
**P.Little** : Globin gene organization.
**M.Ptashne** (Harvard): Lambda phage regulatory signals.
**A.Rich** (Cambridge,MA): Different theories on the selective recognition of nucleic acids by protein. Gene 5 protein interaction with deoxynucleotides.
**R.J.Roberts** (Cold Spring Harbor): Adenovirus 2: gene organization and processing. Restriction enzymes.
**S.Steitz** : Ribosomal RNA precursors.
**T.Steitz** : Crystallographic methods with application to the structure of the repressor.
**V.Sgaramella** (Pavia): Transposons - expression of E.coli genes in B.subtilis. DNA ligases.
**G.Stöffler** (Berlin): Ribosome architecture.
**C.Weissmann** (Zürich): Vectors - plasmids, lambda, cosmids, yeast, SV40. Interaction of Qß replicase with Qß RNA.
**H.G.Zachau** (München): Chromatin domains. Repetitive DNA sequences.

SPETSAI 1980

GENOME ORGANIZATION AND FUNCTION
August 30 – September 12

**Organizers**  H.G.Zachau (München)(chairman),  A.E.Evangelopoulos (Athens),
H.Feldmann (München).

## Lecturers and Lectures

**G.Bernardi** (Paris): Sequence organisation of the eukaryotic genome.
**M.L.Birnstiel** (Zürich): Surrogate genetics in the Xenopus oocyte. Promoter
elements of tRNA and 5S RNA genes. Promoter elements of mRNA genes.
**B.F.C.Clark** (Aarhus):  General features of protein biosynthesis.  The
ribosome. Decoding and miscoding.
**F.H.C.Crick** (La Jolla): Supercoiled DNA. Selfish DNA.
**M.Eigen** (Göttingen):  Criteria  for  determining  evolutionary  kinship
relations. Mechanisms of template instructed polymerization. An attempt to
reconstruct early evolutionary events from present experimental facts.
**H.Feldmann** (München): Organization and expression of tRNA genes. Movable
genetic elements. What is interesting in yeast?
**G.Felsenfeld** (Bethesda):  Probing nucleosome structure,  the forces of
stabilisation. Structure of transcriptionally active chromatin. The higher
order structure of chromatin: physiochemical studies.
**R.A.Flavell** (London):  The structure and sequence organization of globin
genes. Expression of globin genes in vitro.
**H.M.Goodman** (San Francisco):  Gene isolation,  sequencing,  and structure.
Evolution of mammalian polypeptide hormon genes. Introduction of new genes
into mammalian cells.
**A.Klug** (Cambridge):  Chromatin I:  nucleosome structure.  Chromatin II:
higher  order  structures.   Three-dimensional  structure  determination  of
biological  macromolecules  by  electon  microscopy.   The  structure  and
mechanism of assembly
of tobacco mosaic virus.
**B.Müller–Hill** (Cologne):  Point mutations and the functional analysis of
proteins.   Gene  fusions  and  the  functional  analysis  of  proteins.
Construction  of  point  mutations,   deletions  and  gene  fusions  by
manipulation of DNA in vitro.
**K.Murray** (Heidelberg):  Comparative aspects of prokaryotic vectors.  DNA
transfer in eukaryotic cells. Hepatitis B virus and its molecular biology.
**N.Murray** (Heidelberg):  Bacteriphage lambda.  Applications of recombinant
DNA technology to the molecular genetics of prokaryotes.
**V.Pirotta** (Heidelberg):  Drosophila chromosomes:  genome organization –
development.
**P.Reichard** (Stockholm): DNA replication.
**R.Roberts** (Cold Spring Harbor):  DNA sequence analysis.  Restriction
endonucleases. Adenvirus 2: gene organization and RNA processing.
**A.Tait** (Edinburgh):  Organisms,  disease and the application of basic
biological investigation.  The genome of parasites.  Gene regulation and
organization.
**H.G.Zachau** (München):  Some facts on repetitive DNA and heterochromatin.
Special aspects of chromatin. Immunoglobulin genes.

## Participants
O.Amster (Rehovot), V.Amzel (Baltimore), N.A.van Arkel (Eindhoven), E.Back
(Freiburg),  B.W.Baer (Stanford),  H.Bagci (Ankara),  M.Bienz (Zürich),
D.Blaas (Vienna),  W.Block (Hannover),  M.Blumenfeld (St.Paul Minn.),
M.G.Bovida (Lisboa), S.Bodary (Geneva), R.K.Booth (Uppsala), D.Bramhill
(Newcastle-upon- Tyne),  A.M.C.Brown (Edinburgh),  D.Büchel (Cologne),
K.J.Burger (Würzburg),  M.Bywater (Uppsala),  J.Caraindros (Patras),
Z.Carvalho (Oeiras),  P.Charnay (Paris),  M.Choder (Rehovot),  Y.W.Chooi

(Bloomington Ind.), E.Ceglarz (Poznan), R.Clerc (Zürich), H.Cooke (Edinburgh), K.R.Dahlberg (Lincoln Neb.), A.Deggerdal (Oslo), A.Dollery (Salisbury), J.Drouin (San Francisco), C.Elsner (Berlin), M.Enigün (Istanbul), B.Felber (Bern), R.Frunzio (Naples), L.C.Fuith (Innsbruck), H.Gadler (Stockholm), A.Garel (Villeurbanne), N.Geisler (Göttingen), E.Gilson (Vincennes), A.Goldfarb (München), F.Gossard (Montreal), E.G.Gounaris (Thessaloniki), M.Grez (Berlin), R.Grosskopf (München), L.Hall (Glasgow), A.Haslinger (Vienna), H.Haymerle (Cambridge), R.Heiermann (Bochum), G.Heinrich (Basle), N.Heisterkamp (Groningen), J.Höchtl (München), E.Hornes (Oslo), V.von Hoyningen (Giessen), M.A.Innis (Wilmington Del.), J.Jongstra (Basel), A.H.Jung (Cologne), A.Kalogeropoulos (Orsay), O.Kalogeropoulos (Gif-sur-Yvette), B.Kan (Istanbul), S.Kearsey (Oxford), H.S.Khan (Ankara), D.Kioussis (London), B.Kirdar (Istanbul), A.J.Kool (Amsterdam), K.K.Kotinis (Thessaloniki)`, J.Kurjan (Eugene Ore.), P.Labhart (Zürich), O.de Lapeyriere (Marseille), A.G.Larsen (Oslo), A.Lassar (St.Louis Miss.), A.La Volpe (Naples), M.Leitner (Rehovot), P.Levantis (London), A.Levy (Basel), I.Di Liegro (Palermo), W.Lindenmaier (Berlin), E.Livneh (Rehovot), B.Löwkvist (Lund), R.H.Lovell-Badge (Cambridge), Luo Zhong Xun (Rome), C.A.Maack (London), D.P.Matthopoulos (Ioannina), S.Mellon-Nussbaum (New-York), K.R.Mitchelson (Portsmouth), G.L.Norman (Los Angeles), G.J.B. van Ommen (Amsterdam), B.A.Oostra (Groningen), M.L.Osorio Almeida (Olivais), F.Perrin (Strasbourg), F.Peters (Nijmegen), M.Philip (Basel), J.M.Pipas (Baltimore), L.Rat (Paris), B.Rexer (Tutzing), T.J.Richmond (Cambridge), I.Riede (Tübingen), H.Roiha (London), R.M.Ruiz-Vazquez (Berlin), S.Ruszoni (Zürich), M.Schäfer-Ridder (München), J.Schröder (Cologne), H.Schröter (Braunschweig), K.I.Seferiadis (Ioannina), C.K.Shewmaker (London), R.J.Shott (Glasgow), C.Sinogas (Oeiras), G.R.Smith (Eugene Ore.), G.Somme (Paris), J. Sonnenbichler (München), D.H.Spathas (Patras), L.Stefani (Glasgow), P.B.Suhr-Jessen (Copenhagen), G.Symonds (Rehovot), L.Thelander (Stockholm), S.H.Thorbjarnadottir (Reykjavik), S.T.Tjia (Cologne), A.Ullrich (San Francisco), A.J.Vandenberghe (Antwerpen), H.P.Vogt (Nijmegen), Y.Wang (München), A.Weydert (Paris), G.Widera (Braunschweig), R.Winkler-Ostwatitsch (Göttingen), D.E.Woods (London), J.M.Wower (Berlin), T.A.Zarvalis (Thessaloniki).


## SPETSAI 1981

### CONTROL AND PROCESSING IN THE BIOSYNTHESIS OF MACROMOLECULES
August 30 - September 12

**Organizers** B.F.C.Clark (Aarhus)(chairman), A.E.Evangelopoulos (Athens), N.O.Kjeldgaard (Aarhus), A.E.Smith (London).

### Lecturers and Lectures

**H.Bujard** (Heidelberg): Control of transcription in prokaryotes.
**B.F.C.Clark** (Aarhus): Elongation and termination of protein biosynthesis. Ribosome structure.
**P.Cohen** (Dundee): The neutral and hormonal control of glycogen metabolism. Protein phosphorylation and the co-ordinated control of cellular activities by hormones.
**J.G.Demaille** (Montpellier): Phosphorylation and contractility- (i) Phosphorylation in the control of Ca++ fluxes through plasma and reticular membranes. (ii) Ca++/ calmodulin - and cAMP-dependent phosphorylation of contractile and regulatory proteins.
**B.Dobberstein** (Heidelberg): Co- and post-translational modifications of proteins.
**R.L.Erikson** (Denver): RNA tumor viruses, cell tranformation and protein phosphorylation.

H.**Feldmann** (München): Structure and expression of mitochondrial DNA.
W.**Fiers** (Gent): RNA bacteriophage as model system in molecular biology. Expression of cloned genes in E.coli. The human fibroblast interferon gene.
M.**Grunberg-Manago** (Paris): Initiation and regulation of prokaryotic protein synthesis. Initiation of mammelian protein synthesis. Translational control of gene expression in prokaryotes.
R.**Kamen** (London): Trancription in eukaryotes.
N.O.**Kjeldgaard** (Aarhus): The transcription and its control in prokaryotes. The molecular biology of retroviruses.
A.**Klug** (Cambridge): The structure of the nucleosome and the folding of chromatin. Chromatin and DNA structure probed by nuclease digestion.
P.**Kourilsky** (Paris): Structural organization of eukaryotic genes.
A.J.**Levine** (Stony Brook): Viral transformation - DNA tumor viruses.
G.**Marbaix** (Rhodes-St-Genese): Studies on the translation of mRNAs microinjected into Xenopus oocytes.
R.**Palmiter** (Seattle): Molecular aspects of hormone action.
D.A.**Peattie** (London): Chemically probing RNA sequences, conformations, and intermolecular interactions - experimental results and methodology.
T.E.**Petersen** (Aarhus): Proteolysis in biological regulation.
G.**Schatz** (Basle): Synthesis and transport of mitochondrial proteins.
A.E.**Smith** (London): Eukaryotic viruses as models for cellular regulation.
H.**Soreq** (Rehovot): The use of microinjected Xenopus oocytes for the expression of scarce mRNA species directing the synthesis of biologically active proteins.
G.**Tocchini-Valentini** (Rome): Transcription of eukaryotic tRNA genes.
C.**Weissmann** (Zürich): Isolation and characterization of eukaryotic genes. Investigation of structure-function relationships by reverse genetics.


**Participants**
E.Ackerman (Cambridge), A.Adams (London), A.Aitken (Dundee), S.Alonso (Paris), P.Andersen (Aarhus), N.Axelsen (Copenhagen), R.E.Baker (München), M.Bernäng (Lund), J.Biard (Grenoble), J.Bingdong (Cold Spring Harbor), C.Birchmeier (Zürich), C.R.Birkett (Canterbury), E.Bock (Copenhagen), H.J.Brüning (Braunschweig), J.Buchanan (London), M.N.Carin (Istanbul), I.N.Clarke (Coventry), S.Compere (Seattle), K.G.Cook (Newcastle upon Tyne), N.Courty (Athens), H.H.Dahl (London), O.Danos (Paris), K.Dawidowitz (Caracas), G.Dimitriadis (Athens), A.Dmochowska (Warsaw), H.Domdey (Lausanne), M.Dumont (New Haven), N.Eberhardt (San Francisco), P.Favrizio (Rome), J.Favaloro (London), D.Ferbus (Paris), L.Frretti (Pavia), P.Filetici (Basel), A.Fjose (Bergen), G.Galli (Zürich), A.Garcia (Paris), I.Garcia (Paris), D.Gardner (San Francisco), T.Giannakouros (Thessaloniki), P.Giardina (Naples), K.Giesecke (Berlin), J.Giraudat (Paris), S.Goelz (Basel), G.Görtz (München), J.Gould (Edinburgh), E.Gounaris (Thessaloniki), A.Gozdzicka-Jozefiak (Poznan),R.Harvey (London), N.Hay (Rehovot), N.Hernandez (Heidelberg), B.van Heuverswyn (Bussels), P.Hjorth (Aarhus), A.Hradilek (Prague), T.Ingebritsen (Dundee), M.Innis (Berkeley), K.D.Jentsch (Aarhus), S.J.Johnson (Calgary), O.Kämpe (Uppsala), D.Kalderon (London), N.Kalkkinen (Helsinki), K.Kannan (London), L.Kayrin (Ankara), N.Kecskemethy (Bochum), R.Koren (Rehovot), K.Kotinis (Thessaloniki), G.I.Kristjansson (Uppsala), A.Lamouroux (Orsay), N.Lan (San Francisco), A.Laskaridou (Athens), E.Lazar (Strasbourg), A.R.Leal Lino (Lisboa), M.A.A.Lebrero (Madrid), A.Levi (Rehovot), A.Levinson (San Francisco), H.Lorberboum (Jerusalem), T.Maassen (Leiden), G.Marbaix (Rhodes- St-Genese), F.Maschat (Paris), G.J.Matlashewski (Ontario), P.Matthias (Heidelberg), D.P.Matthopoulos (Ioannina), M.Mazon (Madrid), J.McVey (Glasgow), S.Merten (Strasbourg), M.Michelinakis (Patras), F.D.Miller (Alberta), J.A.Mitlin (London), R.Monk (Philadelphia), M.D.Morch (Paris), J.Morris ( Cambridge), E.Müller (Vienna), L.Nelles (Wilrijk), M.H.M.Noteborn (Groningen), C.O'Reilly (Dublin), E.M.Palla

(Pavia), J.Panourgias (Athens), A.Papahajopoulou (Patras), M.Patrinou-Georgollas (Athens), D.Peattie (London), T.E.Petersen (Aarhus), M.Podravec (Zagreb), S.Raftopoulos (Patras), T.Rapoport (Berlin), C.Rodrigues-Pousada (Oeiras), J.C.L.Sanjuan (Madrid), T.Sclaviadis (Thessaloniki), Z.Scouras (Thessaloniki), K.Seferiadis (Ioannina), E.Sheybani (Geneva), A.L.Sivertsen (Copenhagen), H.Slegers (Wilrijk), T.Smith (Amersham), H.Soreq (Rehovot), M.Streiff (Basel), M.Stupar (Futog), D.Stüber (Heidelberg), C.Swimmer (Stony Brook), U.Szybiak (Poznan), K.Szyfter (Poznan), M.Tenniswood (London), R.Terracol (Paris), J.Thompson (Amherst Mass.), K.K.Thomsen (Aarhus), T.Trangas (Athens), A.S.Tsaftaris (Thessaloniki), M.Urban (Cologne), C.Urbanke (Hannover), M.Varsanyi (Bochum), M.R.Vasconcelos (Lisboa), T.M.Veldman (Amsterdam), C.Watson (London), I.Weygand (Zagreb), B.Will (Edinburgh), H.Witkiewicz (Innsbruck), W.Wolf (Tutzing), J.Zarkadis (Patras), D.Zevin-Sonkin (Rehovot).


## SPETSAI 1982

## REGULATION OF GENE EXPRESSION IN PROKARYOTES AND EUKARYOTES
August 30 - September 11

**Organizers** M.Grunberg-Manago (Paris)(chairman), A.L.Haenni (Paris), E.Brody (Paris), A.E.Evangelopoulos (Athens).

## Lecturers and Lectures

**J.Abelson** (San Diego): A role for the intervening sequence in the biosynthesis of yeast tyrosine tRNA. Mechanism of ligation during tRNA splicing.
**D.Baltimore** (Cambridge Mass.): Decision making in the immune system. Viral oncogenes and cell transformation.
**F.Blasi** (Naples): Regulation by attenuation in bacteria.
**E.Brody** (Paris): Bacteriophage-coded regulators of transcription. Self-recognition and protection of bacteriophage DNA.
**T.Caskey** (Houston): Phenotypic reversion at the HPGRT locus as a consequence of gene amplification. Molecular alterations in the HPGRT locus of rodents and man.
**B.F.C.Clark** (Aarhus): The elongation step of protein biosynthesis.
**D.Dixon** (Brighton): Regulation of the expression of the nitrogen fixation (Nif) operons.
**J.P.Ebel** (Strasbourg): The structure of transfer RNA and its interaction with elements of the translation machinery. Structure and function of ribosomes.
**R.A.Flavell** (London): Structure and expression of haemoglobin genes. Structure and function of the genes encoding the major histocompatibility complex (MHC) of the mouse.
**R.Gesteland** (Salt Lake City): The role of tRNA in the regulation of gene expression.
**M.Grunberg-Manago** (Paris): Regulation of gene expression of translational components of bacteria.
**R.Gumport** (Urbana Ill.): Site-specific mutagenesis.
**A.Haenni** (Paris): Plant viral genomes.
**T.Hunt** (Cambridge): Protein synthesis and its control in reticulocyte lysates.
**T.Igo-Kemenes** (München): Structure of chromatin and chromosomes. Active chromatin.
**W.Keller** (Heidelberg): Transcription and RNA processing in eukaryotes. Transcription and RNA processing by the tumor viruses.
**U.Z.Littauer** (Rehovot): Control of the diversity of micotubule proteins during nerve outgrowth and differentiation.

254

**M.Rosenberg** (Bethesda): Protein-nucleic acids interactions involved in transcription activation. Using the E.coli galactokinase gene to study prokaryotic and eukaryotic gene control elements.
**M.Schwartz** (Paris): Protein export in bacteria.
**P.Slonimski** (Gif-sur-Yvette): The interplay of maturases and signal sequences of split mitochondrial genes.
**A.Ullmann** (Paris): Regulation of catabolic operons in E.coli.
**C.Weissmann** (Zürich): Functional analysis of eukaryotic genes by reversed genetics.

## Participants
J.Ackroyd (Brighton), H.D.Andersen (Aarhus), A.Athanassiadou (Patras), R.Bernander (Lund), E.Boel (Baksvaerd), U.Bonas (Cologne), S.Bray (Cambridge), P.Britton (Cambridge), G.Bugaisky (Paris), L.Bugaisky (Paris), D.W.Burt (Leicester), A.Cano (Madrid), O.Capasso (Naples), A.Carrasco (Basel), M.Cazillis (Orsay), C.Chapon (Paris), S.Chladek (Detroit), N.Coppard (Aarhus), F.Creusot (Paris), R.Crkvenjakov (Belgrade), M.David (Castanet-Tolosan), A.M.Degener (Rome), D.Denicourt (Rhodes St.Genese), C.Edelist (Orsay), P.Ferrara (Paris), J.Finidori (Creteil), E.Firpo (Buenos Aires), S.Forss (Heidelberg), M.Frangou Lazaridis (Ioannina), A.Frisch (Tel Aviv), J.Frydenberg (Aarhus), H.Galski (Jerusalem), T.Geller (Rehovot), R.Gentz (Heidelberg), R.Gesteland (Salt Lake City), P.Giacomoni (Villejuif), E.Grelland (Oslo), L.Guang-Di (München), M.L.Guerinot (East Lansing), A.Guialis (Athens), R.Gumport (Urbana), J.F.Hasson (Paris), P.Haliotis (Kingston Can.), L.Hellman (Uppsala), M.Henry (Grenoble), C.Hughes (Würzburg), I.Ioannou (Athens), T.Jakubowicz (Lublin), D.J.Jeenes (Gwynedd), J.Jensen (Brighton), L.K.Jolliffe (Piscataway NJ), D.J.Joslyn (Camden), M.Joannaud (Paris), S.Kagan (Norwich), C.A.Komly (Joinville), Z.Koukolikova-Nicola (Basle), M.Kröger (Freiburg), G.Langsley (Edinburgh), I.Lazaridis (Ioannina), K.A.W.Lee (Montreal), C.Levy (Basle), M.Lewis (Cambridge), M.Lipoldova (Prague), I.Malec (Salzburg), G.de Martinoff (Bussels), J.McCafferty (Glasgow), P.Melancon (Madison), D.Melloul (Rehovot), H.I.Miller (Rockville), S.Morley (Cambridge), N.Moschonas (Athens), K.Nagai (Cambridge), Y.Ohyama (Tokyo), C.Oker-Blom (Helsinki), H.Pannekoek (Amsterdam), G.Pande (Rome), C.Parsot (Paris), E.Pashnina (Rehovot), J.C.Patte (Orsay), E.Patzelt (Vienna), A.J.Perlman (New York), H.U.Petersen (Aarhus), J.Piette (Brussels), F.Poirier (Orsay), D.Przybyl (Freiburg), R.Rappuoli (Siena), S.M.H.Richardson (Dundee), H.Riedel (Heidelberg), F.Robbiati (Milan), A.Rosenthal (Jerusalem), V.Russo (Berlin), T.R.Rutherford (Oxford), C.Sacerdot (Paris), Santos Lemos (Porto), T.Samuelsson (Göteborg), H.Schnabel (München), R.Schnabel (München), U.Schomburg (Hamburg), M.Sikorski (Berlin), V.de Simone (Naples), J.Skouv (Copenhagen), D.Smith (Cambridge), M.Soria (Pavia), R.Sprengel (Heidelberg), P.Stanssens (Gent), A.Steinmetz (Strasbourg), P.Stiegler (Strasbourg), N.Strandberg Pedersen (Copenhagen), M.Streuli (Zürich), L.Symington (Glasgow), D.Synetos (Patras), P.Szafranski (Warsaw), P.Terpstra (Rotterdam), J.J.Toulme (Paris), R.Villaroel-Mandolia (Gent), D.Vogel (Berlin), D.Weichenhan (Oldenburg), W.Werr (Cologne), S.W.White (Berlin), S.Whitehouse (London), S.Wright (London), Z.Xian-Yang (München), P.Zabel (Wageningen), T.A.Zarualis (Thessaloniki).

## SPETSAI 1983

### MOLECULAR BIOLOGY OF ANIMAL CELLS
August 28 - September 10

**Organizers** C.T.Caskey (Houston)(chairman), J.Hershey (Davis Cal.), J.Paul (Glasgow), A.E.Evangelopoulos (Athens).

## Lecturers and Lectures

**P.Borst** (Amsterdam): The mechanism of antigenic variation in trypanosomes. Gene rearrangements controlling differentiation.

**E.M.Bradbury** (Davis Cal.): Chromatin and chromosome structure.

**M.E.Buckingham** (Paris): Skeletal muscle myogenesis as a model system for the study of terminal differentiation. Actin and myosin coding sequences: their genomic structure, organisation, and expression.

**M.R.Capecchi** (Salt Lake City): Studies of recombination between DNA molecules microinjected into cultured mammalian cells. Informational suppressors in cultured mammalian cells.

**C.T.Caskey** (Houston): Selection and characterization of induced mutations and gene amplification in cultured cells. The HPRT locus- structure, mutations, and gene transfer.

**A.Efstratiadis** (New York): The structure, evolution and expression of preproinsulin genes. DNA conformation of eukaryotic promoters.

**H.Feldmann** (München): Structure and expression of mitochondrial DNA. Organizational patterns and expression of eukaryotic tRNA genes.

**W.W.Franke** (Heidelberg): Cytoskeleton – the insoluble proteinaceous architectual framework of the mammalian cell. The desmosome – an example of a specific membrane-filament complex.

**J.W.B.Hershey** (Davis Cal.): The mechanism and control of protein synthesis in mammalian cells. The role of initiation factors in translational control.

**L.H.Hood** (Pasadena): Antibody genes. Genes of the major histocompatibility complex.

**K.Illmensee** (Geneva): Mammalian chimeras. Experimental genetics of the mouse embryo.

**C.Kedinger** (Strasbourg): Eukaryotic promoter control. Sequence elements controlling the transcription of adenovirus protein coding genes.

**C.J.Marshall** (London): The transformed cell. Cellular oncogenes studied by DNA mediated gene transfer.

**R.D.Palmiter** (Seattle): Transcriptional control in mammalian systems. Gene transfer into mammals.

**J.Paul** (Glasgow): Unorthodox aspects of globin gene transcription. Normal and abnormal gene expression.

**L.Philipson** (Heidelberg): Control of adenovirus gene expression. Region E1A mediated control of early adenovirus gene expression.

**R.G.Roeder** (New York): In vitro analysis of the mechanism and regulation of transcription of class III genes. In vitro analysis of the mechanism and regulation of transcription of class II genes.

**W.Schaffner** (Zürich): Enhancer sequences.

**D.S.Secher** (Cambridge): Monoclonal antibodies. Interferon purification and assay.

**C.Weissmann** (Zürich): RNA splicing – an overview. RNA splicing – analysis by classical and reverse genetics.

**K.Willecke** (Essen): Molecular biology of gap junctions – organelles for intercellular communication. Oncogenes derived from normal and tumorigenic cells.

## Participants

R.Abulafia (Rehovot), C.Agelidis (Ioannina), R.L.Allgren (Stanford), G.Almis-Kanigur (Istanbul), J.Alves (Hannover), A.Anagnostopoulou (Patras), A.M.Aragay (Barcelona), J.Arendes (Mainz), A.Ashworth (London), V.Attadia (Naples), E.Barbanti (Naples), H.Z.Barbera-Saldana (Strasbourg), G.Barsh (Seattle), P.J.R.Barton (Paris), D.P.Bazett-Jones (La Jolla), G.Bengha (Cluj-Napoca), F.Broders (Paris), M.L.Cardoso de Almeida (Cambridge), S.L.Carrol (Houston), W.K.Cavenee (Salt Lake City), M.Ciangriglia (Siena), W.Craigen (Houston), U.Danesch (Heidelberg), B.R.Davis (El Toro Cal.), B.Devaux (Paris), P.de Vilee (Leiden), L.Dirckx (Leuven), B.Distel (Amsterdam), B.Ek (Uppsala), J.Engelbrecht

(Copenhagen), S.Eriksson (Stockholm), J.Feigon (Cambridge Mass), J.P.Gagner (Montreal), A.Gal (Gif-sur-Yvette), R.Galler (Heidelberg), T.Gerster (Heidelberg), J.A.Grifo (Cleveland), R.A.Grymes (Stanford), K.Gustafsson (Uppsala), K,Hammarstrom (Uppsala), O.C.Hansen (Copenhagen), D.Hartley (London), Y.Hassan (Beer-Sheva), A.Hatzopoulos (Evanston Ill.), M.M.Heck (Baltimore), I.A.Hope (Edinburgh), G.Huber (Basle), M.Hulethel (Beer-Sheva), E.Ioakimidis (Berlin), D.A.Jackson (Oxford), H.Jacobsen (Heidelberg), S.B.Jakowilew (Strasbourg), A.Jalanko (Helsinki), M.W.Kilimann (München), T.Kleinberger (Rehovot), A.Kramer (Heidelberg), E.M.Lafer (Cambridge Mass.), C.Lasserre (Paris), R.Lathe (Strasbourg), M.Laughrea (Montreal), R.Layden (London), T.Leandersson (Basle), G.Levi (Rehovot), M.Levi-Strauss (Paris), R.Lewis (New Haven), A,Lopo (Davis Cal.), A.Freytag von Loringhoven (München), K.H.Lundström (Helsinki), G.Macgregor (Brighton), F.McCormick (Berkeley), D.Majiecko (Warsaw), R.Male (Bergen), C.Malet (Kremlin-Bicetre), J.Marynski (Brussels), G.Mengod (Basle), R.Mortara (Cambridge), N.R.Movva (Geneva), A.S.Muel (Paris), M.Müller (Cologne), S.Müller (Strasbourg), E.J.Murray (London), J.L.Nahon (Gif-sur-Yvette), S.Nasi (Rome), M.Ozguc (Ankara), C.Pangiotidis (Thessaloniki), A.Papachatopoulou (Patras), E.Paulssen (Oslo), L.Peltonen (Oulu), B.Pentecost (Calgary Alb.), G.de Petro (Pavia), D.Phillips (London), B.Piechulla (Göttingen), T.Pieler (Berlin), L.Pitto (Pisa), S.E.Plon (Cambridge Mass.), G.A.Rappold (Heidelberg), J.K.V.Reichardt (Stanford), K.Robson (Oxford), U.B.Rosenberg (Tübingen), U.Rüther (Cologne), M.Schmidt (Würzburg), C.Schneider (London), D.L.Simmons (Edinburgh), D.Spathas (Patras), H.Storchova-Grunnerova (Prague), K.Strub (Zürich), V.Subramanian (Hyderabad), K.Takkinen (Helsinki), J.Taljanidis (Budapest), V.M.Taylor (Cambridge), C.Tyler-Smith ( Edinburgh), A.Utterlinden (Rijswyk), N.V.Ursini (Naples), K.Vass (Glasgow), M.Verhoyen (Gent), T.F.Vogt (Philadelphia), U.H.Weidle (Tutzing), T.Williams (London), A.P.Wolffe (London), L.Yaneva (Sofia).


## SPETSAI 1984

### GENOME ORGANIZATION AND FUNCTION
August 30 - September 12

**Organizers:** H.G.Zachau (München)(chairman), A.E.Evangelopoulos (Athens), H.Feldmann (München), T.Igo-Kemenes (München).

### Lecturers and Lectures

**D.Baltimore** (Cambridge Mass.): Immunoglobulin genes. Viral oncogenes.
**P.Borst** (Amsterdam): Gene amplification and other forms of gene rearrangements. Gene rearrangements and modification controlling antigenic variation in trypanosomes.
**B.Daneholt** (Stockholm): Higher order structures in chromatin. Formation and transport of hnRNP particles. The Balbiani ring genes - an example of satellite-like evolution.
**M.Eigen** (Göttingen): The optimization of gene and their products.
**H.Feldmann** (München): Organizational patterns and expression of eukaryotic tRNA genes. Organization and function of mitochondrial genomes.
**G.R.Fink** (Cambridge Mass.): Cis and trans acting elements in HIS4 regulation in yeast.
**M.Grunberg-Manago** (Paris): Translational factors. Regulation of gene expression of translational components in prokaryotes.
**P.Gruss** (Heidelberg): Enhancers as transcriptional control elements. Transcriptional aspects of the differentiation of terato-carcinoma stem cells.
**T.Igo-Kemenes** (München): The structure of chromatin - Nucleosome phasing. Active chromatin.

**F.Kafatos** (Harvard): The chorion of insects as a model system for the study of cell differentiation and molecular evolution.

**S.McKnight** (Seattle): Expression of genes in transgenetic mice. Mechanisms of stroid hormone action.

**F.Melchers** (Basle): T Lymphocyte-dependent and independent B lymphocyte activation to proliferation and immunoglobulin secretion. Development of B cells from stem cells.

**K.Murray** (Edinburgh): The present generation of phage lambda vectors which allow direct selection of recombinants. Molecular biological approaches to the study of viral hepatitis. Genetic manipulation in industrial biology.

**R.Roberts** (Cold Spring Harbor): Application of computers to research on nucleic acids. Structure and function of the adenovirus 2 genome. Structure and function of restriction enzyme genes.

**G.M.Rubin** (Bethesda): Transposable elements in Drosophila. Genetic transformation of Drosophila. Molecular genetics of the white locus.

**B.Shilo** (Rehovot): Viral and cellular oncogenes. Human tumor oncogenes. Drosophila cellular oncogenes.

**M.Steinmetz** (Basle): Molecular genetics of the major histocompattibility complex.

**C.Weissmann** (Zürich): RNA splicing. Interferon genes and their expression.

**R.Winkler** (Göttingen): Comparative sequence analysis.

**H.G.Zachau** (München): Repetitive DNA. Structure and expression of immunoglobulin genes.


### Participants

D.Aghib (Milan), M.D.Amaral (Lisboa), P.Angel (Karlsruhe), A.Avni (Rehovot), T.Barlow (Stockholm), A.Bartoszek (Gdansk), K.Beucamp (Tutzing), C.Berger (Basle), M.Bergman (Stockholm), J.Bernheim (Brussels), B.Berse (Warsaw), B.Bienz (Rehovot), M.Bourouis (Strasbourg), S.Burckhardt (Cologne), N.Busso (Villejuif), Shing Chang (Emeryville Ca), T.Christensen (Aarhus), M.H.Citterich (Rome), P.M.Couble (Villeurbanne), D.H.Coucheron (Trondheim), N.C.P.Cross (Cambridge), I.Davidson (Glasgow), J.L.Degen (Basle), S.J.Degen (Basle), J.L.N.Derbyshire (Cambridge), J.J.Devlin (Cambridge), H.Dinter (Braunschweig), R.M.Duvoisin (Geneva), P.Eckes (Cologne), L.Ettinger (Jerusalem), A.Fontaine (Geneva), M.D.Ford (London), F.V.Fuller-Pace (Edinburgh), M.L.P.Galego (Oeiras), . C.Galup (Nice), A.Ganguly (Calcutta), I.Garner (Paris), S.M.Gasser (Geneva), A.Gautier (Geneva), P.Ghazal (Edinburgh), M.Gessler (Giessen), M.I.S.Gomes (Lisboa), D.R.Greaves (London), P.Hantzopoulos (New York), B.Herrmann (Heidelberg), A.Hill (Oxford), C.S.L.Höög (Stockholm), A.L.Jörgensen (Aarhus), C.Kessler (Tutzing), E.Kjeldsen (Aarhus), W.Klump (München), B.König (Frankfurt), B.Korczak (Toronto), A.Koromilas (Thessaloniki), A.Kretsovali (Paris), G.Krupp (New Haven Con), I.Kryspin-Sörensen (Gentofte), A.Kumar (Washington), S.Lambert (Cambridge), P.Landolt (Fribourg), B.Lapeyre (Toulouse), F.E.Leichtfried (Vienna), L.Lemieux (Nice), Y.Lemoine (Strasbourg), E.Lötscher (München), D.E.Lohr (Tempe Arz.), C.Lougovoi (Thessaloniki), D.G.Lowe (Toronto), B.Lüscher (Zürich), S.Luria (Rehovot), V.Maksimovic (Belgrade), R.S.Mann (Cambridge Mass), A.Maouri (Thessaloniki), A.Marmenout (Gent), M.C.Martinez Gomes (Barcelona), I.Mason (London), J.F.Mastronardy (New Jersey), M.J.McLean (Birmingham), B.Mertens (Leuven), H.Meunier (London), S.E.Millar (London), K.Moses (Cambridge), C.R.Müller (Würzburg), B.S.Munro (Cambridge), S.Muyldermans (Sint-Genesius-Rode), I.Nikas (Glasgow), G.Nilsson (Umea), A.Özer (Istanbul), K.E.O'Neill (Cold Spring Harbor), C.A.Pangiotidis (New Haven), K.Paulsen (Freiburg), P.Pfeiffer (Strasbourg), S.P.Mungio (Barcelona), I.Puff (Würzburg), V.Raymond (Oxford), U.Rdest (Würzburg), P.Ricciardi-Castagnoli (Milan), A.Riccio (Naples), A.Di Rienzo (Rome), R.Rigler (Stockholm), M.J.Rogers (Birmingham), S.Rosahl (Cologne), P.Ruzicska (Szeged), S.Schießl (München), C.Schmutzler (Würzburg), B.J.Scholte

(Amsterdam), Z.Scouras (Thessaloniki), K.Sege (Uppsala), A.Sette (Rome), A.Sidoli (Oxford), A.Simeone (Naples), F.Sinangil (Omaha Ne), C.Sorbas (München), A.Spandidos (Glasgow), G.Spyrou (Stockholm), C.Stein (Rehovot), M.D.Strathearn (Urbana), B.Straubinger (München), K.Strebel (Heidelberg), B.Suri (Basle), D.Tautz (Cambridge), B.Timmerman (Gent), K.Tyc (Warsaw), M.Uhlen (Stockholm), E.Vakalopoulou (München), C.L.Verweij (Amsterdam), H.Vogelsang-Wenke (München), N.Warburton (High Wycombe), F.Weber (Zürich), U.H.Weidle (Tutzing), G.Weisinger (Rehovot), M.Wende (Berlin), R.F.J.de Winter (Portsmouth), H.Wolfes (Hannover), B.Wulff (Copenhagen), Y.Yarden (Rehovot), D.Zajchowski (Strasbourg), I.Zarkadis (Patras), T.Zarvalis (Thessaloniki).


SPETSAI 1985

MATURATION AND MIGRATION OF PROTEINS
September 1 - September 14

Organizers: B.F.C.Clark (Aarhus)(chairman), A.E.Evangelopoulos (Athens), G.Schatz (Basle).

Lecturers and Lectures

B.F.C.Clark (Aarhus): Protein biosynthesis.
J.Davis (Geneva): Biotechnology.
R.L.Erikson (Cambridge, Mass.): Functional significance of post-translation modification of proteins. Does reversible protein phosphorylation play a role in the regulation of cell growth or malignant transformation?
A.E.Evangelopoulos (Athens): Control of cellular activity by protein phosphorylation- dephosphorylation processes.
M.Grunberg-Manago (Paris): Regulation of gene expression of translational components in prokaryotes. Aminoacyl-tRNA synthetases.
R.Henderson (Cambridge):
G.Koehler (Freiburg): Monoclonal antibodies.
C.Kurland (Uppsala): Tuning the ribosome.
D.Louvard (Paris): Development of cell polarity using an established epithelial cell-line in vitro. Organelle-specific antibodies as probe to study intracellular transport.
D.Meyer (Heidelberg): Translocation of nascent peptides across the membrane of the endoplasmic reticulum: Step one in intracellular transport. Cotranslational events in peptide secretion: the translocation system of rough endoplasmic reticulum.
N.Nelson (Nutley, NJ.): Biogenesis and assembly of protein complexes in energy-transducing membranes.
E.de Robertis (Basle) :
J.-D.Rochaix (Geneva): Biosynthesis and assembly of chloroplast proteins involved in photosynthesis. Synthesis, assembly and properties of two important chloroplast polypeptides.
J.Sambrook (Cold Spring Harbor): Transport of viral proteins in mammalian cells.
G.Schatz (Basle): The biogenesis of mitochondria: Background. The biogenesis of mitochondria: How proteins are imported into specific intra-mitochondrial locations.
M.Schwartz (Paris): The secretion of proteins by bacteria. The use of Petri dishes and mice to study the structure of a membrane protein.
T.J.Silhavy (Princeton): The use of gene fusions to study protein localization. Intragenic information specifying export of lamb to the outer membrane of E.coli. Genetic identification of cellular components of the export machinery.

**K.Simons** (Heidelberg): The life cycle of an enveloped virus in its host cell. The use of enveloped viruses to study the biogenesis of cell surface polarity in epithelial cells.
**W.Tanner** (Regensburg): Protein glycosylation: (1) The pathways of protein N- and O-glycosylation, (2) Possible role of carbohydrate moieties of glycoproteins.
**B.Wickner** (Los Angeles): Mechanisms of bacterial membrane assembly.


**Participants**
S.Alexson (Stockholm), P.Andersen (Aarhus), J.Armstrong (Warwick), P.Baeuerle (Martinsried), G.Banting (London), J.Barbas (Madrid), A.Baroin (Gif-sur- Yvette), M.Berger (Giessen), Y.Berko-Flint ( Rehovot), N.Bilgin-Aktar (Istanbul), J.Blenis (Cambridge, Mass), O.Blingsmo (Oslo), A.Boukla (Thessaloniki), G.van den Broeck (Gent), T.Buerglin (Basle), N.Campos (Barcelona), J.Cavallius (Aarhus), C.Chia (East Lansing, Michigan), V.Citovski (Jerusalem), M.Courtney (Strasbourg), S.Cox (Canterbury), A.Delahodde (Gif-sur-Yvette), I.Diamantis (Basle), G.Dinos (Patras), G.Durso (Seattle), P.Ferreira (Porto), V.Feys (Gent), P.Fragapane (Rome), G.Fumagalli (Milan), A.Futerman (Rehovot), M.Gallagher (Dundee), P.Gallagher (Cold Spring Harbor), R.Gentz (Basle), T.Giannakouros (Thessaloniki), Y.Gibson (Leicester), P.Gluschankof (Paris), E.Gormley (Dublin), C.Halpin (Warwick), E.Hatzivassiliou (Thessaloniki), R.Hengge (Konstanz), D.Henning (Giessen), D.Henrique (Oeiras), M.Herruer (Amsterdam), M.Hortsch (Heidelberg), L.Honberg (Valby), M.Ioannou (Patras), E.Jackson (Valby), G.S.Jensen (Aarhus), J.Lipp (Heidelberg), M.E.Lloyd (Edinburgh), C.Lougovoi (Thessaloniki), L.R.Lund (Copenhagen), S.Macintyre (Tübingen), B.Maier (Freiburg), A.Maouri (Thessaloniki), P.Mariottini (Rome), P.Martinez (Barcelona), C.Martins de Sa (Paris), N.Matta (Kurukshetra), A.Mitraki (Orsay), J.Mulholland (Aachen), N.Nikolakaki (Thessaloniki), L.Nilsson (Stockholm), O.Nybroe (Copenhagen), P.Oudshoorn (Amsterdam), A.Pantazaki (Thessaloniki), H.Pahverk (Uppsala), G.Patey (Gif- sur-Yvette), E.Paulssen (Oslo), H.van Pelt-Heerschap (Leiden), B.Petridou (Paris), N.Pfanner (Munich), L.Popolo (Milan), C.Prody (Rehovot), S.K.Rhee (Seoul), J.Roitelman (Tel Aviv), I.Sadler (Munich), F.Schirmaier (Basle), H.Semb (Umea), L.Severinsson (Uppsala), D.Simmons (Edinburgh), D.Sofianos (Athens), U.Stochaj (Cologne), J.Struck (Berlin), I.Suominen (Turku), S.Tapio (Uppsala), J.Tavernier (Gent), R.Tipirdamaz (Ankara), R.Valle (Paris), U.Weidle (Tutzing), H.Wikström (Uppsala), C.Witte (Basle), T.Zarvalis (Thessaloniki), J.Zeleny (Prague).


**SPETSAI 1986**

**MOLECULAR GENETICS OF MICROORGANISMS**
August 31 - September 13

**Organizers** M.Grunberg-Manago (Paris)(chairman), A.E.Evangelopoulos (Athens), J.P.Lecocq (Strasbourg), M.Schwartz (Paris).

**Lecturers and Lectures**

**W.Boos** (Konstanz): Active transport system in bacteria.
**P.Boquet** (Paris): Microbial toxins.
**P.Borst** (Amsterdam): 1.Discontinuous transcription of protein-coding genes in trypanosomes and related protozoa. 2.Gene rearrangements and gene modification controlling antigenic variation in trypanosomes and in bacteria.
**L.Bosch** (Leiden): Gene organization and expression of prokaryotic and eukaryotic elongation and termination factors.

**E.Brody** (Paris): 1.Yeast nuclear pro-mRNA splicing. 2.The yeast spliceosome and its components.

**B.Chassy** (Bethesda): 1.Molecular biology of lactic acid bacteria. 2.Application of lactic bacteria in food technology, agriculture and industry.

**J.Davies** (Paris): Protein migration and maturation in biotechnology.

**H.Feldmann** (Munich): 1.Eukaryotic tRNA genes as a model for expression and its control. 2.Repetitive, transposable elements in yeast: organisation and function.

**C.Georgopoulos** (Geneve): 1.Bacteriophage lambda - E.coli interactions. 2.The heat shock response in E.coli.

**M.Grunberg-Manago** (Paris): 1.Translational autoregulation of threonyl-tRNA synthetase expression. 2.Mechanism of regulation of translational initiation factors expression.

**J.W.B.Hershey** (Davis Ca.): 1.Regulation of prokaryotic translation and the synthesis of ribosomes. 2.Translational control in eukaryotic cells.

**M.Holland** (Davis Ca.): 1.Regulation of mRNA and rRNA synthesis in Saccharomyces cerevisiae. 2.Regulation of expression of yeast glycolytic genes.

**F.Imamoto** (Wako Jap.): Molecular structure and transcriptional control of the two tRNA (fMet)

**D.Kaiser** (Stanford): Social cells and intercellular communications.

**J.-P.Lecocq** (Strasbourg): 1.Prokaryotic and eukaryotic vectors for gene expression: which one and when? 2.Molecular biology and rabies: a step to a "comeback" of vaccinia virus among vaccines.

**R.Losick** (Cambridge Mass.): RNA polymerase heterogeneity. 2.Genetics of endospore formation.

**M.van Montagu** (Ghent): Regulation of gene expression in plants.

**W.Neupert** (Munich): How mitochondria import proteins from the cytoplasm and how these proteins are correctly distributed within mitochondria.

**T.Platt** (Rochester): 1.Gene regulation by transcription termination in prokaryotes. 2.Gene regulation by transcription termination in yeast and higher eukaryotes.

**M.Schwartz** (Paris): 1.Protein export. 2.The secretion of proteins by bacteria.

**P.Slonimski** (Gif-sur-Yvette): Intron encoded proteins from mitochondria: key elements of gene expression and genomic evolution.

**W.Szybalski** (Madison): Antitermination of transcription and novel regulatory circuits.

**A.Toussaint** (Rhode-St.-Genese): 1.Mobile genetic elements of gram⁻ bacteria and their mechanism of transpositions. 2.Mu and transposons as genetic tools in gram⁻ bacteria.

**A.Ullmann** (Paris): 1.Cyclic AMP in bacteria: catabolite repression. 2.Cyclic AMP and bacterial virulence.

## Participants

L.B.Abrahmsen (Stockholm), P.F.de Almeida (Oeiras), P.Ameloot (Gent), C.R.de Andrade (Recife), P.P.de Andrade (Recife), E.Arikan (Diyarbakir), M.C.Arikan (Ankara), A.Banerjee (New-Dehli), I.G.A.M.Bakkeren (Basel), K.Bauer (Konstanz), P.Bertin (Louvain-la-Neuve), C.Bibus (Basel), C.Bourgouin (Paris), A.Brändli (Basel), Y.Brun (Ste-Foy C.), P.Bruneau (Charbonnieres), S.Brunet (Aulnay-sous-Bois), S.Butler (Rochester), I.M.C.Calvente de Barahona (Oeiras), A.Campbell (London), A.Carattoli (Rome), M.Cardarelli (Rome), P.Caspers (Basel), M.Chiurazzi (Naples), B.Cirakoglu (Gebze), K.Clare (London), T.Collet (Sommerville Ma.), J.Conley (New Haven Conn.), K.D.Cromie (Brighton), H.S.Cummings (Davis Ca.), M.K.Dahl (Konstanz), W.Dalemans (Leuven), M.G.Dominguez (Leon), M.Dziegiel (Kalundborg), C.d'Enfert (Paris), N.Ekaterinaki (Glasgow), J.Ericson (Umea), I.Fijalkownska (Warsaw), D.Fox (New York), T.Gaal (Budapest), J.Gebert (Konstanz), D.Gennimata (Thessaloniki), F.Habib Shah (Bangi), J.Haarmans-Stoorvogel (Leiden), K.Harmak (Aarhus), C.Hatt

(Southampton), M.Hentze (Bethesda), P.-P.C.Henze (Berlin), N.Homatidis (Thessaloniki), W.Hönerlage (Hamburg), J.Houmard (Paris), P.Huber (Zürich), H.Hug (Zürich), L.Isaksson (Uppsala), A.Khan (Edinburgh), M.Kiremitci (Ankara), K.Korpela (Espoo), H.M.Krisch (Geneva), W.M.A.von Krüger (Rio de Janeiro), J.Kulik (Warsaw), M.Latta (Joinville-le-Pont), Y.Lemoine (Strasbourg), M.van der Linden (Groningen), D.d.S.Lobo (Rio de Janeiro), B.Lapeyre (Toulouse), M.Lobocka (Warsaw), M.Lynch (Edinburgh), W.Meijer (Groningen), R.Medici (Naples), S.Metzger (Jerusalem), U.Michelsen (Erlangen), S.Milburn (Davis Ca.), P.Moller Sorensen (Paris), B.Müller (Zürich), W.Musters (Amsterdam), S.Mazan (Toulouse), J.R.McCormick (Rochester), H.Neuhaus (Freiburg), H.Nielsen (Copenhagen), I.d.S.Nogueira (Oeiras), S.Nyström (Umea), M.L.Osorio-Almeida (Monte da Caparica), M.Ouellette (Boul Can.), U.Patel (Portsmouth), B.Pavlovic (München), C.Persson (Uppsala), B.Pisan (Basel), I.Plavec (Zürich), P.Poulsen (Copenhagen), T.Raabe (Basel), P.Raeymaekers (Wilrijk), A.Richter (Uppsala), J.Robert-Baudouy (Villeurbanne), S.Rokem (Jerusalem), L.Ruohonen (Helsinki), M.Ryden (Uppsala), K.Schörgendorfer (Graz), H.Schwelberger (Graz), A.Spalding (Canterbury), H.Sychrova (Prague), N.Samuelov (Jerusalem), B.Sandahl Sorensen (Aarhus), K.Schnetz (Fribourg), A.Senelonge (Lyon), P.Shiels (Glasgow), M.Skrzypek (Warsaw), J.Sobczak (Paris), J.Soppa (Martinsried), S.Sperka (Tutzing), J.Stockhaus (Köln), A.Tanuri (Rio de Janeiro), C.Taschke (Heidelberg), M.Todorova (Sofia), M.-P.E.Tranchant (Madrid), A.Trivedi (London Can.), S.Tsirka (Thessaloniki), S.Vamvakas (Würzburg), S.Vanhanen (Espoo), G.Vancanneyt (Cologne), E.Vanmechelen (Antwerpen), A.C.de Vera (Sevilla), A.C.P.Vicente (Rio de Janeiro), Y.Vizirianakis (Thessaloniki), H.Vogelsang-Wenke (Martinsried), A.Vonshak (Sede Boqer Campus), G.Wagner (Uppsala), K.Wang (Gent), M.Wehrmann (Hannover), T.Wells (London), K.C.Williams (London), S.Williamson (Midlothian), C.M.Williamson (Berkshire), O.Yarden (Rehovot), D.Zerbib (Toulouse), T.Zimmermann (Visp), T.Zusman (Tel Aviv).


## SPETSAI 1987

## MOLECULAR BIOLOGY OF DEVELOPMENT
August 30 - September 12

**Organizers** C.T.Caskey (Houston)(chairman), A.E.Evangelopoulos (Athens), W.J.Gehring (Basel), J.Hershey (Davis).

## Lecturers and Lectures

**S.Artavanis-Tsakonas** (New Haven): 1.Neurogenesis in Drosophila: a genetic and molecular approach I. 2. Neurogenesis in Drosophila: a genetic and molecular approach II.
**A.Bradley** (Houston): Manipulation of the mouse genome via ES cells modified in culture.
**C.T.Caskey** (Houston): 1.Gene Replacement. 2.Antisense gene regulation.
**V.N.Chapman** (Buffalo): 1.X-chromosome regulation in mammalian development. 2.Gametic imprinting and its effects on embryonic gene development.
**T.E.Evangelopoulos** (Athens): 1.Control of cellular activity by protein phosphorylation-dephophorylation processes. 2.Glycogen metabolism in smooth muscle-chicken gizzard phosphorylase kinase.
**W.J.Gehring** (Basel): 1.Homeotic genes, the homeobox and the control of development. 2.Designing and redesigning of the fruitfly.
**H.M.Goodman** (Boston): 1.Arabidopsis thaliana as a model for the study of plant development. 2.Isolation of developmentally important genes from Arabidopsis thaliana.
**P.Gruss** (Göttingen): 1.Strategies for the identification of genes controlling mammalian differentiation and development. 2.Enhancers as tissue-specific transcriptional control elements.


262

**J.W.B.Hershey** (Davis): 1.The molecular biology of sea urchin early development. 2.Translational control of gene expression during sea urchin development.

**M.Lenardo** (Cambridge Ma.): 1.Leukemogenesis by the abl gene. 2.Control of antibody gene transcription.

**A.L.McLaren** (London): 1.Gonadal sex determination in mammals. 2.Differentiation of germ cells in mice.

**G.Morata** (Madrid): 1.Genetic specification of the body plan of Drosophila. 2.Genetic structure of the bithorax complex of Drosophila.

**P.A.Overbeek** (Houston): 1.Alterations in development and tissue-specific oncogenesis in transgenic mice. 2.Dominant and recessive developmental disorders in transgenetic mice: Cataracts and fused phalanges.

**R.A.Palmiter** (Seattle): 1.Tissue-specific gene expression in transgenic mice. 2.Transgenic oncogenesis.

**R.A.Phillips** (Toronto): 1.Genetic manipulation of hematopoietic stem cells. 2.Retinoblastoma as a model for recessive oncogenes and their role in differentiation.

**D.A.Sassoon** (Paris): 1.The actin and myosin multigene families. 2.Actin and myosin gene regulation during myogenesis.

**W.Schaffner** (Zürich): 1.Enhancers and viral gene expression. 2.The modular structure of DNA sequences controlling cell type-specific transcription.

**P.Soriano** (Cambridge Ma.): 1.Retrovirus and insertional mutagenesis. 2.Retroviruses as chromosomal markers and as insertional mutagens.

**J.Sulston** (Cambridge U.K.): 1.The cell lineage of Caenorhabditis. 2.Construction of a genome map for Caenorhabditis.

**E.F.Wagner** (Heidelberg): 1.Gene transfer with retroviral vectors into embryonic and hemopoietic stem cells of the mouse. 2.Oncogene expression in transgenic mice.

**D.J.Wolgemuth** (New York): 1.Development of the mammalian germ cell lineage. 2.Development stage specific expression of cellular oncogenes and homeo box-containing genes in mammalian germ cells.

## Participants

A.Aysel (Antalya), E.Agar (Atakum), M.Adamczewski (Freiburg), M.Al-Ubaidi (Houston), H.Amerein (Zürich), A.J.Baekgaard (Aarhus), A.Barberis (Zürich), H.A.Barrera-Saldana (Monterrey), M.M.R.Baroso (Oieras), K.Basler (Zürich), S.Bhandari (Cambridge), M.Bodner (La Jolla), M.Boissinot (Quebec), I.Busseau (Aubiere), P.Callaerts (Leuven), P.S.L.B.Campo (Lisboa), B.Cardinali (Rome), M.Caserta (Rome), C.Cerni (Vienna), J.Chester (Glasgow), B.F.C.Clark (Risskov), V.Colot (Cambridge), F.Cutruzzola (Rome), R.L.Davis (Seattle), A.de Basturia (Madrid), M.de Lorenzi (Zürich), F.J.de Mayo (Houston), A.Derventzi (Thessaloniki), V.de Simone (Heidelberg), Z.Dominski (Warsaw), P.Einat (Rehovot), M.Fabre (Castanet-Tolosan), A.M.Ford (London), C.J.Gallo (Davis), J.Gautier (Toulouse), S.Gautron (Paris), A.Giangrande (Strasbourg), A.Graham (London), H.Gram (Basel), E.Gratwohl (Basel), K.J.P.Griffin (London), A.H.Handyside (London), C.Henchcliffe (Oxford), M.Henriksson (Stockholm), D.Henrique (Oeiras), F.Hilberg (Hamburg), H.Hirt (Vienna), R.C.Hoeben (Leiden), H.J.Holtke (Tutzing), K.Hooper (Oxford), R.H.Jackson (Buckinghamshire), M.Jamrich (Bethesda), L.Jeannotte (Montreal), P.Jones (Oxford), S.R.Kala (Umea), C.Kalogera (Ioannina), M.Keil (Berlin), S.Khochbin (Grenoble), K.Kohrer (Martinsried), P.Koopman (London), R.Krieg (München), F.-D.Kuhl (Zürich), D.Levanon (Rehovot), D.Loncar (Stockholm), S.G.Mackie (Cambridge), L.H.Y.Madsen (Aarhus), I.G.D.F.Maessen (Leiden), K.A.Mahone (Bethesda), G.M.Maniatis (Patras), E.Martin (Madrid), D.Meijer (Rotterdam), F.Melin (Villejuif), I.Miltner (Ulm), A.Molven (Bergen), K.Mooslehner (Hamburg), I.M.Morgan (Glasgow), A.Negro (Siena), C.Niehrs (Heidelberg), P.R.Njolstad (Bergen), V.Orlando (Rome), V.Orphanos (Patras), N.Papalopulu (London), I.S.Pappas (Thessaloniki), F.Payre (Toulouse), J.A.Pearlman (Houston), M.Pettersson (Zürich), D.B.Pilgrim (Seattle), R.Possenti (Rome), E.Pringault (Paris), A.Puschel (Göttingen),

S.I.S.Rattan (Aarhus), A.D.Reith (London), P.Remy (Strasbourg), D.Resnitzky (Rehovot), M.Roth (Baltimore), S.Ruppert (Heidelberg), S.Ryser (Basel), L.Sastre (Madrid), E.Schejter (Rehovot), S.Scherer (Houston), S.Sesodia (Paris), K.L.Signorelli (New York), A.Simeone (Naples), P.Sorernsen (Aarhus), E.Spanopoulou (London), M.-L.Steen (Uppsala), V.Stiefel (Barcelona), B.Stein (Karlsruhe), K.Tasiouka (Thessaloniki), H.-J.Thiesen (Heidelberg), B.Thisse (Strasbourg), J.Thompson (London), T.Unger (Rehovot), G.Veres (Szeged), I.Vernos (Madrid), G.Vezina (Quebec), C.M.Viviano (New York), G.Wasner (Salzburg), S.E.Wedden (Boston), M.van Wiles (London), N.C.Wrighton (London), C.Yanicostas (Paris), A.Yarden (Rehovot), M.Zernicka (Warsaw).

## SPETSAI 1988

## ORGANIZATION AND FUNCTION OF THE EUKARYOTIC GENOME
September 1 - 10, 1988

**Organizers** H.G. Zachau (München)(chairman), A.E.Evangelopoulos (Athens), H.Feldmann (München).

## Lecturers and Lectures

**M.Beato** (Marburg): 1.Regulation of transcription in animal cells, a survey. 2.Gene regulation by steroid hormones.
**A.Bird** (Wien): 1.DNA methylation and differential gene expression. 2.CpG islands as gene markers in the vertebrate nucleus.
**P.Borst** (Amsterdam): 1.Antigenic variation and discontinuous mRNA synthesis in Trypanosomes. 2.Amplified genes involved in multi-drug resistance in cancer cells.
**M.Eigen** (Göttingen): 1.Theory, experiment and reality of molecular evolution. 2.Evolution of sequences.
**H.Feldmann** (München): 1.Gene regulation in yeast. 2.Mobile elements in lower and higher eukaryotes.
**R.Gallo** (Bethesda): 1.Human lymphotropic viruses. 2.AIDS.
**S.E.Humphries** (London): 1.Molecular biology techniques in the analysis of multifactorial diseases. 2.Structure and function of the apolipoprotein gene families.
**H.Jäckle** (München): 1.Drosophila development, homeotic genes. 2.Molecular analysis of segmentation in Drosophila.
**F.Melchers** (Basel): 1.Development of lymphocytes from stem cells. 2.Cell co-operations in the immune response.
**W.Neupert** (München): 1.Intracellular protein sorting: how proteins are targeted to the various cell organelles. 2.How proteins are sorted into the various mitochondrial subcompartments.
**L.Orgel** (San Diego): 1.Biochemistry of prebiotic evolution, a survey. 2.Non-enzymatic transcription of oligonucleotides.
**S.Ottolenghi** (Milan): 1.Structure and function of human globulin genes. 2.Inherited defects of hemoglobin synthesis.
**A.Rich** (Cambridge Ma.): 1.Structural polymorphism in DNA. 2.Z-DNA distribution in chromatin and its role in homologous recombination.
**C.Sander** (Heidelberg): 1.Sequence analysis by computer. 2.Studies in protein design.
**D.Stehelin** (Lille): 1.Oncogenes, an overview. 2.Oncogenes and the genetic dissection of cancer.
**Ch.Weissmann** (Zürich): 1.Mechanism of splicing. 2.The molecular biology of scrapie, a slow, transmissable disease of the nervous system.
**H.G.Zachau** (München): 1.Immunoglobulin and T-cell receptor genes. 2.The human immunoglobulin loci and their rearrangements.

264

## Participants

K.Alexciev (Sofia), S.Altiok (Istanbul), C.Angelidis (Ioannina), I.Arzimanoglou (Athens), C.Bagni (Rome), K.Beaucamp (Tutzing), J.Bernues (Barcelona), J.Bogerd (Amsterdam), P.Bonaldo (Aviano), F.Brombacher (Freiburg), P.Bull (London), D.Buonamassa (Siena), D.Christodoulou (Thessaloniki), A.Ciccodicola (Naples), M.Classon (Stockholm), R.C.Ferreira (Oeiras), W.Coppieters (Merelbeke), W.Dalemans (Strasbourg), J.M.Darby (Bucks), P.Dincer (Ankara), T.Dobner (München), H.Erdem (Ankara), G.Fourel (Paris), M.Garcia-Ramirez (Barcelona), R.Gargouri (Paris), T.Geladopoulos (Athens), G.Gibson (Basel), D.Ginsberg (Rehovot), V.Giossi (Athens), M.Gomperts (Cambridge), P.Gregor (Rehovot), G.Hansson (London), M.Hartl (München), E.Heard (London), K.Helin (Copenhagen), A.Hemming (Göteborg), P.Henttu (Oulu), C.H.Sanchez (Genève), L.A.Herzenberg (Stanford), R.Hochstenbach (Nijmegen), R.Höfgen (Berlin), L.Höglund (Stockholm), D.Hoffman (Warsaw), C.Huber (München), O.P.L.Hughes (San Francisco), S.Humphries (London), M.Ioannou (Patras), P.Jackson (Cambridge, Ma.), K.Jacobsen (Aarhus), L.Janson (Uppsala), M.Jung (Warsaw), D.Katcoff (Ramat Gan), A.A.Katz (Rehovot), I.Kemler (Zürich), D.Kipling (Oxford), A.Krzyzaniak (Cambridge, Ma), P.Kylsten (Stockholm), E.Lagasse (Basel), G.Lang (London), M.Lazard (Gif sur Yvette), D.Leiss (Martinsried), P.Lemaire (Heidelberg), G.Lennon (Philadelphia), C.Le Van Kim (Paris), M.Ligtenberg (Amsterdam), J.Lorens (Bergen), B.Lutz (Zürich), B.Majello (Naples), B.Marciniak (Warsaw), B.Meier (Basel), C.Melani (Genova), J.Mertsching (Martinsried), M.Mieszczak (Strasbourg), A.E.Morris (New York), U.Mortensen (Aarhus), A.Mouzaki (Genève), S.Munemitsu (San Francisco), T.Munoz-Antonia (New Haven), J.Muser (Basel), A.Nazeem (Zürich), G.P.Nolan (Stanford), M.Nordling (Göteborg), A.Nygren (Stockholm), R.Oakey (Oxford), F.Palla (Palermo), I.Palmero (Madrid), L.Parreira (Lisbon), C.Patriotis (Sofia), T.Perlmann (Stockholm), U.Pfeffer (Genova), M.P.Valles (Barcelona), H.-J.Pucher (Braunschweig), G.Radziwill (Heidelberg), V.L.R.Marques (Oeiras), J.Rich (Madison), D.Rotem (Tübingen), I.Rubelj (Zagreb), F.Ruder (Tübingen), A.Salzberg (Haifa), C.Sander (Heidelberg), D.Schaller (Fribourg), M.Schlaeppi (Basel), U.Schlokat (San Francisco), P.Schmid (Ulm), B.Schnierle (München), H.Schwelberger (Graz), R.M.Schwendenwein (Wien), J.Serth (Hannover), R.Simon (Basel), A.Sinclair (Heidelberg), M.P.Somma (Rome), I.Stefanov (Szeged), S.L.Sturley (Madison), D.J.Talbot (London), M.Todorova (Szeged), F.Tronche (Paris), A.Tsamadou (Thessaloniki), B.Turan (Ankara), P.Vankan (Basel), A.R.Venkitaraman (London), V.Wallet (Paris), F.Weih (Heidelberg), R.A.Wells (Oxford), S.Wendel (Neuherberg), M.Werner (Göttingen), A.Wilson (London), C.-Y.Yu (Cambridge), N.Zander (Bochum), T.A.Zarvalis (Thessaloniki), F.-J.Zimmer (München), J.Zhu (Shanghai), F.Zwartkruis (Utrecht).

(Reprinted from Nature, Vol. 220, no 5174, pp. 1275-1276, December 28, 1968)

## ON RUNNING A SUMMER SCHOOL

by
FRANCIS CRICK
MRC Laboratory of Molecular Biology,
Hills Road, Cambridge

*Would-be organizers of summer schools are offered some advice on how to increase the efficiency of communication between participants*

The sort of "summer" school I have in mind can occur at any time of the year, lasts for any period from a few days to two or three weeks (the commonest time is one week), consists wholly of lectures and discussions, without practical classes, and is attended by anything from fifty to a few hundred postgraduate or postdoctoral students. There may be some older people from adjacent fields, or some who want to change their field entirely, but most of the audience will already have had a grounding in at least a part of the subject to be covered. My own experience has been limited to schools in molecular biology. I have not been to many of these, but I have noticed that they all have certain problems in common. With a little care some of these difficulties could be avoided.

All the schools I have attended have coped very well with the more obvious problems of organization. They have been held in pleasant places, for everybody knows that scientists are reluctant to go to meetings unless they are held half way up Mont Blanc, or on a Greek island, or in some similar place. (I notice that scientists have not yet achieved a really peak position in our affluent society. If they meet in expensive places, they do so slightly out of season, when the rates are appreciably cheaper.) The mechanics of housing people, getting them to and from the airport or the railway station, the arrangement of the lecture room and other such matters are usually done rather well. And, although I have not been privy to the financial side, the lecturers have been tempted with a modest honorarium in addition to their expenses, and many of the students assisted with grants from one source or another. Of course, there can be hazards even here. One organizer told me that they had omitted to ask, on the application form, for the sex of the applicants, thinking that they could deduce this from the first name (the students were sharing two to a room), but they were stumped in several cases by such names as Leslie. Another organizer told me that for their meeting they had fortunately remembered to put the entry "sex" on the application form. "Of course", he said, "a few people came up with the old joke, and simply put "yes", but they always turned out to be male".

Nor have the difficulties come because the lecturers were inexperienced. It is not generally realized by the world at large just how accustomed scientists are to talking for fifty minutes, at least by the time they are in mid-career. It is possible for a novelist to win the Nobel Prize for Literature without ever having made a public speech, but I doubt if any comparable scientific figure could be inexperienced in this way. Even a taciturn scientist can often give a good lecture. Moreover, scientists nowadays mix fairly easily, because, contrary to popular belief, scientific research is often a rather gregarious activity, unlike writing, which is usually rather solitary. Of course, some scientists give poor lectures, but they are unlikely to be asked to speak at summer schools.

The organizers usually make it quite clear in their advertisements what the course will cover, but they often omit to say at what level it will be given and exactly what background is expected of the student. It is well worth considering whether the first few lectures should not be entirely introductory, to bring everybody up to the same level. This is especially important for new subjects which fall between two already established ones. In such cases the initial lectures might be run in parallel. For example, if the summer school is designed to lure recruits to the new and fascinating field of, say, astrobotany, then the astronomers might get an introductory day on botany, while at the same time the botanists could be brushed up on the solar system and the structure of the galaxy. At the very least one or two suitable text-books could be indicated, the contents of which the students were supposed to master beforehand, though whether in fact they would do this only experience can show.

This uncertainty about background also affects the speakers, for, curiously enough, the organizers often fail to brief the speakers on this point. I have heard a really excellent lecture given on "work in progress in my lab" when what I thought the audience needed was a broad review of the field. The organizers, because they select the students, usually know a fair amount about their background, but this useful information they often keep to themselves. Sometimes they volunteer it after the lecturers have arrived, but this is usually too late for the lecturers to alter their talks very much, if only because they may not have brought the right slides. For all I know the organizers of the meetings I attended may have considered all these points but have decided that old so-and-so will always give the same lecture, no matter what he is asked to talk about, or, alternatively, that it doesn't matter what he talks about because the students never really understand it, although he makes it all sound fascinating. But I believe most lecturers would welcome a little guidance from the organizers and would in most cases adjust their lectures to meet the occasion.

A much more difficult point concerns written lecture notes. Most lecturers nowadays have so many calls on their time that they will not accept an invitation to a summer school if they are expected to provide a manuscript for publication. Of course, the organizers may try to bribe them with a very large honorarium (to be paid when the manuscript is delivered!), but this is expensive and even then will not always work. One school I attended taped all the lectures. From these tapes one of the organizers made lengthy summaries, which were then mimeographed by a very efficient secretarial staff and distributed to the students, but this puts a very heavy strain on the persons concerned and inevitably rather a lot of mistakes creep in. (Nevertheless, Volker Kasche and his colleagues (see footnote, page 1276) consider that these summaries were very valuable and strongly recommend that such papers be prepared in future courses.)

On the other hand, if nothing is provided the students may have a very hard time taking useful notes. A good compromise is to persuade the lecturers to provide an outline or a summary, together with a few key references (too many is worse than too few). This could even be done after the lecturer had arrived, though it is obviously better to have the summaries available before the lecture is delivered. Most speakers lecture from notes and it is not a great deal of trouble to transform these into an outline of the lecture, provided no attempt is made to adapt such notes for more formal publication. (The difficulties which this produces have been discussed by Bragg[1]).

The principal fault, however, of almost all schools is not the quality of the lectures but the quantity. One organizer of a school admitted to me that they had too many speakers. "We asked a few more than we needed", he told me, "feeling certain that one or two would refuse, but to our surprise they all accepted!" Whatever the reason the organizers almost always arrange too many lectures for the time available. At one time I thought that my fatigue was just another distressing symptom of my advancing years, but I have made enquiries among my younger colleagues and (with only one exception) they have all complained that before the end of the meeting they were satiated and could no longer take things in. Most organizers know that they should leave at least part of the day completely free ("for discussion", they usually say, but a walk or a swim would be better). Even so, they often manage to programme as much as six or seven hours of lectures a day. I think this is too much (except perhaps for a meeting lasting only one day) though what the optimum is I am not sure. I suspect four hours of listening a day is enough, plus perhaps an hour or so for informal discussion. Moreover, even at this rate, if the meeting lasts any time at all there should be complete days (or half-days) left free, at about the rate of one day off for every three or four days on. The wise organizer would even arrange more free time than this, to allow for the demand, which usually arises if the school is a good one, for "a little extra meeting this afternoon to hear about such-and-such". It is these little extra meetings, like those little extra drinks, which are really killing. I would very much like to know the opinions of students on their powers of assimilation. Good data on this would be invaluable to the organizers of all sorts of meetings. I am assuming here that the school is such that almost all the students will want to attend all the lectures.

It follows that because the total time available for the school is usually fixed in advance the organizers must resist the temptation to try to cover too much, and must design the content of the programme with great care. Experience shows that this is more easily said than done.

But suppose that the organizers have solved all these problems, that the students have all acquired a suitable background, that the lecturers, being well briefed, have all delivered model lectures and that the programme is well chosen and not overcrowded.

There is one outstanding problem—the discussion ; or, in more general terms, the reciprocal interaction between lecturer and student. There should always be some discussion in the lecture room at the end of each lecture or groups of lectures. Sometimes the students are shy and the questions get asked only by the other lecturers. They have usually heard the lecture several times already in other parts of the world, but out of politeness they may ask leading questions, or even occasionally attack the speaker. Such a conflict is not necessarily a bad thing. I have noticed that students like nothing better than seeing a couple of established workers going at each other hammer and tongs, provided it is done with good humour. But if, as the meeting goes on, the students themselves are not asking the questions, something is wrong.

And this brings me to the informal discussion. Students want very much to be able to meet the speakers, not only to ask them the more technical questions, but also to get some impression of the way their minds work. This cannot be done merely by having the lecturers and students eat together, because the audience for each speaker is then too small and too frozen. Someone down the table, who desperately wants to hear some topic discussed, may be just too far away.

An excellent solution, used by one school I attended, was to make the morning's speakers available, after lunch, on the terrace when coffee was being served. A small group would form round each speaker (sometimes round a pair of speakers) and the questioning could go on for perhaps an hour or so. Nobody (except the lecturers !) was compelled to stay and eventually the talk would gradually fade away. The amount of ground covered in this way can be astonishing. Not only can misunderstandings be quickly cleared up, but much detailed technical information can be conveyed rather rapidly to the more advanced students. Moreover, nothing removes the students' shyness so quickly, and this will usually solve the difficulty of getting a good discussion immediately after the lectures and also give them enough courage to talk freely to the lecturers when they come across them at meals or in the bar. Such conversations can be invaluable to the students. For this reason lecturers should always be encouraged to stay at the school for as long as possible, preferably for all of the time if they can spare it. They will find that if they do so the students will be very appreciative. Whether they should be expected to listen to all the other lecturers is another matter !

Such informal contacts also allow the lecturers and the organizers to get some impression as to how the school is going, but I suspect that if one really wants to know what the students feel some more organized feed-back is necessary. I would suggest that something along the following lines might be tried. Let the students be put into groups of ten or a dozen, each with a spokesman. The groups, as far as possible, should be homogeneous. Either all botanists, or all Frenchmen, or all postdocs, or, better still, all three. The spokesmen should be articulate and liked by their fellows. Each day the groups should meet informally for a few minutes (no need to spend too much time on it) and discuss their reactions to the lectures of the day and the way the school is going. Then the group leaders and the organizers should meet collectively (again, for a brief period) to pass on this information. Lecturers, as far as possbile, should not be included, so that everybody can be frank I Even if this is not done every day, it should be done at least soon after the start of the course and again at the end. It would not be a bad thing if, after the course, students were asked to comment on it in writing if they felt they had something useful to say*. It seems to me that without this kind of feedback it is going to be impossible to improve the organization of summer schools.

Finally, I should like to make a personal plea to the organizers on two minor points. It really does help to have name tags (and legible name tags that one can read without peering) together with a list of the names of addresses of all lecturers and students. This is usually quite easy to compile because the organizers have all the data passing through their hands. The other point concerns the social life. A good school has a convenient bar, where people can relax over drinks and in some schools dances are organized or trips to see the local night life. Personally, I love dancing in rather dimly lit cellars, but I find that I cannot stay up till all hours of the morning and be fresh and receptive next day. Age again, you will say, but I believe that only the very young can do this. It makes life much less of a strain if the late nights are organized on the evenings before a totally free morning. I know this sounds elementary, but it never seems to happen.

Looking back on what I have written, I see that my chief theme has been efficient communication. Research workers are quite experienced at communicating with other research workers. If they teach, they acquire, over the years, some experience in instructing a recurrent set of undergraduates. But the summer school is often a one-shot teaching situation, involving rather more sophisticated students. If we are to become professional at it, and not just play it by ear in an amateur fashion, we must take special steps to see that this communication of knowledge, experience and enthusiasm is both efficient and enjoyable.

*This was done at one school I attended. A group of Swedish students, headed by Volker Kasche, produced a very valuable critique of the school. They particularly stressed the advantages of having a high lecturer/student ratio. I should like to thank them for making their comments available to me.

[1] Bragg, L., *Science*, 154, 1613 (1966).

*Printed in Great Britain by Fisher, Knight and Co., Ltd., St. Albans.*

CONTRIBUTORS

Dr. John ABELSON
California Institute of Technology
Division of Biology 147-75
Pasadena, California 91125
U.S.A.

Dr. Giorgio BERNARDI
Laboratoire de Génétique Molé-
culaire
Institut Jacques Monod
Université Paris VII - CNRS
Tour 43 - 2 place Jussieu
75251 PARIS CEDEX 05
FRANCE

Dr. Piet BORST
Het Nederlands Kankerinstitut
Antoni van Leeuwenhoekhuis
plesmanlaan 121
1066 CX AMSTERDAM
THE NETHERLANDS

Dr. Edward BRODY
LP 2420 - Bâtiment 24
Centre de Génétique Moléculaire
du C.N.R.S.
91190 GIF-SUR-YVETTE
FRANCE

Dr. Charles CANTOR
Dept. of Genetics and Development
College of Physicians and Surgeons
of Columbia University
701 West 168th Street
New York, N.Y. 10032
U.S.A.

Dr. Thomas CASKEY
Baylor College of Medicine
Texas Medical Center
Institute for Molecular Genetics
One Baylor Plaza
Houston, Texas 77030
U.S.A.

Dr. Brian F.C. CLARK
Dept. of Chemistry
Division of Biostructural Chemistry
Kemisk Institut
Aarhus University
Langelandsgade 140
8000 AARHUS C
DENMARK

Dr. Jean-Pierre EBEL
Institut de Biologie Moléculaire
et Cellulaire du C.N.R.S.
15 rue René Descartes
67084 STRASBOURG CEDEX
FRANCE

Dr. Manfred EIGEN
Max-Planck-Institut für biophysi-
kalische Chemie
Karl-Friedrich-Bonhoeffer-Institut
Abteilung biochemische Kinetik
Am Fassberg - Postfach 2841
3400 GÖTTINGEN
FEDERAL REPUBLIC OF GERMANY

Dr. Athanasios EVANGELOPOULOS
National Hellenic Research Foundation
Biological Research Center
48. Vas. Constantinou Avenue
ATHENS 116 35
GREECE

Dr. Horst FELDMANN
Institut für physiologische Chemie,
physikalische Biochemie und Zellbiologie der Universität
Goethestrasse 33
8000 MÜNCHEN 2
FEDERAL REPUBLIC OF GERMANY

Dr. Marianne GRUNBERG MANAGO
Institut de Biologie Physico-
Chimique
13 rue Pierre et Marie Curie
75005 PARIS
FRANCE

Dr. Anne-Lise HAENNI
Laboratoire de Biochimie du
Développement
Institut Jacques Monod
Université Paris VII - CNRS
Tour 43 - 2 place Jussieu
75251 PARIS CEDEX 05
FRANCE

Dr. John HERSHEY
University of California Davis
School of Medicine
Dept. of Biol. Chem.
Davis, California 95616
U.S.A.

Dr. Niels KJELDGAARD
Institute for Molecular Biology
and Plant Physiology
Aarhus University
C.F. Møllers Allé 130
8000 AARHUS C
DENMARK

Dr. Leslie ORGEL
The Salk Institute
P.O. Box 85800
San Diego, California 92138-9216
U.S.A.

Dr. Lennart PHILIPSON
E.M.B.L.
Postfach 10.2209
Meyerhofstrasse 1
6900 HEIDELBERG
FEDERAL REPUBLIC OF GERMANY

Dr. Alexander RICH
Department of Biology
Massachusetts Institute of
Technology
Cambridge, Massachusetts 02139
U.S.A.

Dr. Gottfried SCHATZ
Biozentrum der Universität Basel
Abteilung Biochemie
Klingelbergstrasse 70
4056 BASEL
SWITZERLAND

Dr. Maxine SINGER
Carnegie Institution of Washington
1530 P. Street, Northwest
Washinton, D.C. 20005-1910
U.S.A.

Dr. Hans ZACHAU
Institut für physiologische Chemie,
physikalische Biochemie und Zellbiologie der Universität
Goethestrasse 33
8000 MÜNCHEN 2
FEDERAL REPUBLIC OF GERMANY

A-DNA (see also DNA), 5, 7

B-DNA (see also DNA), 2, 7

Cell cycle, 101, 106
  negative regulation of growth,
    101-108
  proliferation, 101
  quiescent state, 101
  transformed cells, 101, 104
Chloroplast, 73
Chromatin
  nucleosome, 84
  organization, 83
  structure, 84
  Z DNA binding proteins in, 11
Cytochrome-oxidase
  bc1 complex, 76
  subunit IV precursor, 73
Dihydrofolate-reductase, 74, 170
DNA
  A form, 5, 7
  B form, 2, 7
  Z form, 2, 3, 7, 11
    binding proteins, 11
    genetic recombination, 14
    immunogenicity of, 9
    negative supercoiling, 9
  conformational equilibrium, 5, 10
  conformational polymorphism, 1
  synthesis by resting cell mRNA, 106

Elongation factor Tu, 41
  as GTP-binding protein, 41, 43
  domains, 40
  expression of G-domain, 46
  site-directed mutagenesis of, 45
Exon duplication, 210

Gene
  activation, 13
  expression, 17, 143, 148
    in E.coli, 17
    translational control, 17

Genome
  mammalian, 49, 150
  megabase perspective, 143
  polymorphisms, 1, 145
Glycolysis, 163, 171
  evolution of, 169
  in cytosol, 163, 167
  in glycosome, 163
    origin of, 170
  in Tripanosoma brucei, 163
    membrane, 165
    intermediates of, 163
    p-glycerate-kinase, 164
Growth arrest specific genes (gas),
    102-104
Growth factors, 102, 108

hnRNPs, 194
Human immunoglobin K, 111
  cosmid clones, 113
  duplication of, 116
  evolution, 111, 115
  locus organisation, 112
  transcriptional polarity, 116

Intron (see also Splicing)
  group I and II, 193, 203
  secondary and tertiary
  structures, 193
  splicing of, 208

Kinase C, as $Ca^{2+}$-calmodulin-
dependent kinase, 59, 63
  activators and inhibitors, 61
  bovine, 65
  location of $Ca^{2+}$ binding site, 64
  $Zn^{2+}$ binding site, 64

Line-1 sequences, 155

Mammalian initiation factors, 49
  binary and ternary complexes, 50
  DAI (eIF2 α kinase), 53
  HCR inhibition, 51

Mammalian initiation factors
    (continued)
  initiation pathway, 49
  phosphorylation, 49, 51, 54
  translational control, 49
Mitochondria, 71
  contact sites, 73
  DNA, 71
  extramitochondrial cytoplasm, 71
  intramitochondrial compartment, 71
  matrix, 71
    localized processing protease,
      75
  oxydative phosphorylation system
      in, 167
  precursor proteins, 75
  targeting signal, 73
mRNA
  inhibition of DNA synthesis by,
    105
Murine ornithine transcarbamylase,
    121
  activity, 121
  gene structure, 122
  in transgenic mice, 126
  promoter, 125

Nucleic acids
  conformational changes, 1
  inheritent self-production, 227
  Nucleotide synthesis, 216

Oxidative phosphorylation, 71

Phosphorylase kinase, 59
  activators and inhibitors, 61
  endogenous and exogenous, 64
  proteolysis activation of, 60
Plant RNA viruses, 175
  (+) RNA
    "picornia-like" group, 178
    "sindbis-like" group, 178
    functional domains in proteins
      coded by, 178
    NTP-binding domains, 180
  DI particles, 177, 185
  evolution of, 176
    modular evolution hypothesis,
      180
  mutation rate, 176
  recombination, 175, 176, 177, 181
    examples of, 182, 183, 184
    homologous vs non-homologous,
      185
    in functional domains, 177
Proteins
  allostery, 1
  GTP binding, 46
  nuclear, SEF1, 88

kinase, 59
  nucleic acid binding, 155
  precursors, 71, 73
    $Ca^{2+}$/ATPase function, 63
    phospholipid activated, 65
  ras p21, 40
  ribonucleic, 193, 194
  thermal stability of, 140
Pseudogenes, 114
  orphons, 117

Quasispecies, 225, 227
  molecular, 228

ras p21 protein, 40
Retroviruses, 157, 160, 228
  as quasispecies behaviour model,
    228
  similarities with class I
Retrotransposons, 157
RNA
  ancestors of, 215, 218
  replication theory, 220, 223
  synthesis, 216
    enantiomeric cross-inhibition
      of, 219
    inhibitors of, 217
RNA polymerase III, 84

Sequence space 225, 226
  binary sequence, 226
  features of , 227
  quaternary digits, 226
  statistical geometry in, 228
snRNPs, 193
Splicing
  accuracy of, 206
  alternative to tissue-specific,
    203, 209
  consensus sequence, 193, 206
  in Yeast, 193, 206
    pre-mRNA mutant, 193, 195
    extragenic suppressors, 200
    localization and role, 196
  mechanism of, 193, 203
  RNA structure in, 209
    stem-loop inhibition, 210
  site, 193, 208
  thermodynamic differences, 208
Spliceosome, 194, 206, 208
SV40
  core motif, 88
  minichromosomes, 11

Threonyl-tRNA synthetase, 17
  activity, 24
  domains, 19, 24
    regulatory, 21
  isoacceptors, 25

Threonyl-tRNA synthetase (continued)
  mRNA of, 17
    anticodon-like region, 21, 23
    binding, 21, 24
    feedback control, 21
    structure, 19
  mutants, 24
  operator, 19
  SD sequence, 19
  translation repression, 24
Transcription,
  cis- and trans-acting elements,
      79
  enhancers, 88
  factors, 79, 84
  rate of, 80
Translation initiation factor, 49
Translocation, 71, 138
Transposable elements, 155
  DNA-mediated transpositions, 158
  human, 155
  in Drosophila melanogaster, 155
  retrotransposons, 155, 160
    class I and II, 155, 156, 158
tRNA
  D and T-loop, 29, 79
  genes, 79
    A and B boxes, 79
    expression, 79
    intron-like sequence, 81
    organization, 79, 80
    promoters, 84
  primary structure of, 29
TYMV RNA, 29
  tRNA-like end of, 29
  recognition by the synthetase, 36
  structural organization
      of, 29, 30, 34
  valylation activity of, 34, 37

U3 region
  deletion analysis of, 91
  mutagenesis, 89
  role and pathogenicity and tissue
      specificity, 89

Vertebrate genomes, 133
  base substitution process, 136
  evolution, 134, 141
    conservative mode of, 141
  GC levels in, 133
    in homologous coding sequences,
      136
  introns, 140
  isochores, 133, 136, 148

Yeast pre-mRNA splicing mutants,
  193

Z-DNA (see also DNA), 2, 3, 7, 11
  binding proteins, 11